Nonlinear Systems and Complexity

Volume 38

Nonlinear Systems and Complexity provides a place to systematically summarize recent developments, applications, and overall advance in all aspects of nonlinearity, chaos, and complexity as part of the established research literature, beyond the novel and recent findings published in primary journals. The aims of the book series are to publish theories and techniques in nonlinear systems and complexity; stimulate more research interest on nonlinearity, synchronization, and complexity in nonlinear science; and fast-scatter the new knowledge to scientists, engineers, and students in the corresponding fields. Books in this series will focus on the recent developments, findings and progress on theories, principles, methodology, computational techniques in nonlinear systems and mathematics with engineering applications. The Series establishes highly relevant monographs on wide ranging topics covering fundamental advances and new applications in the field. Topical areas include, but are not limited to: Nonlinear dynamics Complexity, nonlinearity, and chaos Computational methods for nonlinear systems Stability, bifurcation, chaos and fractals in engineering Nonlinear chemical and biological phenomena Fractional dynamics and applications Discontinuity, synchronization and control.

Carla M. A. Pinto • Clara Mihaela Ionescu
Editors

Computational and Mathematical Models in Biology

Springer

Editors
Carla M. A. Pinto
School of Engineering, Center of Mathematics
Polytechnic of Porto
Porto, Portugal

Clara Mihaela Ionescu
Faculty of Engineering and Architecture
Ghent University
Ghent, Belgium

ISSN 2195-9994 ISSN 2196-0003 (electronic)
Nonlinear Systems and Complexity
ISBN 978-3-031-42688-9 ISBN 978-3-031-42689-6 (eBook)
https://doi.org/10.1007/978-3-031-42689-6

This Springer imprint is published by the registered company Springer Nature Switzerland AG
The registered company address is: Gewerbestrasse 11, 6330 Cham, Switzerland

Paper in this product is recyclable.

Contents

Chapter 1
Numerical Simulations for Viscous Reactive Micropolar Real Gas Flow

Angela Bašić-Šiško and Ivan Dražić

1.1 Micropolar Fluids

One of the best-known mathematical models is the classical flow model represented by the Navier-Stokes equations, which is used in fluid mechanics to study the static and dynamic properties of fluids. Due to the neglect of certain material properties, this model, although robust, is not sufficient to describe some specific phenomena, e.g., the effect of erythrocytes on blood flow in capillaries or, more generally, the effect of local motions of solid particles dispersed in a fluid on the flow with a small characteristic dimension [34, 36].

The shortcomings of the classical model in such situations can be reduced by adding new variables describing microeffects in the continuum. The model of the micropolar fluid presented by Ahmed Cemal Eringen in the 1964 paper [30] stands out. This non-Newtonian fluid represents a viscous medium in which there are randomly scattered particles whose deformations are neglected [36].

The idea of considering microstructure in material modeling existed before Eringen introduced the micropolar continuum. The first microcontinuum model was described by the Cosserat brothers [17, 62]. It contains nine additional degrees of freedom compared to the classical fluid model, which makes the mathematical analysis very complicated and almost impossible. For this very reason, such models have been neglected in mathematics and applications for a long time. The Eringen model takes into account only the rotation of the particles, which leads to the addition of an extra field with three degrees of freedom representing the microrotation rate. This reduction of the newly introduced parameters made

A. Bašić-Šiško · I. Dražić (✉)
Faculty of Engineering, University of Rijeka, Rijeka, Croatia
e-mail: abasicsisko@riteh.hr; idrazic@riteh.hr

C. M. A. Pinto, C. M. Ionescu (eds.), *Computational and Mathematical Models in Biology*, Nonlinear Systems and Complexity 38,
https://doi.org/10.1007/978-3-031-42689-6_1

this model the most significant and widely used generalization of the classical model for describing fluid flows with a microstructure, since it allows all the usual mathematical analyses with a reasonable additional computational effort. The effort invested in a more complex analysis is justified since this model has been shown to describe much better the behavior of fluids at the microscale [54, 67].

Since microinertia is present in the micropolar continuum, in contrast to the classical fluid, another tensor must be introduced in addition to the normal stress tensor, the so-called contact pair tensor, for the modeling of stresses due to the interactions between the particles. Here, the normal stress tensor is not symmetric, which, together with the intrinsic angular momentum originating from the contact stress, leads to the fact that the conservation law for the total angular momentum is not trivial. In view of this, the micropolar model has an additional equation compared to the classical model. The laws of conservation of mass and momentum have the same tensor form as in the classical model, while the law of conservation of energy changes due to the kinetic energy caused by microrotation, i.e., the work done by the residual stress and torsion [36]. In addition to the mathematical theory of fluid mechanics, the model is also of great interest to other disciplines, to which the accelerated development of nanotechnology in recent years has contributed in particular. The model has been successfully applied in mechanical engineering for lubricant research [8, 11], in biomedical engineering for modeling biological fluids [33, 39], and in modeling geophysical processes [1].

1.1.1 Generalization of the Model in the Thermodynamic Sense

In addition to the introduction of microstructure, the classical models of fluid mechanics can be generalized in a thermodynamic sense by reexamining the concepts of the equation of state describing the interdependence of temperature, volume, and pressure. In modeling compressible fluids, the idealized model of Clapeyron is usually used, which is characterized by an equation of state in which pressure is proportional to the product of temperature and mass density. This model is considered an acceptable approximation for describing the behavior of gasses, especially at low density [14, 16].

However, in many cases, such as near critical points or at extreme values of system variables (low temperatures, high density, or high pressure), the behavior of gasses deviates significantly from that predicted by the idealized model. Therefore, efforts are being made to find new models that better describe gasses in a broader range of system states, but many proposals have proven impractical due to the complexity of the expressions and the large number of parameters [13]. One of the better known generalizations is the class of cubic equations of state, which have been shown to model real gasses more faithfully [13, 55].

Feireisl and Novotný proposed a generalized equation of state in which the pressure is proportional to the product of the absolute temperature and a power of the mass density, and which, according to the authors, should better describe the behavior of real gasses once the temperature is not close to absolute zero [31]. In the case of a classical fluid, this model, although more complex, turned out to allow analysis by adapting the existing and introducing some new techniques [20, 35, 59, 63].

As mentioned above, the deviation from the idealized model becomes particularly important under extreme conditions, so it is useful to apply it in the study of the model of flow and thermal explosion of a reactive fluid describing the behavior of a gas mixture during a chemical reaction [9, 35, 56]. The associated system results from the extension of the classical system by introducing a new variable representing the mass fraction of the fuel and new equations describing the dynamics of the chemical reaction, and it is necessary to adapt the law of conservation of energy to take into account also the temperature change during the combustion [9, 56, 60].

1.1.2 Literature Review

Originally, the analysis of micropolar fluid focused mainly on the stationary and incompressible case [36], and it was not until the end of the twentieth century that Nermina Mujaković began the mathematical analysis of nonstationary compressible micropolar fluid. In her first paper [41], she showed that the one-dimensional flow model of an ideal micropolar gas has a generalized solution locally in time and that it is unique. Later, she also proved its global existence [40] and some additional properties such as regularity and stabilization [42, 43]. The numerical method to solve the problem is presented in [44, 47], and in [48, 51–53], the Cauchy problem is analyzed. Numerical analysis based on the Faedo-Galerkin method and the finite difference method is compared in [18, 19].

The three-dimensional model was analyzed under the assumptions of spherical [23–25, 27, 45, 46, 49] and cylindrical symmetry [21, 26, 28, 32, 50]. The shear flow problem was analyzed in [29, 61].

There are also researches on some special classes of micropolar fluids, e.g., magneto-micropolar fluids [68], and generalizations such as fluids with nonconstant (micro)viscosity coefficients [15].

The generalized equation of state of the gas was applied to the model of classical [20, 38, 58, 59, 63–66] and reactive gas [35, 57, 59]. In these studies, the global existence and asymptotic behavior of the solution were studied.

The one-dimensional flow of a compressible viscous micropolar real gas, with the additional assumption that the gas is thermodynamically polytropic, was analyzed in [3, 4, 7, 22]. In these works, the numerical solution of the model, the local and global existence of the generalized solution, and the uniqueness of the generalized solution were analyzed. The same model for the reactive fluid was considered in

[5, 6], where the local existence of the solution and the numerical solution based on the Faedo-Galerkin method were given.

1.2 Mathematical Model for Micropolar Reactive Fluid

The micropolar fluid model describes the behavior of a fluid with microstructure in such a way that, in addition to the classical hydrodynamic and thermodynamic variables of mass density, velocity, and absolute temperature, the microrotation of particles is also taken into account. A comprehensive overview of the mathematical model is given in [36], and the general model as well as most of the notations used in this work were taken from there.

Flow of micropolar fluid on the domain $Q_T = \Omega\times]0,T[$, where $T > 0$ and $\Omega \subset \mathbf{R}^3$, is described by conservation laws, as follows:

- by the law of conservation of mass

$$\dot{\rho} = -\rho\nabla\cdot\mathbf{v}, \tag{1.1}$$

- by the law of conservation of momentum

$$\rho\dot{\mathbf{v}} = \nabla\cdot\mathbf{T} + \rho\mathbf{f}, \tag{1.2}$$

- by the law of conservation of angular momentum

$$\rho\dot{\mathbf{l}} = \nabla\cdot\mathbf{C} + \mathbf{T}_x + \rho\mathbf{g}, \tag{1.3}$$

- by the law of energy conservation

$$\dot{E} = -\nabla\cdot\mathbf{q} + \mathbf{T} : \nabla\mathbf{v} + \mathbf{C} : \nabla\boldsymbol{\omega} - \mathbf{T}_x\cdot\boldsymbol{\omega}, \tag{1.4}$$

where ρ and E are scalar fields that represent mass density and internal energy, and $\mathbf{v}$ and $\boldsymbol{\omega}$ are vector fields of velocity and microrotational velocity, respectively.

$\mathbf{T} = [\mathbf{T}_{ij}]$ and $\mathbf{C} = [\mathbf{C}_{ij}]$ are the symbols for the stress tensor and the tensor of the contact pair. The function $\mathbf{f}$ is the volume force density, and $\mathbf{g}$ is the volume pair density. By $\mathbf{l}$, we denote the specific spin, and by $\mathbf{q}$ the heat flux.

The dot differential operator on the left side of the Eqs. (1.1)–(1.4) denotes the material derivative and acts on the vector field $\mathbf{b} = \mathbf{b}(\mathbf{x},t) = [b_i(\mathbf{x},t)]_i$ as

$$\dot{\mathbf{b}} = \partial_t\mathbf{b} + (\nabla\mathbf{b})\frac{d\mathbf{x}}{dt}, \tag{1.5}$$

where $\mathbf{x} = \mathbf{x}(t)$ is the position of the particle at time t, $\partial_t\mathbf{b} = [\partial_t b_i]_i$, $\nabla\mathbf{b} = [\partial_j b_i]_{ij}$ and $\partial_j = \partial_{x_j}$, and $\frac{d\mathbf{x}}{dt}$ is the flow velocity. In the case of a scalar field $b = b(\mathbf{x},t)$ the material derivative is given by

$$\dot{b} = \partial_t b + (\nabla) \cdot \frac{d\mathbf{x}}{dt}, \tag{1.6}$$

where $\nabla b = [\partial_i b]_i$.

The vector $\mathbf{T}_x$ represents an axial vector and its components are given by

$$\mathbf{T}_x = [\varepsilon_{ijk}\mathbf{T}_{jk}]_i, \tag{1.7}$$

where ε_{ijk} is the Levi-Civita symbol. Here and in the following, we use Einstein notation for the summation. By :, we denote the scalar product of the tensor, in particular we have

$$\mathbf{T} : \nabla \mathbf{v} = \mathbf{T}_{ji}\partial_j v_i. \tag{1.8}$$

The specific spin of a micropolar fluid is given by

$$\mathbf{l} = \mathbf{J}\boldsymbol{\omega}, \tag{1.9}$$

where $\mathbf{J}$ is the microinertial tensor. In this work, we consider the case of a homogeneous isotropic micropolar fluid, which means that the microinertia tensor has the following form

$$\mathbf{J} = j_I \mathbf{I}, \tag{1.10}$$

where j_I $\left[\mathrm{m}^2\right]$ is a positive constant that we call micro-inertia density, and $\mathbf{I}$ is a unit tensor.

We assume that the fluid is polytropic in the thermodynamic sense, i.e., that the internal energy is given by s

$$E = c_v \theta, \tag{1.11}$$

where θ is a scalar field and represents the absolute temperature of the fluid, and the positive constant c_v $\left[\mathrm{m}^2\mathrm{s}^{-2}\mathrm{K}^{-1}\right]$ is the specific heat at constant volume.

In addition, we assume that the fluid conducts heat, i.e., that Fourier's law applies to the heat flux

$$\mathbf{q} = -\kappa \nabla \theta, \tag{1.12}$$

where the positive constant κ $\left[\mathrm{kg\ m\ s}^{-3}\mathrm{K}^{-1}\right]$ is the thermal conductivity coefficient.

The constitutive equations of the micropolar fluid are given by components s

$$\mathbf{T}_{ij} = (-P + \lambda \partial_k v_k)\delta_{ij} + \mu \left(\partial_i v_j + \partial_j v_i\right) + \mu_r \left(\partial_i v_j - \partial_j v_i\right) - 2\mu_r \varepsilon_{mij}\omega_m, \tag{1.13}$$

$$\mathbf{C}_{ij} = c_0 \partial_k \omega_k \delta_{ij} + c_d \left(\partial_i \omega_j + \partial_j \omega_i\right) + c_a \left(\partial_i \omega_j - \partial_j \omega_i\right), \tag{1.14}$$

where $P = P(\rho, \theta)$ is the pressure, δ_{ij} is the Kronecker symbol, λ $\left[\mathrm{kg\,m^{-1}s^{-1}}\right]$ i μ $\left[\mathrm{kg\,m^{-1}s^{-1}}\right]$ viscosity coefficients, and μ_r $\left[\mathrm{kg\,m^{-1}s^{-1}}\right]$, c_0 $\left[\mathrm{kg\,m\,s^{-1}}\right]$, c_d $\left[\mathrm{kg\,m\,s^{-1}}\right]$ i c_a $\left[\mathrm{kg\,m\,s^{-1}}\right]$ microviscosity coefficients for which the Clausius-Duhamel inequalities apply

$$\begin{aligned} \mu \geq 0, \qquad 3\lambda + 2\mu \geq 0, \qquad \mu_r \geq 0, \\ c_d \geq 0, \quad 3c_0 + 2c_d \geq 0, \quad |c_d - c_a| \leq c_d + c_a. \end{aligned} \tag{1.15}$$

Note that the Eqs. (1.1)–(1.4) reduce to conservation laws for the classical fluid model if we set $\boldsymbol{\omega} = 0$ and $\mu_r = 0$. We would like to emphasize that unlike the classical fluid, the stress tensor $\mathbf{T}$ of the micropolar fluid is not symmetric, which allows modeling phenomena beyond the scope of the classical model. It is important to point out that the symmetry of the stress tensor is the reason that in the classical Navier-Stokes equations, the conservation of angular momentum is derived from the conservation of mass and momentum. This is not the case in the micropolar model.

For the sake of simplicity, we also assume in this work that the following applies

$$\mathbf{f} = \mathbf{g} = 0. \tag{1.16}$$

1.2.1 Micropolar Real Gas

The model of compressible micropolar fluid has been considered so far exclusively with the Clapeyron equation of state of an ideal gas, given as follows

$$P(\rho, \theta) = R\rho\theta, \tag{1.17}$$

where R $\left[\mathrm{m^2K^{-1}s^{-2}}\right]$ is the specific gas constant. In this work, we depart from the assumption of ideality and consider the equation of state of a real gas, whose general form is

$$P(\rho, \theta) = f(\rho) + \theta g(\rho), \tag{1.18}$$

where the function f represents the external pressure and g the internal pressure. A generalized equation of state of this form was first introduced in [31] and is intended to describe the behavior of gasses in more detail. In the real gas model we consider, we ignore the external pressure and assume that the internal pressure is proportional to a power of the mass density, i.e.,

$$g(\rho) = R\rho^p, \tag{1.19}$$

where $p \geq 1$ is a dimensionless constant that we call the pressure exponent, and $R\ \left[\mathrm{m}^{3p-1}\ \mathrm{K}^{-1}\ \mathrm{s}^{-2}\ \mathrm{kg}^{1-p}\right]$ a positive constant. In other words, we consider the following equation of state

$$P = R\rho^p\theta. \tag{1.20}$$

Note that for $p = 1$ (1.20), it reduces to the equation of state of an ideal gas. We would like to emphasize that R is no longer a specific gas constant once p is not equal to 1, and that its dimension (unit of measure) depends on p.

Taking into account all the previously mentioned assumptions, we obtain

$$\begin{aligned}
\mathbf{T} =& (-R\rho^p\theta + \lambda\nabla\cdot v)\mathbf{I} + \mu\left((\nabla\mathbf{v})^T + \nabla\mathbf{v}\right) + \mu_r\left((\nabla\mathbf{v})^T - \nabla\mathbf{v}\right) \\
&- 2\mu_r\begin{bmatrix} 0 & \omega_3 & -\omega_2 \\ -\omega_3 & 0 & \omega_1 \\ \omega_2 & -\omega_1 & 0 \end{bmatrix}, && (1.21) \\
\nabla\cdot\mathbf{T} =& - R\nabla(\rho^p\theta) + (\lambda+\mu-\mu_r)\nabla(\nabla\cdot\mathbf{v}) + (\mu+\mu_r)\Delta\mathbf{v} + 2\mu_r\nabla\times\boldsymbol{\omega}, && (1.22) \\
\mathbf{T}:\nabla\mathbf{v} =& - R\rho^p\theta\nabla\cdot\mathbf{v} + \lambda(\nabla\cdot\mathbf{v})^2 + \mu((\nabla\mathbf{v})^T + \nabla\mathbf{v}):\nabla\mathbf{v} + \mu_r(\nabla\times\mathbf{v})^2 \\
&- 2\mu_r\boldsymbol{\omega}\cdot(\nabla\times\mathbf{v}), && (1.23) \\
\mathbf{T}_x =& 2\mu_r\nabla\times\mathbf{v} - 4\mu_r\boldsymbol{\omega}, && (1.24) \\
\mathbf{C} =& (c_0\nabla\cdot\boldsymbol{\omega})\mathbf{I} + c_d\left((\nabla\boldsymbol{\omega})^T + \nabla\boldsymbol{\omega}\right) + c_a\left((\nabla\boldsymbol{\omega})^T - \nabla\boldsymbol{\omega}\right), && (1.25) \\
\nabla\cdot\mathbf{C} =& (c_0 + c_d - c_a)\nabla(\nabla\cdot\boldsymbol{\omega}) + (c_d + c_a)\Delta\boldsymbol{\omega}, && (1.26) \\
\mathbf{C}:\nabla\boldsymbol{\omega} =& c_0(\nabla\cdot\boldsymbol{\omega})^2 + c_d((\nabla\boldsymbol{\omega})^T + \nabla\boldsymbol{\omega}):\nabla\boldsymbol{\omega} + c_a((\nabla\boldsymbol{\omega})^T - \nabla\boldsymbol{\omega}):\nabla\boldsymbol{\omega}, && (1.27) \\
\nabla\cdot\mathbf{q} =& - \kappa\Delta\theta, && (1.28)
\end{aligned}$$

therefore, the general three-dimensional model of a real micropolar gas reads:

$$\begin{aligned}
\partial_t\rho =& - (\nabla\rho)\cdot\mathbf{v} - \rho\nabla\cdot\mathbf{v}, && (1.29) \\
\rho\partial_t\mathbf{v} =& - \rho(\nabla\mathbf{v})\cdot\mathbf{v} - R\nabla(\rho^p\theta) + (\lambda+\mu-\mu_r)\nabla(\nabla\cdot\mathbf{v}) \\
&+ (\mu+\mu_r)\Delta\mathbf{v} + 2\mu_r\nabla\times\boldsymbol{\omega}, && (1.30) \\
j_I\rho\partial_t\boldsymbol{\omega} =& - j_I\rho(\nabla\boldsymbol{\omega})\cdot\mathbf{v} + (c_0 + c_d - c_a)\nabla(\nabla\cdot\boldsymbol{\omega}) \\
&+ (c_d + c_a)\Delta\boldsymbol{\omega} + 2\mu_r(\nabla\times\mathbf{v} - 2\boldsymbol{\omega}) && (1.31) \\
c_v\rho\partial_t\theta =& - c_v\rho(\nabla\theta)\cdot\mathbf{v} + \kappa\Delta\theta - R\rho^p\theta\nabla\cdot\mathbf{v} + \lambda(\nabla\cdot\mathbf{v})\cdot(\nabla\cdot\mathbf{v})
\end{aligned}$$

$$+\frac{1}{2}\mu(\nabla\mathbf{v}+\nabla\mathbf{v}^T):(\nabla\mathbf{v}+\nabla\mathbf{v}^T)$$
$$+4\mu_r\left(\frac{1}{2}\nabla\times\mathbf{v}-\boldsymbol{\omega}\right)\cdot\left(\frac{1}{2}\nabla\times\mathbf{v}-\boldsymbol{\omega}\right)+c_0(\nabla\cdot\boldsymbol{\omega})\cdot(\nabla\cdot\boldsymbol{\omega})$$
$$+(c_d-c_a)\nabla\boldsymbol{\omega}:\nabla\boldsymbol{\omega}+(c_d+c_a)(\nabla\boldsymbol{\omega})^T:\nabla\boldsymbol{\omega}. \tag{1.32}$$

We add suitable initial conditions to the model

$$\rho(\mathbf{x},0)=\rho_0(\mathbf{x}),\quad \mathbf{v}(\mathbf{x},0)=\mathbf{v}_0(\mathbf{x}),\quad \boldsymbol{\omega}(\mathbf{x},0)=\boldsymbol{\omega}_0(\mathbf{x}),\quad \theta(\mathbf{x},0)=\theta_0(\mathbf{x}), \tag{1.33}$$

as well as homogeneous boundary conditions

$$\mathbf{v}|_{\partial\Omega}=0,\quad \boldsymbol{\omega}|_{\partial\Omega}=0,\quad \nabla\theta|_{\partial\Omega}=0. \tag{1.34}$$

Under the chosen boundary conditions for $\mathbf{v}$ and θ, the initial boundary problem (1.29)–(1.34) models the flow of a micropolar real gas through a pipe with solid and thermally isolated walls. On the other hand, there is still some disagreement about the correct choice of boundary condition for $\boldsymbol{\omega}$. The homogeneous Dirichlet boundary condition used here is the most commonly used in the literature [10, 36, 37]. An example of a different boundary condition for $\boldsymbol{\omega}$ can be found in [12].

For the model to be physically meaningful, the initial mass density and absolute temperature should be nonnegative, i.e.,

$$\rho_0(\mathbf{x})\geq 0,\quad \theta_0(\mathbf{x})\geq 0,\quad \mathbf{x}\in\Omega. \tag{1.35}$$

Thus, in the most general case, vacuum and absolute zero are allowed, i.e., $\rho_0(\mathbf{x})=0$, for some $\mathbf{x}\in\Omega$ and $\theta_0(\mathbf{x})=0$, for some $\mathbf{x}\in\Omega$. In this paper, however, it is additionally assumed that the pipe is completely filled with liquid at the initial time, i.e., there is no vacuum and the absolute temperature is strictly positive.

1.2.2 One Dimensional Model

In this work, we study the one-dimensional flow of a micropolar real fluid. We obtain the corresponding initial boundary problem from the general model (1.29)–(1.34) by introducing additional assumptions

$$\rho(\mathbf{x},t)=\rho(x,t), \tag{1.36}$$
$$\mathbf{v}(\mathbf{x},t)=(v(x,t),0,0), \tag{1.37}$$
$$\boldsymbol{\omega}(\mathbf{x},t)=(\omega(x,t),\omega_2(x,t),\omega_3(x,t)), \tag{1.38}$$
$$\theta(\mathbf{x},t)=\theta(x,t), \tag{1.39}$$

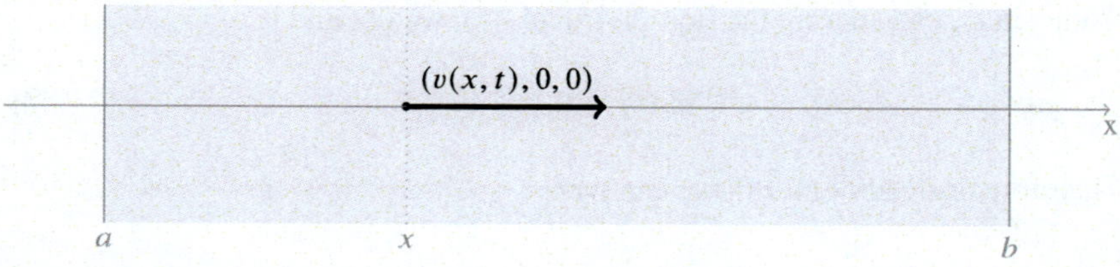

Fig. 1.1 One-dimensional flow model: geometry and velocity direction

for $(x,t) \in]a,b[\times]0,T[$, where $a < b$.

Figure 1.1 shows the geometry of the one-dimensional flow and the direction of the velocity.

Substituting (1.36)–(1.39) into (1.29)–(1.32) and dividing by the components of the vector fields, we obtain

$$
\begin{aligned}
\partial_t \rho =& - v\partial_x \rho - \rho \partial_x v, && (1.40)\\
\rho\partial_t v =& - \rho v \partial_x v - R\partial_x(\rho^p\theta) + (\lambda + 2\mu)\partial_{xx} v, && (1.41)\\
0 =& - 2\mu_r \partial_x \omega_3, && (1.42)\\
0 =& 2\mu_r \partial_x \omega_2, && (1.43)\\
j_I \rho \partial_t \omega =& - j_I \rho v \partial_x \omega + (c_0 + 2c_d)\partial_{xx}\omega - 4\mu_r \omega, && (1.44)\\
j_I \rho \partial_t \omega_2 =& - j_I \rho v \partial_x \omega_2 + (c_d + c_a)\partial_{xx}\omega_2 - 4\mu_r \omega_2, && (1.45)\\
j_I \rho \partial_t \omega_3 =& - j_I \rho v \partial_x \omega_3 + (c_d + c_a)\partial_{xx}\omega_3 - 4\mu_r \omega_3, && (1.46)\\
c_v \rho \partial_t \theta =& - c_v \rho v \partial_x \theta + \kappa \partial_{xx}\theta - R\rho^p\theta\partial_x v + (\lambda + 2\mu)(\partial_x v)^2 \\
& + (c_0 + 2c_d)(\partial_x \omega)^2 + (c_d + c_a)\big((\partial_x \omega_2)^2 + (\partial_x \omega_3)^2\big) && (1.47)\\
& + 4\mu_r(\omega^2 + \omega_2^2 + \omega_3^2).
\end{aligned}
$$

We introduce an additional assumption on the microviscosity coefficient μ_r, which is generally nonnegative (see (1.15)). From now on, we will assume that it is

$$\mu_r > 0. \tag{1.48}$$

Model-wise, this assumption is justified since in the case of $\mu_r = 0$ the stress tensor (1.13) becomes symmetric and independent of the microrotation, and our goal in this work is to investigate the full effect of micropolarity.

From (1.42)–(1.43) and (1.48), we get

$$\omega_2(x,t) = \omega_2(t), \quad \omega_3(x,t) = \omega_3(t), \tag{1.49}$$

from where, considering the Eqs. (1.45)–(1.46), we obtain

$$\omega_i(t) = 0, \quad i = 2, 3. \tag{1.50}$$

Therefore, instead of (1.38), we can write

$$\boldsymbol{\omega}(\mathbf{x}, t) = (\omega(x, t), 0, 0). \tag{1.51}$$

Substituting (1.50) into (1.40)–(1.47), we obtain a system of differential equations describing the one-dimensional flow of a real micropolar fluid in Euler coordinates

$$\partial_t \rho = - v \partial_x \rho - \rho \partial_x v, \tag{1.52}$$

$$\rho \partial_t v = - \rho v \partial_x v - R \partial_x (\rho^p \theta) + (\lambda + 2\mu) \partial_{xx} v, \tag{1.53}$$

$$j_I \rho \partial_t \omega = - j_I \rho v \partial_x \omega + (c_0 + 2c_d) \partial_{xx} \omega - 4\mu_r \omega, \tag{1.54}$$

$$\begin{aligned} c_v \rho \partial_t \theta = & - c_v \rho v \partial_x \theta + \kappa \partial_{xx} \theta - R \rho^p \theta \partial_x v + (\lambda + 2\mu)(\partial_x v)^2 \\ & + (c_0 + 2c_d)(\partial_x \omega)^2 + 4\mu_r \omega^2, \end{aligned} \tag{1.55}$$

for $(x, t) \in]a, b[\times]0, T[$. The corresponding initial conditions become

$$\rho(x, 0) = \rho_0(x), \quad v(x, 0) = v_0(x), \quad \omega(x, 0) = \omega_0(x), \quad \theta(x, 0) = \theta_0(x), \tag{1.56}$$

for $x \in]a, b[$, and homogeneous boundary conditions are

$$v(a, t) = v(b, t) = 0, \quad \omega(a, t) = \omega(b, t) = 0, \quad \partial_x \theta(a, t) = \partial_x \theta(b, t) = 0, \tag{1.57}$$

for $t \in]0, T[$.

The resulting system consists of four quasi-linear partial equations and is of the mixed parabolic-hyperbolic type. To simplify the analysis of the model, we translate the equations from the Eulerian to the Lagrangian description. Namely, in this way, we achieve the parabolicity of the system, which allows the use of many already known approaches for the analysis of such systems.

We observe a fluid particle with an initial position at point ξ moving through point x at time $t > 0$. We call the point ξ the Lagrangian or material coordinate and x the Euler coordinate or position of the observed particle [2, 36]. It holds

$$\frac{d}{dt} x(\xi, t) = v(x(\xi, t), t) = v(x, t), \qquad x(\xi, 0) = \xi, \tag{1.58}$$

where ξ is a parameter, that is,

$$x(\xi, t) = \xi + \int_0^t \tilde{v}(\xi, \tau) d\tau \tag{1.59}$$

where

$$\tilde{v}(\xi, t) = v(x(\xi, t), t). \tag{1.60}$$

In the following, for any differentiable function $\phi = \phi(x, t)$, we denote by $\tilde{\phi}(\xi, t) = \phi(x(\xi, t), t)$ the corresponding function in the Lagrangian description. To transfer the problem (1.52)–(1.57) into a Lagrangian description, we define an invertible mapping

$$\mathcal{M}_t = x(\cdot, t) : \xi \mapsto x, \tag{1.61}$$

whose Jacobian is

$$J(\xi, t) = \frac{dx}{d\xi}. \tag{1.62}$$

The partial derivative of the Jacobian (1.62) in time, after substitution (1.58), can be written as

$$\partial_t J = J \partial_x v. \tag{1.63}$$

For mass density $\tilde{\rho}(\xi, t) = \rho(x(\xi, t), t)$, after we apply (1.58) on (1.52), we get

$$\partial_t \tilde{\rho} = \partial_x \rho \cdot \partial_t x + \partial_t \rho = v \partial_x \rho + \partial_t \rho = -\rho \partial_x v = -\tilde{\rho} \partial_x v. \tag{1.64}$$

From (1.63) and (1.64) follows

$$\frac{1}{\tilde{\rho}} \partial_t \tilde{\rho} + \frac{1}{J} \partial_t J = 0, \tag{1.65}$$

hence

$$\partial_t \ln(\tilde{\rho} J) = 0. \tag{1.66}$$

Integrating (1.66) over $[0, t]$, since for $t = 0$ we have $\xi = x$ and $J = 1$, we obtain

$$\tilde{\rho} J = \rho_0. \tag{1.67}$$

Applying the relation (1.62) to the differentiable function ϕ, we obtain

$$\partial_x \phi = \frac{\tilde{\rho}}{\rho_0} \partial_\xi \tilde{\phi}, \tag{1.68}$$

$$\partial_{xx} \phi = \frac{\tilde{\rho}}{\rho_0} \partial_\xi \left(\frac{\tilde{\rho}}{\rho_0} \partial_\xi \tilde{\phi} \right), \tag{1.69}$$

$$\partial_t \tilde{\phi} = \partial_t \phi + v\,\partial_x \phi. \tag{1.70}$$

If we put the above relations into (1.52)–(1.57), we get the same system in the Lagrangian description

$$\partial_t \tilde{\rho} = -\frac{\tilde{\rho}^2}{\rho_0}\partial_\xi \tilde{v}, \tag{1.71}$$

$$\partial_t \tilde{v} = -\frac{R}{\rho_0}\partial_\xi\left(\tilde{\rho}^p \tilde{\theta}\right) + \frac{\lambda + 2\mu}{\rho_0}\partial_\xi\left(\frac{\tilde{\rho}}{\rho_0}\partial_\xi \tilde{v}\right), \tag{1.72}$$

$$j_I \partial_t \tilde{\omega} = \frac{c_0 + 2c_d}{\rho_0}\partial_\xi\left(\frac{\tilde{\rho}}{\rho_0}\partial_\xi \tilde{\omega}\right) - 4\mu_r\frac{\tilde{\omega}}{\tilde{\rho}}, \tag{1.73}$$

$$\begin{aligned} c_v \partial_t \tilde{\theta} = &\frac{\kappa}{\rho_0}\partial_\xi\left(\frac{\tilde{\rho}}{\rho_0}\partial_\xi \tilde{\theta}\right) - \frac{R}{\rho_0}\tilde{\rho}^p\tilde{\theta}\partial_\xi \tilde{v} + \frac{\lambda + 2\mu}{\rho_0^2}\tilde{\rho}\left(\partial_\xi \tilde{v}\right)^2 \\ &+ \frac{c_0 + 2c_d}{\rho_0^2}\tilde{\rho}\left(\partial_\xi \tilde{\omega}\right)^2 + 4\mu_r\frac{\tilde{\omega}^2}{\tilde{\rho}}, \end{aligned} \tag{1.74}$$

for $(\xi, t) \in]a, b[\times]0, T[$,

$$\tilde{\rho}(\xi, 0) = \rho_0(\xi), \quad \tilde{v}(\xi, 0) = v_0(\xi), \quad \tilde{\omega}(\xi, 0) = \omega_0(\xi), \quad \tilde{\theta}(\xi, 0) = \theta_0(\xi), \tag{1.75}$$

for $\xi \in [a, b]$,

$$\tilde{v}(a, t) = \tilde{v}(b, t) = 0, \quad \tilde{\omega}(a, t) = \tilde{\omega}(b, t) = 0, \quad \partial_x \tilde{\theta}(a, t) = \partial_x \tilde{\theta}(b, t) = 0, \tag{1.76}$$

for $t \in [0, T]$.

The obtained system consists of four quasi-linear partial differential equations and is of parabolic type. Note that the first Eq. (1.71) has a simpler form than the corresponding equation in Euler's description (1.52), in which the spatial derivative of the function ρ appears in addition to time derivative.

We simplify the system further by switching to Lagrangian mass coordinates [2]. We define the function η by

$$\eta(\xi) = \int_a^\xi \rho_0(s)ds. \tag{1.77}$$

Because of (1.67), we have

$$\eta(\xi) = \int_a^{x(\xi,t)} \rho(s, t)ds. \tag{1.78}$$

Since we have assumed that at the initial time the entire domain is filled with fluid, i.e., that we do not allow a vacuum at the initial time, the following applies

$$\inf_{x\in[a,b]} \rho_0(x) > 0. \tag{1.79}$$

From (1.79), it follows that η is invertible, which allows the following definition of a new dimensionless spatial coordinate

$$q = \frac{1}{L}\eta(\xi), \tag{1.80}$$

where

$$L = \int_a^b \rho_0(s)ds > 0 \tag{1.81}$$

is the characteristic length ($L\ [\mathrm{kg\ m^{-2}}]$). We call the points (q,t) the Lagrangian mass coordinates, that take values on the set

$$Q_T =]0,1[\times]0,T[. \tag{1.82}$$

For a differentiable function, ϕ, from (1.77) and (1.80), we get

$$\partial_\xi \tilde{\phi} = \frac{1}{L}\rho_0 \partial_q \hat{\phi}, \tag{1.83}$$

where $\hat{\phi}(q,t) = \tilde{\phi}(\xi,t)$.

We write the initial boundary value problem (1.71)–(1.76) in the new coordinates (q,t), writing (q,t) instead of (x,t) for simplicity and omitting the labels $\hat{\cdot}$ and $\tilde{\cdot}$

$$\partial_t \rho = -\frac{1}{L}\rho^2 \partial_x v, \tag{1.84}$$

$$\partial_t v = -\frac{R}{L}\partial_x(\rho^p \theta) + \frac{\lambda + 2\mu}{L^2}\partial_x(\rho \partial_x v), \tag{1.85}$$

$$j_I \partial_t \omega = \frac{c_0 + 2c_d}{L^2}\partial_x(\rho \partial_x \omega) - 4\mu_r \frac{\omega}{\rho}, \tag{1.86}$$

$$\begin{aligned} c_v \partial_t \theta =& \frac{\kappa}{L^2}\partial_x(\rho \partial_x \theta) - \frac{R}{L}\rho^p \theta\, \partial_x v + \frac{\lambda + 2\mu}{L^2}\rho(\partial_x v)^2 \\ &+ \frac{c_0 + 2c_d}{L^2}\rho(\partial_x \omega)^2 + 4\mu_r \frac{\omega^2}{\rho}, \end{aligned} \tag{1.87}$$

for $(x,t) \in]0,1[\times]0,T[$,

$$\rho(x,0) = \rho_0(x), \quad v(x,0) = v_0(x), \quad \omega(x,0) = \omega_0(x), \quad \theta(x,0) = \theta_0(x), \tag{1.88}$$

for $x \in [0,1]$,

$$v(0,t)=v(1,t)=0,\quad \omega(0,t)=\omega(1,t)=0,\quad \partial_x\theta(0,t)=\partial_x\theta(1,t)=0, \tag{1.89}$$

for $t \in [0, T]$.

This one-dimensional model was derived in [7].

1.2.3 Reactive Gas

One of the main reasons for introducing the generalized equation of state (1.20) is the expectation that the fluid model will more accurately describe the behavior of gasses under some extreme conditions, such as during a chemical reaction. In this section, we apply the micropolar real gas model to the reactive fluid model that describes the fluid during the dynamic combustion process. We consider a simple model in which a gas mixture has a single molecular mass and a single diffusion coefficient. This type of model is described in the context of the classical fluid model in [35, 56, 59, 60], from which the labels, nomenclature, and basic form of the model used in this work are taken.

We observe a one-step irreversible reaction in which the reactants transform into products in the presence of oxidizing agents

$$\text{fuel} + \text{oxidant} \rightarrow \text{products}. \tag{1.90}$$

The dynamics of the chemical reaction is described by the following equation

$$\rho\dot{z} = \sigma\nabla\cdot(\rho\nabla z) - \rho r(\rho,\theta,z), \tag{1.91}$$

where z is the fraction of unburned fuel in the gas mixture and a positive constant σ $\left[\mathrm{m}^2\mathrm{s}^{-1}\right]$ is diffusion coefficient. Function r $\left[\mathrm{s}^{-1}\right]$ describes the intensity of the reaction and is most often stated in the form of the Arrhenius law

$$r(\rho,\theta,z) = \epsilon\rho^{m-1}z^m \exp\frac{\theta-1}{\epsilon\theta}, \tag{1.92}$$

where $m \geq 1$ is an integer equal to the sum of the reaction orders of the reactants, and the positive constant ϵ is the activation energy.

Introducing a new process into the model affects other aspects of the model. In the case of a reactive real micropolar fluid, the expression of the law of conservation of energy changes compared to the model without a chemical reaction. Namely, the law of conservation of energy for a reactive micropolar gas is

$$\rho\dot{E} = -\nabla\cdot\mathbf{q} + \mathbf{T}:\nabla\mathbf{v} + \mathbf{C}:\nabla\boldsymbol{\omega} - \mathbf{T}_x\cdot\boldsymbol{\omega} + \delta\rho r(\rho,\theta,z), \tag{1.93}$$

where the additional term in relation (1.4) is a consequence of the change in internal energy caused by combustion. Here the positive constant δ $\left[\mathrm{m}^2\mathrm{s}^{-2}\right]$ describes the reaction rate.

By respecting assumptions (1.5)–(1.16), (1.20) in Eqs. (1.91) and (1.93), we get

$$
\begin{aligned}
c_v\rho\partial_t\theta = & -c_v\rho(\nabla\theta)\cdot\mathbf{v}+\kappa\Delta\theta-R\rho^p\theta\nabla\cdot\mathbf{v}+\lambda(\nabla\cdot\mathbf{v})^2+c_0(\nabla\cdot\boldsymbol{\omega})\cdot(\nabla\cdot\boldsymbol{\omega})\\
&+\frac{1}{2}\mu(\nabla\mathbf{v}+\nabla\mathbf{v}^T):(\nabla\mathbf{v}+\nabla\mathbf{v}^T)\\
&+4\mu_r\left(\frac{1}{2}\nabla\times\mathbf{v}-\boldsymbol{\omega}\right)\cdot\left(\frac{1}{2}\nabla\times\mathbf{v}-\boldsymbol{\omega}\right)\\
&+(c_d-c_a)\nabla\boldsymbol{\omega}:\nabla\boldsymbol{\omega}+(c_d+c_a)(\nabla\boldsymbol{\omega})^T:\nabla\boldsymbol{\omega}+\delta\rho f(\rho,\theta,z)
\end{aligned}
\tag{1.94}
$$

$$
\rho\partial_t z=-\rho(\nabla z)\cdot\mathbf{v}+\sigma\nabla\cdot(\rho\nabla z)-\rho f(\rho,\theta,z), \tag{1.95}
$$

$$
\mathbf{v}|_{\partial\Omega}=0,\quad \boldsymbol{\omega}|_{\partial\Omega}=0,\quad \nabla\theta|_{\partial\Omega}=0,\quad \nabla z|_{\partial\Omega}=0. \tag{1.96}
$$

For the function z, we use the initial condition $z(\mathbf{x},0) = z_0(\mathbf{x})$, as well as homogeneous Neumann boundary conditions, according to [35, 56, 60].

For the model to be consistent with its physical interpretation, the proportion of fuel in the mixture should meet the following conditions

$$
0\le z_0(\mathbf{x})\le 1,\quad x\in\Omega. \tag{1.97}
$$

Using the procedure described earlier, we write these equations in one-dimensional form and translate them into Lagrangian mass coordinates. Finally, we obtain the system

$$
\partial_t\rho=-\frac{1}{L}\rho^2\partial_x v, \tag{1.98}
$$

$$
\partial_t v=-\frac{R}{L}\partial_x(\rho^p\theta)+\frac{\lambda+2\mu}{L^2}\partial_x(\rho\partial_x v), \tag{1.99}
$$

$$
j_I\partial_t\omega=\frac{c_0+2c_d}{L^2}\partial_x(\rho\partial_x\omega)-4\mu_r\frac{\omega}{\rho}, \tag{1.100}
$$

$$
\begin{aligned}
c_v\partial_t\theta = & \frac{\kappa}{L^2}\partial_x(\rho\partial_x\theta)-\frac{R}{L}\rho^p\theta\,\partial_x v+\frac{\lambda+2\mu}{L^2}\rho(\partial_x v)^2+\frac{c_0+2c_d}{L^2}\rho(\partial_x\omega)^2\\
&+4\mu_r\frac{\omega^2}{\rho}+\delta r(\rho,\theta,z),
\end{aligned}
\tag{1.101}
$$

$$
\partial_t z=\frac{\sigma}{L^2}\partial_x\left(\rho^2\partial_x z\right)-r(\rho,\theta,z), \tag{1.102}
$$

for $(x,t)\in]0,1[\times]0,T[$,

$$
\begin{aligned}
&\rho(x,0)=\rho_0(x),\ v(x,0)=v_0(x),\ \omega(x,0)=\omega_0(x),\\
&\theta(x,0)=\theta_0(x),\ z(x,0)=z_0(x),
\end{aligned}
\tag{1.103}
$$

for $x \in [0, 1]$,

$$\begin{aligned} v(0,t) = v(1,t) = 0, \quad \omega(0,t) = \omega(1,t) = 0, \\ \partial_x\theta(0,t) = \partial_x\theta(1,t) = 0, \quad \partial_x z(0,t) = \partial_x z(1,t) = 0, \end{aligned} \tag{1.104}$$

for $t \in [0, T]$.

This one-dimensional model was derived in [6].

1.3 Local Existence of Generalized Solution

Let us now define what we consider by the solution to the problem (1.98)–(1.104) [5].

Definition 1 A generalized solution of the problem (1.98)–(1.104) in $Q_T =]0,1[\times]0,T[$, for some $T > 0$, is a function

$$(x,t) \mapsto (\rho, v, \omega, \theta, z)(x,t), \quad (x,t) \in Q_T$$

such that

$$\begin{aligned} \rho \in L^\infty\big(0,T; H^1(]0,1[)\big) \cap H^1(Q_T), \quad \inf_{Q_T} \rho > 0 \\ v, \omega, \theta, z \in L^\infty\big(0,T; H^1(]0,1[)\big) \cap H^1(Q_T) \cap L^2\big(0,T; H^2(0,1)\big), \end{aligned} \tag{1.105}$$

whereby ρ, v, ω, θ, and z satisy equations (1.98)–(1.104) a.e. in Q_T, initial conditions (1.88) a.e. in $]0,1[$ and boundary conditions (1.89) in the sense of traces.

Directly from Definition 1, using theorems of embedding for function spaces, we can conclude that the functions of the generalized solution satisfy the following inclusions.

Proposition 1 *Let $(\rho, v, \omega, \theta, z)$ be the generalized solution from Definition 1. Then*

$$\begin{aligned} \rho \in C\big([0,T]; L^2(0,1)\big) \cap L^\infty\big(0,T; C([0,1])\big) \cap C\big(\overline{Q}_T\big), \\ v, \omega, \theta, z \in L^2\big(0,T; C^1([0,1])\big) \cap C\big([0,T]; H^1(]0,1[)\big) \cap C\big(\overline{Q}_T\big). \end{aligned}$$

According to [5] to be able to prove the local existence theorem, some additional assumptions should apply. For initial functions ρ_0, θ_0, and z_0, we suppose that

$$\rho_0(x) \geq m_0, \qquad \theta_0(x) \geq m_0, \qquad \text{for } x \in]0,1[, \tag{1.106}$$

where m_0 is some positive constant, and

$$0 \leq z_0(x) \leq 1, \qquad \text{for } x \in]0, 1[. \tag{1.107}$$

Furthermore, we consider the chemical reaction rate function of the following form

$$r(\rho, \theta, z) = z^m \tilde{r}(\rho, \theta, z), \tag{1.108}$$

where the positive integer exponent m is either odd or equal to 2 and $\tilde{r}$ is a function defined on $[0, +\infty[^3$ such that

$$\tilde{r} \geq 0, \tag{1.109}$$

and for which the following conditions hold:

- $\tilde{r}$ is bounded on each set of the form $[a, b] \times [0, +\infty[\times[0, +\infty[$, (1.110)
- $\tilde{r}$ is continuous with respect to ρ, and globally Lipschitz continuous with respect to θ and z, (1.111)
- $\lim_{\rho \to 0+} \tilde{r}(\rho, \theta, z) = 0$ and $\lim_{\theta \to 0+} \tilde{r}(\rho, \theta, z) = 0.$ (1.112)

This form of the function r is an adaptation of the assumptions used in [35]. Here we wish to note that in the most general case exponent, m can assume any positive integer value, but the assumption given above about the allowable range of values that m may assume proves to be crucial in proving the theorem of local existence. Let us notice that the continuity of the generalized solution asserted in Proposition 1 justifies the requirement of positivity of the infimum of the function ρ in Definition 1.

Now we can state the local existence theorem which is proved in [5].

Theorem 1 *Let the functions* $\rho_0, \theta_0, z_0 \in H^1(]0, 1[)$ *satisfy the conditions* (1.106)–(1.107), $v_0, \omega_0 \in H_0^1(]0, 1[)$. *Let* r *satisfy* (1.108)–(1.112) *and let* m *from* (1.108) *be either an odd integer or equal to 2. There exists* $T_0 \in]0, T]$ *such that the problem* (1.98)–(1.104) *has a generalized solution in* $Q_0 := Q_{T_0}$ *such that*

$$\theta > 0 \quad \text{and} \quad 0 \leq z \leq 1 \quad \text{in} \quad \overline{Q}_0. \tag{1.113}$$

1.4 Approximate Solutions

Here we derive a series of approximate solutions

$$\left(\rho^n, v^n, \omega^n, \theta^n, z^n\right), \quad n = 1, 2, 3, \ldots, \tag{1.114}$$

to our problem using Faedo-Galerkin method introduced in [6]. For more details on the procedure of applying this method, see [7], but the key steps from [6] will be presented here.

We use two systems of $L^2(0, 1)$-orthogonal functions,

$$\{\sin(\pi i x)\}_{i=1}^{\infty} \tag{1.115}$$

and

$$\{\cos(\pi i x)\}_{i=0}^{\infty}. \tag{1.116}$$

Let us define

$$v^n(x,t) = \sum_{i=1}^{n} v_i^n(t)\sin(\pi i x), \tag{1.117}$$

$$\omega^n(x,t) = \sum_{i=1}^{n} \omega_i^n(t)\sin(\pi i x), \tag{1.118}$$

$$\theta^n(x,t) = \sum_{i=0}^{n} \theta_i^n(t)\cos(\pi i x), \tag{1.119}$$

$$z^n(x,t) = \sum_{i=0}^{n} z_i^n(t)\cos(\pi i x), \tag{1.120}$$

$n = 1, 2, 3, \ldots$, where v_i^n, ω_i^n, θ_i^n, and z_i^n are unknown functions. Let us notice that v^n, ω^n, θ^n, z^n satisfy the following boundary conditions

$$v^n(0,t) = v^n(1,t) = 0, \tag{1.121}$$

$$\omega^n(0,t) = \omega^n(1,t) = 0, \tag{1.122}$$

$$\partial_x\theta^n(0,t) = \partial_x\theta^n(1,t) = 0, \tag{1.123}$$

$$\partial_x z^n(0,t) = \partial_x z^n(1,t) = 0. \tag{1.124}$$

According to the Faedo-Galerkin method, approximate solutions should satisfy the following initial conditions

$$v^n(x,0) = \sum_{i=1}^{n} v_{0i}\sin(\pi i x), \tag{1.125}$$

$$\omega^n(x,0) = \sum_{i=1}^{n} \omega_{0i} \sin(\pi i x), \tag{1.126}$$

$$\theta^n(x,0) = \sum_{i=0}^{n} \theta_{0i} \cos(\pi i x), \tag{1.127}$$

$$z^n(x,0) = \sum_{i=0}^{n} z_{0i} \cos(\pi i x), \tag{1.128}$$

where v_{0i}, ω_{0i}, θ_{0i}, z_{0i} are Fourier coefficients of functions v_0, ω_0, θ_0 and z_0, that is,

$$v_{0i} = 2\int_0^1 v_0(x)\sin(\pi i x)dx, \tag{1.129}$$

$$\omega_{0i} = 2\int_0^1 \omega_0(x)\sin(\pi i x)dx, \tag{1.130}$$

$$\theta_{0i} = 2\int_0^1 \theta_0(x)\cos(\pi i x)dx, \tag{1.131}$$

$$z_{0i} = 2\int_0^1 z_0(x)\cos(\pi i x)dx, \tag{1.132}$$

for $i = 1, 2, 3, \ldots$, and

$$\theta_{00} = \int_0^1 \theta_0(x)dx, \quad z_{00} = \int_0^1 z_0(x)dx. \tag{1.133}$$

Let ρ^n be a solution to the initial value problem

$$\partial_t \rho^n + \frac{1}{L}(\rho^n)^2 \partial_x v^n = 0, \tag{1.134}$$

$$\rho^n(x,0) = \rho_0(x). \tag{1.135}$$

From (1.134)–(1.135), it is easy to see that for ρ^n, we have

$$\rho^n(x,t) = \frac{L\rho_0(x)}{L + \rho_0(x)\int_0^t \partial_x v^n(x,\tau)\,d\tau}. \tag{1.136}$$

In accordance with Faedo-Galerkin method, approximate solutions v^n, ω^n, θ^n, z^n satisfy the following system of conditions

$$\int_0^1 \left[\partial_t v^n + \frac{R}{L}\partial_x((\rho^n)^p\theta^n) - \frac{\lambda+2\mu}{L^2}\partial_x(\rho^n\partial_x v^n)\right]\sin(\pi i x)dx = 0, \tag{1.137}$$

$$\int_0^1 \left[\partial_t\omega^n - \frac{c_0+2c_d}{L^2 j_I}\partial_x(\rho^n\partial_x\omega^n) + 4\frac{\mu_r}{j_I}\frac{\omega^n}{\rho^n}\right]\sin(\pi i x)dx = 0, \tag{1.138}$$

$$\int_0^1 \left[\partial_t\theta^n - \frac{\kappa}{L^2 c_v}\partial_x(\rho^n\partial_x\theta^n) + \frac{R}{Lc_v}(\rho^n)^p\theta^n\partial_x v^n - \frac{\lambda+2\mu}{L^2 c_v}\rho^n(\partial_x v^n)^2 \right.$$
$$\left. -4\frac{\mu_r}{c_v}\frac{(\omega^n)^2}{\rho^n} - \frac{c_0+2c_d}{L^2 c_v}\rho^n(\partial_x\omega^n)^2 - \frac{\delta}{c_v}\epsilon(\rho^n)^{m-1}(z^n)^m\exp\frac{\theta^n-1}{\epsilon\theta^n}\right] \tag{1.139}$$
$$\cos(\pi j x)dx = 0,$$

$$\int_0^1 \left[\partial_t z^n - \frac{\sigma}{L^2}\partial_x\left((\rho^n)^2\partial_x z^n\right) + \epsilon(\rho^n)^{m-1}(z^n)^m\exp\frac{\theta^n-1}{\epsilon\theta^n}\right]$$
$$\cos(\pi j x)dx = 0, \tag{1.140}$$

for $i = 1, 2, 3, \ldots$, and $j = 0, 1, 2, \ldots$

Let us introduce the auxiliary functions

$$\zeta_i^n(t) = \int_0^t v_i^n(\tau)d\tau, \quad i = 1, \ldots, n, \tag{1.141}$$

which enable us to derive the following representation for ρ^n

$$\rho^n(x,t) = \frac{L\rho_0(x)}{L + \rho_0(x)\sum_{i=1}^n \zeta_i^n(t) i\pi\cos(\pi i x)}. \tag{1.142}$$

Using relations (1.117)–(1.120), from (1.137)–(1.141), we obtain the following system of ordinary differential equations

$$\dot{v}_i^n(t) = 2\int_0^1 \left[-\frac{R}{L}\partial_x((\rho^n)^p\theta^n) + \frac{\lambda+2\mu}{L^2}\partial_x(\rho^n\partial_x v^n)\right]\sin(\pi i x)dx, \tag{1.143}$$

$$\dot{\omega}_j^n(t) = 2\int_0^1 \left[\frac{c_0+2c_d}{L^2 j_I}\partial_x(\rho^n\partial_x\omega^n) - 4\frac{\mu_r}{j_I}\frac{\omega^n}{\rho^n}\right]\sin(\pi j x)dx, \tag{1.144}$$

$$\dot{\theta}_k^n(t) = \lambda_k\int_0^1 \left[\frac{\kappa}{L^2 c_v}\partial_x(\rho^n\partial_x\theta^n) - \frac{R}{Lc_v}(\rho^n)^p\theta^n\partial_x v^n + \right.$$

$$\frac{\lambda+2\mu}{L^2 c_v}\rho^n(\partial_x v^n)^2 + 4\frac{\mu_r}{c_v}\frac{(\omega^n)^2}{\rho^n} + \frac{c_0+2c_d}{L^2 c_v}\rho^n(\partial_x \omega^n)^2 + \tag{1.145}$$

$$\left.\frac{\delta}{c_v}\epsilon(\rho^n)^{m-1}(z^n)^m \exp\frac{\theta^n-1}{\epsilon\theta^n}\right]\cos(\pi k x)dx,$$

$$\dot{z}_l^n(t) = \lambda_l \int_0^1 \left[\frac{\sigma}{L^2}\partial_x\left((\rho^n)^2\partial_x z^n\right) - \epsilon(\rho^n)^{m-1}(z^n)^m \exp\frac{\theta^n-1}{\epsilon\theta^n}\right] \tag{1.146}$$

$$\cos(\pi l x)dx, \tag{1.147}$$

$$\dot{\zeta}_m^n(t) = v_m^n(t), \tag{1.148}$$

for $i, j, m = 1, \dots, n$ and $k, l = 0, 1, \dots, n$, where λ_k is defined as

$$\lambda_k = \begin{cases} 1, & k = 0 \\ 2, & k = 1, \dots, n \end{cases}.$$

From (1.125)–(1.128), the corresponding initial conditions follow

$$v_i^n(0) = v_{0i}, \quad \omega_j^n(0) = \omega_{0j}, \quad \theta_k^n(0) = \theta_{0k}, \quad z_l^n(0) = z_{0l}, \quad \zeta_m^n(0) = 0, \tag{1.149}$$

for $i, j, m = 1, \dots, n$ and $k, l = 0, 1, \dots, n$.

The Cauchy-Picard theorem implies existence and uniqueness of smooth solutions to the problem (1.143)–(1.149), and therefore, the following proposition is proven.

Proposition 2 *For each $n \in \mathbf{N}$, there exist $T_n \in]0, T]$ and unique functions $(v_i^n, \omega_j^n, \theta_k^n, z_l^n, \zeta_m^n) : i, j, m = 1, \dots, n,\ k, l = 0, 1, \dots, n$ on $[0, T_n]$ such that functions defined by* (1.117)–(1.120), (1.142) *have properties*

$$v^n, \omega^n, \theta^n, z^n \in C^1(\overline{Q}_n), \tag{1.150}$$

$$\rho^n \in C(\overline{Q}_n), \tag{1.151}$$

where $Q_n =]0, 1[\times]0, T_n[$, and satisfy equations (1.137)–(1.140), (1.134) *as well as conditions* (1.125)–(1.128), (1.135).

1.5 Numerical Solution of the Model

In this chapter, we describe a numerical method for solving the initial boundary value problem (1.98)–(1.104) based on the Faedo-Galerkin approximations derived in the previous chapter. Namely, by applying the Faedo-Galerkin projection, we obtain the semi-discretized initial problem (1.143)–(1.149) whose solutions are approximate solutions $(\rho^n, v^n, \omega^n, \theta^n, z^n)$ for the observed initial boundary value problem. The details of this method are described in [6], while here we give only the main ideas and results.

The above initial problems consist of a $5n + 2$ system of ordinary differential equations and the corresponding number of initial conditions. It should be noted that in the context of the proof of the local existence theorem [5], it was proved that the sequences of approximate solutions thus defined converge (on a subsequence) to the generalized solution of the corresponding problem. The complete discretization of the problem is achieved by applying one of the methods for numerical solution of systems of ordinary differential equations.

The described approach has already been applied to the numerical solution of the initial boundary problem for the basic model of the one-dimensional flow of a real micropolar fluid in [7], and it has been used previously for the numerical solution of similar problems, for example, in [19, 28, 29, 47]. In addition, the finite difference method [19, 44–46] and upwind scheme [18] were also applied to similar problems.

The implementation of the numerical problem-solving method was done in the Python programming language using the objects and functions available in its standard library as well as the NumPy, SciPy, and Matplotlib libraries.

The first step to obtain a fully discretized system is to approximate the integral on the right-hand side of the system Eqs. (1.143)–(1.149). For this purpose, we use the Gauss-Legender quadrature formula of order 20, which is given by

$$\int_0^1 f(x)dx \approx \sum_{i=1}^{20} w_i f(x_i), \tag{1.152}$$

whereby w_i, $i = 1, \dots, 20$ are weights, and x_i, $i = 1, \dots, 20$ nodes of the quadrature formula. After applying the chosen quadrature formula, we determine the unknown functions by applying the function for solving ordinary differential equations implemented as part of the SciPy Python library. In the following example, a multistep implicit method based on numerical formulas for approximating derivatives by backward divided differences was used.

As an example, we solve numerically the initial boundary problem (1.98)–(1.104) with the following initial conditions

$$\begin{gathered} \rho_0(x) = 1, \;\; v_0(x) = \sin(\pi x), \\ \omega_0(x) = \sin(2\pi x), \;\; \theta_0(x) = 2 + \cos(\pi x), \;\; z_0(x) = 1. \end{gathered} \tag{1.153}$$

These initial conditions were chosen for simplicity. With this selection of initial functions, the Fourier coefficients can in fact be read directly from their notation, and they are:

$$\begin{aligned} &v_{01} = 1, \quad v_{0i} = 0, \quad i = 2, \dots, n, \\ &\omega_{02} = 1, \quad \omega_{0i} = 0, \quad i = 1, 3, \dots, n, \end{aligned}$$

$$\theta_{00} = 2, \quad \theta_{01} = 1, \quad \theta_{0i} = 0, \quad i = 2, \ldots, n,$$
$$z_{00} = 1, \quad z_{0i} = 0, \quad i = 1, \ldots, n. \tag{1.154}$$

Furthermore, we assume that Arrhenius' law is valid, i.e., that the function of the intensity of the chemical reaction r has the form (1.92). For the system parameters, we take

$$L = 1,\ c_d = 1,\ c_0 = 1,\ c_v = 1,\ R = 1,\ j_I = 1,\ \lambda = -2,$$
$$\mu_r = 1,\ \mu = 3,\ \kappa = 0.024, \tag{1.155}$$

for the pressure exponent, we take $p = 4$, and for the parameters related to the reactive properties of the gas

$$\sigma = 1,\ \delta = 1, \epsilon = 0.2,\ m = 2. \tag{1.156}$$

The numerical solution for the described example is shown in the Figs. 1.2, 1.3, 1.4, 1.5, 1.6 and 1.7.

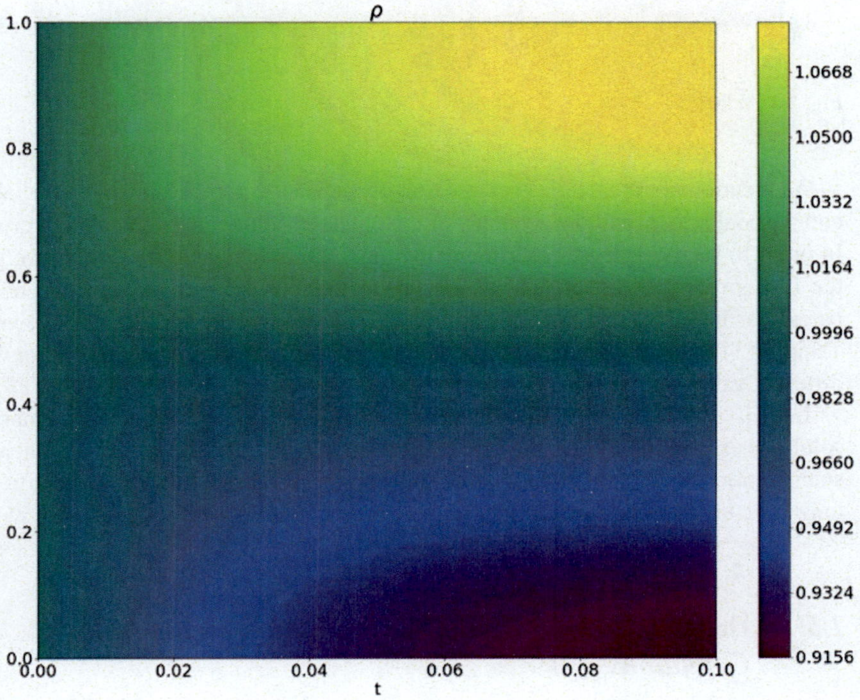

Fig. 1.2 Mass density

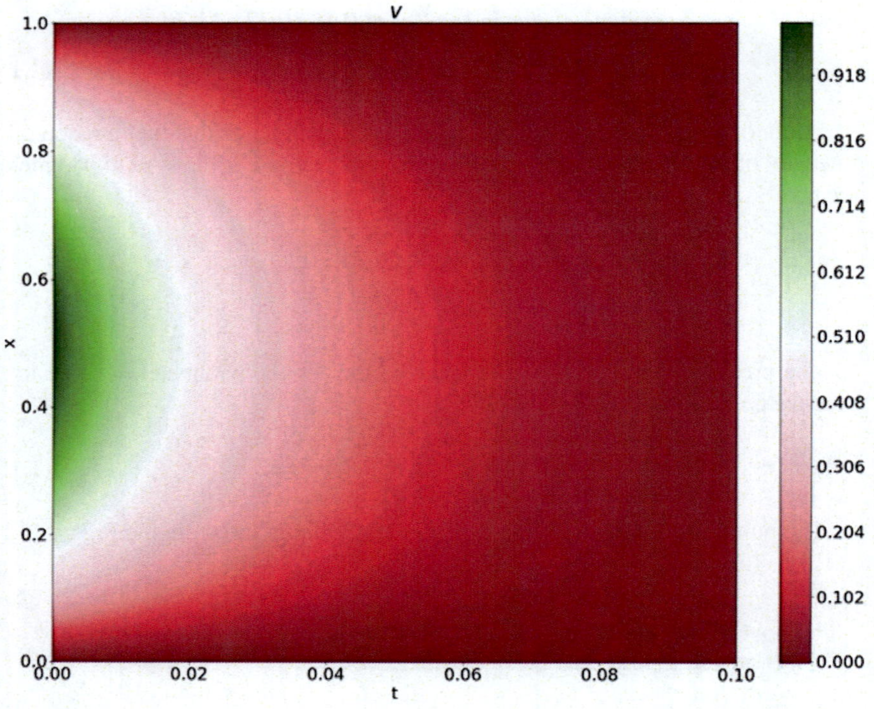

Fig. 1.3 Velocity

As mentioned in [6], the exact solution of the analyzed problem is not available, and we could not verify the accuracy of the obtained solutions by calculating errors. In order to verify, at least experimentally, that the implemented method converges, the L^∞-norm of the differences of the approximations for successive values of the parameter n was observed. The results of the convergence analysis are shown in Table 1.1. The table also shows the CPU times required to compute the solutions for different values of n.

In [6], it was also experimentally confirmed that the calculated numerical solutions exhibit some expected properties, in particular, the property of exponential stabilization toward a stationary solution. The term stabilization here refers to the long-time behavior of the solution, i.e., the limiting behavior when $t \to \infty$.

1.5.1 Dependence of the Numerical Solution on the Initial Conditions

In paper [6], we analyzed only a simple problem in terms of initial conditions with finite Fourier series expansion. Therefore, here we additionally analyze the

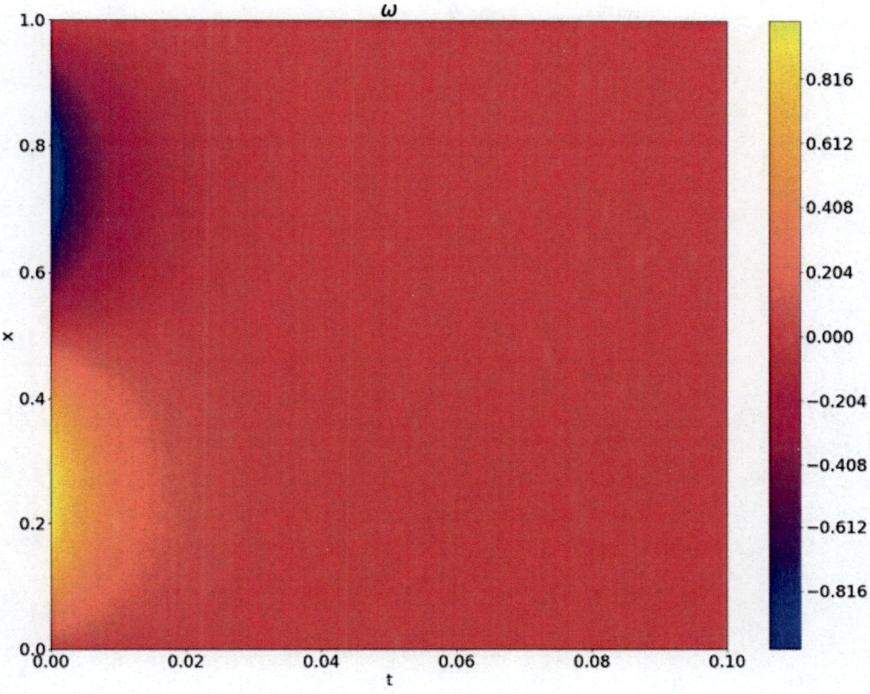

Fig. 1.4 Microrotation

behavior of the described numerical method with respect to two other types of initial conditions.

1. Example 1: Simple initial conditions where the functions have a finite Fourier series expansion:

$$\rho_0(x) = 1, \quad v_0(x) = \sin(\pi x), \quad \omega_0(x) = \sin(2\pi x),$$
$$\theta_0(x) = 2 + \cos(\pi x), \quad z_0(x) = 1. \tag{1.157}$$

2. Example 2: The initial functions are smooth but do not have a finite Fourier expansion, i.e., the corresponding coefficients are approximated numerically:

$$\rho_0(x) = 1,\ v_0(x) = x - x^2,\ \omega_0(x) = x^2 - x^4,$$
$$\theta_0(x) = 2x^3 - 3x^2 + 2,\ z_0(x) = 2x^3 - 3x^2 + 1. \tag{1.158}$$

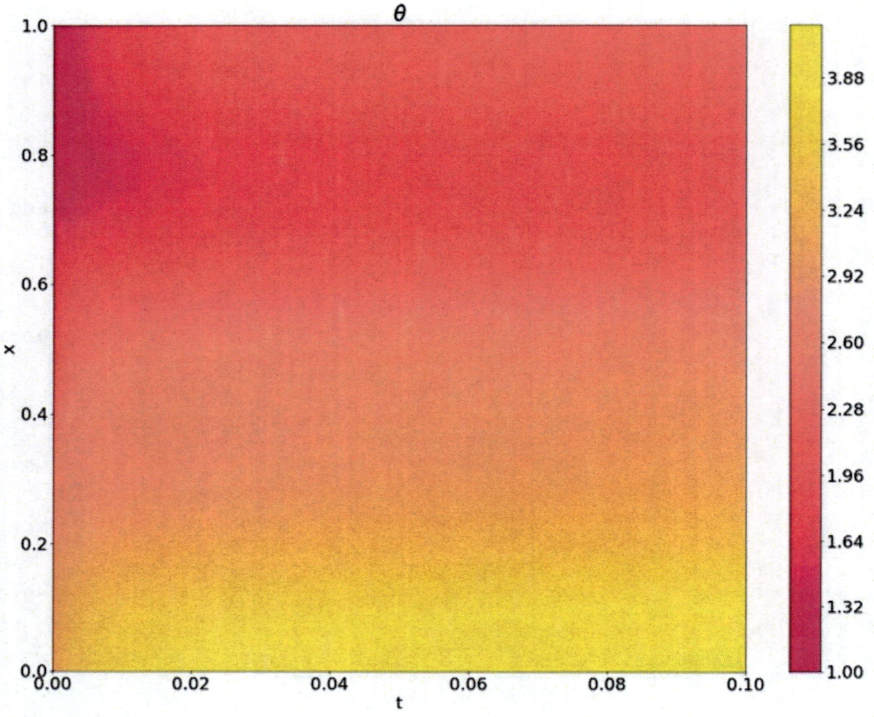

Fig. 1.5 Temperature

3. Example 3: The initial functions are continuous but not smooth:

$$\rho_0(x) = |x^2 - 0.25| + 1,\ v_0(x) = 0,\ \omega_0(x) = 4(x^2 - x^4),\ \theta_0(x) = 0.1,$$

$$z_0(x) = \begin{cases} 1, & x < 0.2 \\ -2x + 1.4, & 0.2 \leq x \leq 0.7 \\ 0, & x > 0.7 \end{cases}. \tag{1.159}$$

Figure 1.8 shows the convergence of the method for the three given examples for the functions ω and z. The test results in Fig. 1.8 (on a logarithmic scale) clearly show that the norms of the differences decrease for higher values of n, but it is also evident that this convergence is slower for more complex initial conditions. CPU times for these calculations are shown in Table 1.2.

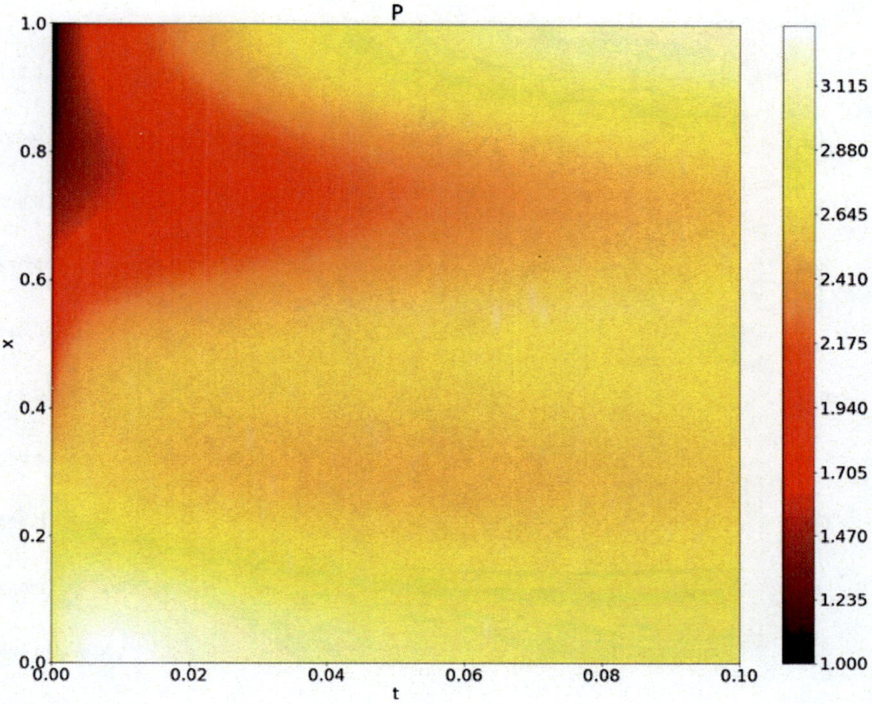

Fig. 1.6 Pressure

1.6 Conclusion

In this work, we have studied a model of a one-dimensional micropolar reactive real gas. The micropolarity is considered by the vector of microrotation velocity, and the gas reactivity is expressed by the mass fraction of unburned fuel and the chemical reaction equation describing the dynamic combustion.

First, we have explained the physical meaning of the micropolar fluid and the historical development of the theory, giving an overview of the existing results in this field. In the next chapter, we have presented the mathematical model of the described fluid and explained the derivation of the associated initial boundary problem. We have also explained the definition of the generalized solution of this problem and established the corresponding existence theorem.

In the following, we explain how to construct approximate solutions of the system using the Faedo-Galerkin method, and the numerical scheme introduced in [6] to solve the problem is explained, giving the main ideas and results.

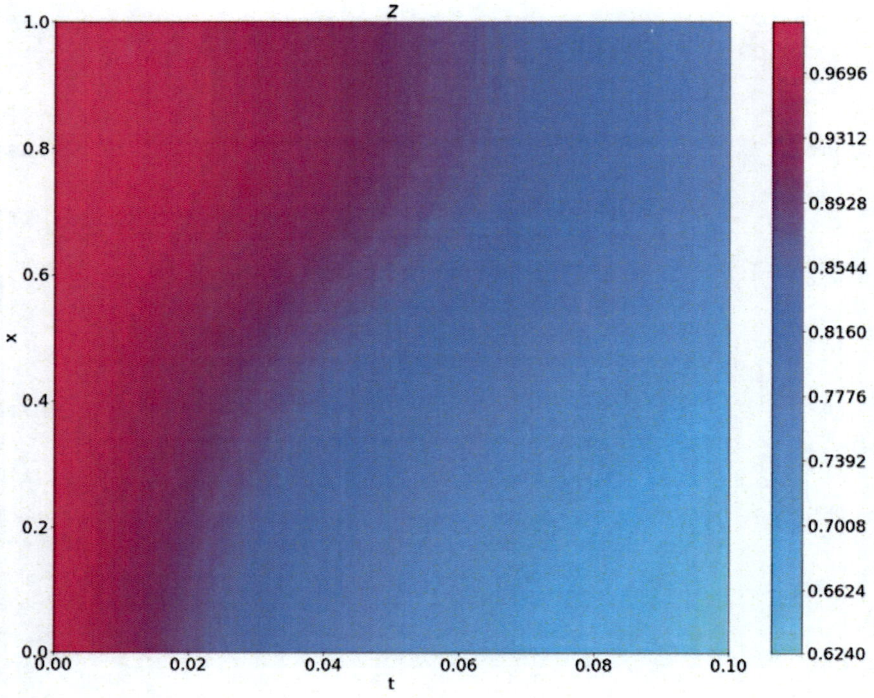

Fig. 1.7 Mass fraction of unburned fuel

Table 1.1 Convergence analysis and CPU times ($T = 1$)

n	$\|\|\rho^n - \rho^{n-1}\|\|_\infty$	$\|\|v^n - v^{n-1}\|\|_\infty$	$\|\|\omega^n - \omega^{n-1}\|\|_\infty$	$\|\|\theta^n - \theta^{n-1}\|\|_\infty$	CPU time (s)
3	$4.9713 \cdot 10^{-3}$	$8.5899 \cdot 10^{-3}$	$7.4429 \cdot 10^{-3}$	$1.0730 \cdot 10^{-2}$	2.9940
4	$1.4356 \cdot 10^{-3}$	$3.8706 \cdot 10^{-3}$	$1.9335 \cdot 10^{-4}$	$4.9781 \cdot 10^{-2}$	4.8381
5	$1.7221 \cdot 10^{-4}$	$2.4116 \cdot 10^{-4}$	$5.5093 \cdot 10^{-5}$	$1.2228 \cdot 10^{-3}$	6.9076
6	$2.2872 \cdot 10^{-5}$	$1.0540 \cdot 10^{-4}$	$5.0302 \cdot 10^{-5}$	$1.0480 \cdot 10^{-4}$	8.8045
7	$1.3044 \cdot 10^{-5}$	$4.1811 \cdot 10^{-5}$	$5.2839 \cdot 10^{-6}$	$6.9502 \cdot 10^{-5}$	9.4298
8	$1.2454 \cdot 10^{-5}$	$2.9395 \cdot 10^{-5}$	$4.4830 \cdot 10^{-6}$	$7.2803 \cdot 10^{-5}$	11.1871

Moreover, we have analyzed here how the numerical scheme behaves under more complex initial conditions, in terms of their more complex expansion in the Fourier series. We have shown experimentally that the method is applicable also in these cases, but it takes more CPU time, and the convergence will be somewhat slower.

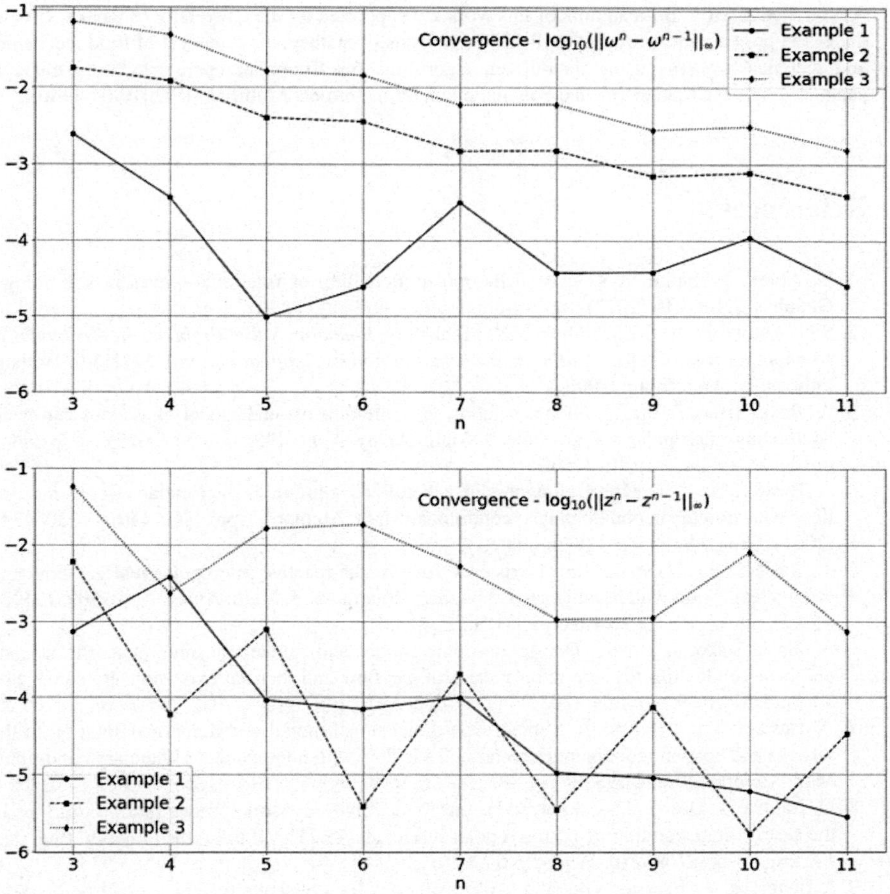

Fig. 1.8 Convergence test

Table 1.2 CPU times (s)

n	Example 1	Example 2	Example 3
3	2.38	2.72	2.74
4	3.19	3.75	5.10
5	4.93	4.13	6.71
6	5.11	5.89	6.47
7	8.12	6.62	7.79
8	8.31	8.23	11.47
9	8.87	8.62	16.44
10	9.70	9.35	18.43
11	9.64	12.90	18.00

Acknowledgments Both authors of this work are supported by the University of Rijeka, Croatia under the project uniri-prirod-18-118-1257 (Analysis of mathematical models of fluid mechanics and technical systems using data-driven algorithms for Koopman operator). Ivan Dražić is supported by the Croatian Science Foundation under the project MultiFM (IP-2019-04-1140).

References

1. R. Abreu, J. Kamm, A.S. Reiss, Micropolar modelling of rotational waves in seismology. Geophys. J. Int. **210** (2017). https://doi.org/10.1093/gji/ggx211
2. S.N. Antontsev, A.V. Kazhikhov, V.N. Monakhov, *Boundary Value Problems in Mechanics of Nonhomogeneous Fluids, Studies in Mathematics and Its Applications*, vol. 22 (North-Holland Publishing, Amsterdam, 1990)
3. A. Bašić-Šiško, I. Dražić, Global solution to a one-dimensional model of viscous and heat-conducting micropolar real gas flow. J. Math. Analy. Appl. **495**, 124690 (2021). https://doi.org/10.1016/j.jmaa.2020.124690
4. A. Bašić-Šiško, I. Dražić, Uniqueness of generalized solution to micropolar viscous real gas flow with homogeneous boundary conditions. Math. Methods Appl. Sci. **44**(6), 4330–4341 (2021). https://doi.org/10.1002/mma.7032
5. A. Bašić-Šiško, I. Dražić, Local existence for viscous reactive micropolar real gas flow and thermal explosion with homogeneous boundary conditions. J. Math. Analy. Appl. **509**, 125988 (2022). https://doi.org/10.1016/j.jmaa.2022.125988
6. A. Bašić-Šiško, I. Dražić, One-dimensional model and numerical solution to the viscous and heat-conducting reactive micropolar real gas flow and thermal explosion. Iranian J. Sci. Technol. Trans. Mech. Eng. (2022). https://doi.org/10.1007/s40997-022-00498-w
7. A. Bašić-Šiško, I. Dražić, L. Simčić, One-dimensional model and numerical solution to the viscous and heat-conducting micropolar real gas flow with homogeneous boundary conditions. Math. Comput. Simul. **195**, 71–87 (2022). https://doi.org/10.1016/j.matcom.2021.12.024
8. U. Baidya, S. Das, S. Das, Analysis of misaligned hydrodynamic porous journal bearings in the steady- state condition with micropolar lubricant. ARCHIVE Proc. Instit. Mech. Eng. Part J J. Eng. Tribol. **208–210**, 1994–1996 (2020)
9. J. Bebernes, A. Bressan, Global a priori estimates for a viscous reactive gas. Proc. R. Soc. Edinburgh Sect. A Math. **101**, 321–333 (1985)
10. M. Benes, I. Pažanin, M. Radulović, Leray's problem for the nonstationary micropolar fluid flow. Mediterran. J. Math. **17** (2020). https://doi.org/10.1007/s00009-020-1493-9
11. B. Bhattacharjee, P. Chakraborti, K. Choudhuri, Evaluation of the performance characteristics of double-layered porous micropolar fluid lubricated journal bearing. Tribol. Int. **138**, 415–423 (2019). https://doi.org/10.1016/j.triboint.2019.06.025
12. M. Bonnivard, I. Pažanin, F.J. Suárez-Grau, Effects of rough boundary and nonzero boundary conditions on the lubrication process with micropolar fluid. Euro. J. Mech. - B/Fluids **72**, 501–518 (2018). https://doi.org/10.1016/j.euromechflu.2018.07.013
13. C. Borgnakke, R.E. Sonntag, *Fundamentals of Thermodynamics*, SI Version, 10th edn. (Wiley, Hoboken, 2019)
14. Y.A. Çengel, R.H. Turner, *Fundamentals of Thermal-Fluid Sciences* (McGraw-Hill Companies, New York City, 2004)
15. Z. Chen, D. Wang, Global stability of rarefaction waves for the 1d compressible micropolar fluid model with density-dependent viscosity and microviscosity coefficients. Nonlinear Analy. Real World Appl. **58**, 103226 (2021). https://doi.org/10.1016/j.nonrwa.2020.103226
16. É. Clapeyron, Mémoire sur la puissance motrice de la chaleur. Journal de l'École Polytechnique, **XIV**(23), 153–191 (1834)

17. E. Cosserat, F. Cosserat, A. Hermann et forie des Corps déformables. A. Hermann et fils, Paris (1909)
18. N. Črnjarić Žic, Upwind numerical approximations of a compressible 1d micropolar fluid flow. J. Comput. Appl. Math. **303**, 81–92 (2016). https://doi.org/10.1016/j.cam.2016.02.022
19. N. Črnjarić-Žic, N. Mujaković, Numerical analysis of the solutions for 1d compressible viscous micropolar fluid flow with different boundary conditions. Math. Comput. Simul. **126**, 45–62 (2016)
20. H. Cui, Z.A. Yao, Asymptotic behavior of compressible p-th power newtonian fluid with large initial data. J. Differ. Equ. **258**(3), 919–953 (2015). https://doi.org/10.1016/j.jde.2014.10.011
21. I. Dražić, Three-dimensional flow of a compressible viscous micropolar fluid with cylindrical symmetry: a global existence theorem. Math. Methods Appl. Sci. **40**(13), 4785–4801 (2017). https://doi.org/10.1002/mma.4344
22. I. Dražić, A. Bašić-Šiško, Local existence theorem for micropolar viscous real gas flow with homogeneous boundary conditions. Math. Methods Appl. Sci. (2022). https://doi.org/10.1002/mma.8841
23. I. Dražić, N. Mujaković, 3-d flow of a compressible viscous micropolar fluid with spherical symmetry: a local existence theorem. Boundary Value Prob. **2012** (2012). https://doi.org/10.1186/1687-2770-2012-69
24. I. Dražić, N. Mujaković, 3-d flow of a compressible viscous micropolar fluid with spherical symmetry: a global existence theorem. Boundary Value Prob. **2015**(98) (2015). https://doi.org/10.1186/s13661-015-0357-x
25. I. Dražić, N. Mujaković, 3-d flow of a compressible viscous micropolar fluid with spherical symmetry: large time behavior of the solution. J. Math. Analy. Appl. **431** (2015). https://doi.org/10.1016/j.jmaa.2015.06.002
26. I. Dražić, N. Mujaković, Local existence of the generalized solution for three-dimensional compressible viscous flow of micropolar fluid with cylindrical symmetry. Boundary Value Prob. **2019** (2019). https://doi.org/10.1186/s13661-019-1131-2
27. I. Dražić, N. Mujaković, L. Simčić, 3-d flow of a compressible viscous micropolar fluid with spherical symmetry: regularity of the solution. Math. Analy. Appl. **438**(1), 162–183 (2016). https://doi.org/10.1016/j.jmaa.2016.01.071
28. I. Dražić, N. Mujaković, N. Črnjarić Žic, Three-dimensional compressible viscous micropolar fluid with cylindrical symmetry: derivation of the model and a numerical solution. Math. Comput. Simul. **140**, 107–124 (2017). https://doi.org/10.1016/j.matcom.2017.03.006
29. I. Dražić, N. Čnjarić Žic, L. Simčić, A shear flow problem for compressible viscous micropolar fluid: derivation of the model and numerical solution. Math. Comput. Simul. **162**, 249–267 (2019). https://doi.org/10.1016/j.matcom.2019.01.013
30. C.A. Eringen, Simple microfluids. Int. J. Eng. Sci. **2**(2), 205–217 (1964). https://doi.org/10.1016/0020-7225(64)90005-9
31. E. Feireisl, A. Novotný, Large time behaviour of flows of compressible, viscous, and heat conducting fluids. Math. Methods Appl. Sci. **29**, 1237–1260 (2006). https://doi.org/10.1002/mma.722
32. L. Huang, I. Dražić, Exponential stability for the compressible micropolar fluid with cylinder symmetry in R^3. J. Math. Phys. **60**, 021507 (2019). https://doi.org/10.1063/1.5017652
33. E.G. Karvelas, A. Tsiantis, T.D. Papathanasiou, Effect of micropolar fluid properties on the hydraulic permeability of fibrous biomaterials. Comput. Methods Program. Biomed. **185**, 105135 (2020). https://doi.org/10.1016/j.cmpb.2019.105135
34. D. Khanukaeva, A. Filippov, Isothermal flows of micropolar liquids: formulation of problems and analytical solutions. Colloid J. **80**, 14–36 (2018). https://doi.org/10.1134/S1061933X18010040
35. M. Lewicka, P. Mucha, On temporal asymptotics for the pth power viscous reactive gas. Nonlinear Analy. Theory, Methods Appl. **57**, 951–969 (2004). https://doi.org/10.1016/j.na.2003.12.001
36. G. Lukaszewicz, *Micropolar Fluids: Theory and Applications*. Modeling and Simulation in Science, Engineering and Technology (Birkhäuser, Boston, 1999)

37. G. Lukaszewicz, I. Pažanin, M. Radulović, Asymptotic analysis of the thermomicropolar fluid flow through a thin channel with cooling. Appl. Analy. **101**, 1–29 (2020). https://doi.org/10.1080/00036811.2020.1836353
38. D. Maltese, M. Michálek, P.B. Mucha, A. Novotný, M. Pokorný, E. Zatorska, Existence of weak solutions for compressible navier-stokes equations with entropy transport. J. Differ. Equ. **261**, 4448–4485 (2016). https://doi.org/10.1016/j.jde.2016.06.029
39. K. Mekheimer, T. Elnaqeeb, M. El Kot, F. Alghamdi, Simultaneous effect of magnetic field and metallic nanoparticles on a micropolar fluid through an overlapping stenotic artery: blood flow model. Phys. Essays **29**, 272–283 (2016). https://doi.org/10.4006/0836-1398-29.2.272
40. N. Mujaković, One-dimensional flow of a compressible viscous micropolar fluid: a global existence theorem. Glasnik matematički **33**(53), 199–208 (1998)
41. N. Mujaković, One-dimensional flow of a compressible viscous micropolar fluid: a local existence theorem. Glas. Mat., III. Ser. **33**(1), 71–91 (1998)
42. N. Mujaković, One-dimensional flow of a compressible viscous micropolar fluid regularity of the solution. Radovi Matematički **10**, 181–193 (2001)
43. N. Mujaković, One-dimensional flow of a compressible viscous micropolar fluid: Stabilization of the solution, in ed. by Drmač, Z., Marušić, M., Tutek, Z., *Proceedings of the Conference on Applied Mathematics and Scientific Computing* (Springer, Dordrecht, 2005). https://doi.org/10.1007/1-4020-3197-1_18
44. N. Mujaković, N. Črnjarić-Žic, Convergent finite difference scheme for 1D flow of compressible micropolar fluid. Int. J. Numer. Anal. Model. **12**(1), 94–124 (2015)
45. N. Mujaković, N. Črnjarić Žic, Convergence of a finite difference scheme for 3d flow of a compressible viscous micropolar heat-conducting fluid with spherical symmetry. Int. J. Numer. Analy. Model. **13**, 705–738 (2016)
46. N. Mujaković, N. Črnjarić Žic, Finite difference formulation for the model of a compressible viscous and heat-conducting micropolar fluid with spherical symmetry, in ed. by S. Pinelas, Z. Došlá, O. Došlý, P.E. Kloeden, *Differential and Difference Equations with Applications*, vol. 164 (Springer, Berlin, 2016), pp. 293–301. https://doi.org/10.1007/978-3-319-32857-7-27
47. N. Mujaković, I. Dražić, Numerical approximations of the solution for one-dimensional compressible viscous micropolar fluid model. Int. J. Pure Appl. Math. **38**, 285–296 (2007)
48. N. Mujaković, I. Dražić, The cauchy problem for one-dimensional flow of a compressible viscous fluid: stabilization of the solution. Glasnik Matematički. Serija III **46** (2011). https://doi.org/10.3336/gm.46.1.16
49. N. Mujaković, I. Dražić, 3-d flow of a compressible viscous micropolar fluid with spherical symmetry: a local existence theorem. Boundary Value Prob. **226**, 1–17 (2014). https://doi.org/10.1186/1687-2770-2012-69
50. N. Mujekaović, I. Dražić, L. Simčić, 3-d flow of a compressible viscous micropolar fluid with cylindrical symmetry: uniqueness of a generalized solution. Math. Methods Appl. Sci. **40**(7), 2686–2701 (2017). https://doi.org/10.1002/mma.4191
51. M. Nermina, One-dimensional flow of a compressible viscous micropolar fluid: the cauchy problem. Math. Commun. **10**, 1–14 (2005)
52. M. Nermina, Uniqueness of a solution of the cauchy problem for one-dimensional compressible viscous micropolar fluid model. Appl. Math. E-Notes **6**, 113–118 (2006)
53. M. Nermina, One-dimensional compressible viscous micropolar fluid model: stabilization of the solution for the cauchy problem. Boundary Value Prob. **2010** (2010). https://doi.org/10.1155/2010/796065
54. I. Papautsky, J. Brazzle, T. Ameel, A. Frazier, Laminar fluid behavior in microchannels using micropolar fluid theory. Sensors Actuat. A Phys. **73**(1), 101–108 (1999). https://doi.org/10.1016/S0924-4247(98)00261-1
55. S. Pimputkar, S. Nakamura, Decomposition of supercritical ammonia and modeling of supercritical ammonia-nitrogen-hydrogen solutions with applicability toward ammonothermal conditions. J. Supercrit. Fluids **107**, 17–30 (2016). https://doi.org/10.1016/j.supflu.2015.07.032

56. J. Poland, D. Kassoy, The induction period of a thermal explosion in a gas between infinite parallel plates. Combust. Flame **50**, 259–274 (1983). https://doi.org/10.1016/0010-2180(83)90069-X
57. Y. Qin, L. Huang, Global existence and exponential stability for the pth power viscous reactive gas. Nonlinear Analy.-Theory Methods Appl. **73**, 2800–2818 (2010). https://doi.org/10.1016/j.na.2010.06.015
58. Y. Qin, L. Huang, Regularity and exponential stability of the pth newtonian fluid in one space dimension. Math. Models Methods Appl. Sci. - M3AS **20** (2010). https://doi.org/10.1142/S0218202510004350
59. Y. Qin, L. Huang, Global *Well-posedness of Nonlinear Parabolic-Hyperbolic Coupled Systems*. Frontiers in Mathematics, 1st edn. (Birkhäuser, Basel, 2012). https://doi.org/10.1007/978-3-0348-0280-2
60. Y. Qin, J. Zhang, X. Su, J. Cao, Global existence and exponential stability of spherically symmetric solutions to a compressible combustion radiative and reactive gas. J. Math. Fluid Mech. **18** (2016). https://doi.org/10.1007/s00021-015-0242-5
61. L. Simčić, A shear flow problem for compressible viscous micropolar fluid: uniqueness of a generalized solution. Math. Methods Appl. Sci. **42**(5), 6358–6368 (2019). https://doi.org/10.1002/mma.5727
62. I. Vardoulakis, *Cosserat Continuum Mechanics: With Applications to Granular Media*. Lecture Notes in Applied and Computational Mechanics, vol. 87 (Springer International Publishing, Cham, 2019)
63. T. Wang, One dimensional p-th power newtonian fluid with temperature-dependent thermal conductivity. Commun. Pure Appl. Analy. **15**, 477–494 (2016). https://doi.org/10.3934/cpaa.2016.15.477
64. L. Wan, T. Wang, Asymptotic behavior for the one-dimensional pth power newtonian fluid in unbounded domains. Math. Methods Appl. Sci. **39** (2015). https://doi.org/10.1002/mma.3539
65. S. Watson, M. Lewicka, J. Watson, Temporal asymptotics for the p'th power newtonian fluid in one space dimension. Zeitschrift für angewandte Mathematik und Physik **54**, 633–651 (2003). https://doi.org/10.1007/s00033-003-1149-1
66. S. Yanagi, Asymptotic stability of the spherically symmetric solutions for a viscous polytropic gas in a field of external forces. Trans. Theory Statist. Phys. **29**, 333–353 (2000). https://doi.org/10.1080/00411450008205878
67. S. Ye, K. Zhu, W. Wang, Laminar flow of micropolar fluid in rectangular microchannels. Acta Mech. Sinica **22**, 403–408 (2006). https://doi.org/10.1007/s10409-006-0023-8
68. P. Zhang, M. Zhu, Global regularity of 3d nonhomogeneous incompressible magneto-micropolar system with the density-dependent viscosity. Comput. Math. Appl. **76** (2018). https://doi.org/10.1016/j.camwa.2018.08.041

Chapter 2
Digital Operators and Discrete Equations as Computational Tools

Alexander Vasilyev, Vladimir Vasilyev, and Anastasia Mashinets

2.1 Introduction

Discrete and difference equations play an important role in different branches of science, particularly in mathematical biology, signal, and image processing [1, 2]. So, for example, a lot of physical and technical processes are described by difference and discrete equations, namely, and continuous models arise as a limit transfer. According to the latter sentence, it is thought that studying of difference and discrete equations is very required.

We will deal with discrete equations related to well-known continuous mathematical objects as pseudo-differential operators and equations [3–6]. There is a series of books devoted to different aspects of the theory of discrete equations and discrete boundary value problems (see, e.g., [7–9]), but as a rule, these methods are developed for partial differential equations only. Also, some authors use projectional and algebraic methods for studying finite approximations for integral and related operator equations [10–12]. Let us remind that the theory of pseudo-differential operators was constructed to join together the theory of differential operators and certain integral ones. Starting from this point of view, we will try to construct a theory of discrete pseudo-differential operators and equations, to study such discrete equations and related discrete boundary value problems, and to verify their approximation properties.

A few years ago, such a work was started, and certain results were obtained. Some of these papers were related to discrete analogues of Calderon–Zygmund operators and corresponding integral equations [13, 14], but latter papers are devoted to studying discrete pseudo-differential equations and discrete boundary value

A. Vasilyev · V. Vasilyev (✉) · A. Mashinets
Belgorod State National Research University, Belgorod, Russia
e-mail: vbv57@inbox.ru

C. M. A. Pinto, C. M. Ionescu (eds.), *Computational and Mathematical Models in Biology*, Nonlinear Systems and Complexity 38,
https://doi.org/10.1007/978-3-031-42689-6_2

problems in a discrete space and a discrete half-space [15–19]. We work with model operators with symbols non-depending on a spatial variable in special canonical domains which are cones in Euclidean space. This methodology is stipulated by a special local principle in the theory of pseudo-differential equations. We will widely use discrete and periodic analogues of classical one-dimensional singular integral operators [20, 21], methods of function theory of many complex variables [22, 23], and the key abstract result from the theory of projectional methods [24].

This paper is devoted to a new concept in the theory of discrete equations and discrete boundary value problems. We will describe here this approach and will present some key results in this direction. Most part of the paper will be related to a plane quadrant, and it is a new type of a conical domain for which we try to expand applicability of the suggested approach.

2.2 Discrete Spaces and Digital Operators

We will use the following notations. Let $\mathbf{T}$ be the segment $[-\pi, \pi]$, $h > 0$, $\hbar = h^{-1}$. We will consider all functions defined in the cube $\mathbf{T}^m$ as periodic functions in $\mathbf{R}^m$ with the same cube of periods.

If $u_d(\tilde{x})$, $\tilde{x} \in h\mathbf{Z}^m$, is a function of a discrete variable, then we call it "discrete function." For such discrete functions, one can define the discrete Fourier transform

$$(F_d u_d)(\xi) \equiv \tilde{u}_d(\xi) = \sum_{\tilde{x}\in h\mathbf{Z}^m} e^{-i\tilde{x}\cdot\xi} u_d(\tilde{x}) h^m, \quad \xi \in \hbar\mathbf{T}^m,$$

if the latter series converges, and the function $\tilde{u}_d(\xi)$ is a periodic function on $\mathbf{R}^m$ with the basic cube of periods $\hbar\mathbf{T}^m$. This discrete Fourier transform preserves basic properties of the integral Fourier transform, particularly the inverse discrete Fourier transform is given by the formula

$$(F_d^{-1}\tilde{u}_d)(\tilde{x}) = \frac{1}{(2\pi)^m} \int_{\hbar\mathbf{T}^m} e^{i\tilde{x}\cdot\xi} \tilde{u}_d(\xi) d\xi, \quad \tilde{x} \in h\mathbf{Z}^m.$$

The discrete Fourier transform is a one-to-one correspondence between the spaces $L_2(h\mathbf{Z}^m)$ and $L_2(\hbar\mathbf{T}^m)$ with norms

$$||u_d||_2 = \left(\sum_{\tilde{x}\in h\mathbf{Z}^m} |u_d(\tilde{x})|^2 h^m\right)^{1/2}$$

and

$$||\tilde{u}_d||_2 = \left(\int\limits_{\xi \in \hbar \mathbf{T}^m} |\tilde{u}_d(\xi)|^2 d\xi \right)^{1/2}.$$

Example 1 Since the definition for Sobolev–Slobodetskii spaces includes partial derivatives, we use their discrete analogue, i.e., divided difference of first order

$$\left(\Delta_k^{(1)} u_d\right)(\tilde{x}) = h^{-1}(u_d(x_1, \ldots, x_k + h, \ldots, x_m) - u_d(x_1, \ldots, x_k, \ldots, x_m)),$$

for which its discrete Fourier transform looks as follows

$$\widetilde{(\Delta_k^{(1)} u_d)}(\xi) = h^{-1}(e^{-ih\cdot\xi_k} - 1)\tilde{u}_d(\xi).$$

Further for the divided difference of second order, we have

$$\left(\Delta_k^{(2)} u_d\right)(\tilde{x}) = h^{-2}(u_d(x_1, \ldots, x_k + 2h, \ldots, x_m)$$
$$- 2u_d(x_1, \ldots, x_k + h, \ldots, x_m) + u_d(x_1, \ldots, x_k, \ldots, x_m))$$

and its discrete Fourier transform

$$\widetilde{(\Delta_k^{(2)} u_d)}(\xi) = h^{-2}(e^{-ih\cdot\xi_k} - 1)^2\tilde{u}_d(\xi).$$

Thus, for the discrete Laplacian, we have

$$(\Delta_d u_d)(\tilde{x}) = \sum_{k=1}^{m} (\Delta_k^{(2)} u_d)(\tilde{x}),$$

so that

$$\widetilde{(\Delta_d u_d)}(\xi) = h^{-2} \sum_{k=1}^{m} (e^{-ih\cdot\xi_k} - 1)^2\tilde{u}_d(\xi).$$

We will use the discrete Fourier transform to introduce special discrete Sobolev–Slobodetskii spaces which are very convenient for studying discrete pseudo-differential operators and related equations.

Now we will introduce the basic space $S(h\mathbf{Z}^m)$ which consists of discrete functions with finite semi-norms

$$|u_d| = \sup_{\tilde{x} \in h\mathbf{Z}^m} (1 + |\tilde{x}|)^l |\Delta^{(\mathbf{k})} u_d(\tilde{x})|$$

for arbitrary $l \in \mathbf{N}$, $\mathbf{k} = (k_1, \ldots, k_m)$, $k_r \in \mathbf{N}$, $r = 1, \ldots, m$, where

$$\Delta^{(\mathbf{k})}u_d(\tilde{x}) = \Delta_1^{k_1}\ldots,\Delta_m^{k_m}u_d(\tilde{x}).$$

In other words, the space $S(h\mathbf{Z}^m)$ is a discrete analogue of the Schwartz space $S(\mathbf{R}^m)$ of infinitely differentiable rapidly decreasing at infinity functions. Usually the space of distributions over the basic space $S(\mathbf{R}^m)$ is denoted by $S'(\mathbf{R}^m)$.

Digital distribution we call an arbitrary linear continuous functional defined on $S(h\mathbf{Z}^m)$. A set of such digital distributions we will denote by $S'(h\mathbf{Z}^m)$, and a value of the functional f_d on the basic function u_d will be denoted by (f_d, u_d).

Together with the space $S(h\mathbf{Z}^m)$, we consider the space $D(h\mathbf{Z}^m)$ consisting of discrete functions with a compact (finite) support. We say that $f_d = 0$ in the discrete domain $M_d \equiv M \cap h\mathbf{Z}^m, M \subset \mathbf{R}^m$, if $(f_d, u_d) = 0, \forall u_d \in D(M_d)$, where $D(M_d) \subset D(h\mathbf{Z}^m)$ consists of discrete functions whose supports belong to M_d. If we will denote $\widetilde{M}_d$, a union of such M_d, where $f_d = 0$ then by definition $supp\ f_d = h\mathbf{Z}^m \setminus \widetilde{M}_d$.

As usual [25], we can define some simplest operations in the space $S'(h\mathbf{Z}^m)$ excluding the differentiation (see below), and a convergence is defined as a weak convergence in the space of functionals $S'(h\mathbf{Z}^m)$.

If $f_d(\tilde{x})$ is a local summable function, then one can define the digital distribution f_d by the formula

$$(f_d, u_d) = \sum_{\tilde{x}\in h\mathbf{Z}^m} f_d(\tilde{x})u_d(\tilde{x})h^m, \quad \forall u_d \in S(h\mathbf{Z}^m).$$

Such distributions we call *regular* digital distributions. But there are so-called *singular* digital distributions like the Dirac mass-function

$$(\delta_d, u_d) = u_d(0),$$

which cannot be represented by the above formula.

Let us denote $\zeta^2 = h^{-2}\sum\limits_{k=1}^{m}(e^{-ih\cdot\xi_k} - 1)^2$ and introduce the following

Definition 1 The space $H^s(h\mathbf{Z}^m)$ is a closure of the space $S(h\mathbf{Z}^m)$ with respect to the norm

$$||u_d||_s = \left(\int\limits_{\hbar\mathbf{T}^m} (1 + |\zeta^2|)^s |\tilde{u}_d(\xi)|^2 d\xi\right)^{1/2}.$$

We would like to note that a lot of properties for such spaces were studied in [26].

Further, let $D \subset \mathbf{R}^m$ be a domain and $D_d = D \cap h\mathbf{Z}^m$ be a discrete domain.

Definition 2 The space $H^s(D_d)$ consists of discrete functions from $H^s(h\mathbf{Z}^m)$ which supports belong to $\overline{D_d}$. A norm in the space $H^s(D_d)$ is induced by a norm of the space $H^s(h\mathbf{Z}^m)$. The space $H_0^s(D_d)$ consists of discrete functions u_d with a support in D_d, and these discrete functions should admit a continuation into the whole $H^s(h\mathbf{Z}^m)$. A norm in the $H_0^s(D_d)$ is given by the formula

$$||u_d||_s^+ = \inf ||\ell u_d||_s,$$

where infimum is taken over all continuations ℓ.

The Fourier image of the space $H^s(D_d)$ will be denoted by $\widetilde{H}^s(D_d)$.

Let $\widetilde{A}_d(\xi)$ be a measurable periodic function in $\mathbf{R}^m$ with the basic cube of periods $\hbar\mathbf{T}^m$. Such functions are called symbols. As usual, we will define a digital pseudo-differential operator by its symbol.

Definition 3 A digital pseudo-differential operator A_d in a discrete domain D_d is called an operator of the following kind

$$(A_d u_d)(\tilde{x}) = \sum_{\tilde{y} \in h\mathbf{Z}^m} \int_{\hbar\mathbf{T}^m} \widetilde{A}_d(\xi) e^{i(\tilde{x}-\tilde{y})\cdot\xi} \tilde{u}_d(\xi) d\xi, \quad \tilde{x} \in D_d.$$

An operator A_d is called an elliptic operator if

$$ess \inf_{\xi \in \hbar\mathbf{T}^m} |\widetilde{A}_d(\xi)| > 0.$$

First as usual, we define the operator A_d on the dense set $S(h\mathbf{Z}^m)$ and then extend it on more general space.

Remark 1 One can introduce the symbol $\widetilde{A}_d(\tilde{x}, \xi)$ depending on a spatial variable $\tilde{x}$ and define a general pseudo-differential operator by the formula

$$(A_d u_d)(\tilde{x}) = \sum_{\tilde{y} \in h\mathbf{Z}^m} \int_{\hbar\mathbf{T}^m} \widetilde{A}_d(\tilde{x}, \xi) e^{i(\tilde{x}-\tilde{y})\cdot\xi} \tilde{u}_d(\xi) d\xi, \quad \tilde{x} \in D_d,$$

For studying such operators and related equations one needs to use more fine and complicated technique.

Definition 4 By definition the class E_α includes symbols satisfying the following condition

$$c_1(1 + |\zeta^2|)^{\alpha/2} \leq |A_d(\xi)| \leq c_2(1 + |\zeta^2|)^{\alpha/2} \tag{2.1}$$

with universal positive constants c_1, c_2 non-depending on h and the symbol $A_d(\xi)$.

The number $\alpha \in \mathbf{R}$ is called an order of a digital pseudo-differential operator A_d.

Obviously that operator A_d satisfying (2.1) is an elliptic operator. Using the last definition one can easily get the following property.

Lemma 1 *A digital pseudo-differential operator $A_d \in E_\alpha$ is a linear bounded operator $H^s(h\mathbf{Z}^m) \rightarrow H^{s-\alpha}(h\mathbf{Z}^m)$ which norm does not depend on h.*

We study the equation

$$(A_d u_d)(\tilde{x}) = v_d(\tilde{x}), \quad \tilde{x} \in D_d, \tag{2.2}$$

assuming that we interested in a solution $u_d \in H^s(D_d)$ taking into account $v_d \in H_0^{s-\alpha}(D_d)$.

Main difficulty for this problem is related to a geometry of the domain D. Indeed, if $D = \mathbf{R}^m$ then the condition (2.1) guarantees the unique solvability for the Eq. (2.2). We will consider here only so-called canonical domains and simplest digital pseudo-differential operators with symbols non-depending on a spatial variable $\tilde{x}$. This fact is dictated by using in the future the local principle. The last asserts that for a Fredholm solvability of the general Eq. (2.2) with symbol $A_d(\tilde{x}, \xi)$ in an arbitrary discrete domain D_d, one needs to obtain invertibility conditions for so-called local representatives of the operator A_d, i.e., for an operator with symbol $A_d(\cdot, \xi)$ in a special canonical domain.

Earlier authors have extracted some canonical domains, namely, $D = \mathbf{R}^m, \mathbf{R}^m_+$, where $\mathbf{R}^m_+ = \{x \in \mathbf{R}^m : x = (x', x_m), x_m > 0\}$.

Everywhere below we study the two-dimensional case for which a domain D is the first quadrant in a plane, $D \equiv K = \{x \in \mathbf{R}^2 : x = (x_1, x_2), x_1 > 0, x_2 > 0\}$. *Moreover, we consider homogeneous equation (2.2) for a simplicity.*

2.3 Solvability of Discrete Equations and Discrete Boundary Value Problems

Let $K_d = K \cap h\mathbf{Z}^2$. We study a solvability of the equation

$$(A_d u_d)(\tilde{x}) = 0, \quad \tilde{x} \in K_d. \tag{2.3}$$

We can describe solvability picture of the Eq. (2.3) if the symbol $\widetilde{A}_d(\xi)$ admits a special representation.

2.3.1 Periodic Wave Factorization

This concept is a periodic analogue of the wave factorization [23]. Some first preliminary considerations and results were described in [15].

We will use certain special domain in two-dimensional complex space $\mathbf{C}^2$. A domain of the type $\mathcal{T}_h(K) = \hbar\mathbf{T}^2 + iK$ is called a tube domain over the quadrant K, and we will consider analytical functions $f(x + i\tau)$ in the domain $\mathcal{T}_h(K) = \hbar\mathbf{T}^2 + iK$.

Definition 5 A periodic wave factorization for the elliptic symbol $A_d(\xi) \in E_\alpha$ is called its representation in the form

$$A_d(\xi) = A_{d,\neq}(\xi) A_{d,=}(\xi),$$

where the factors $A_{d,\neq}(\xi)$, $A_{d,=}(\xi)$ admit analytical continuation into tube domains $\mathcal{T}_h(K)$, $\mathcal{T}_h(-K)$ respectively with estimates

$$c_1(1+|\hat{\zeta}^2|)^{\frac{æ}{2}} \leq |A_{d,\neq}(\xi + i\tau)| \leq c_1'(1+|\hat{\zeta}^2|)^{\frac{æ}{2}},$$

$$c_2(1+|\hat{\zeta}^2|)^{\frac{\alpha-æ}{2}} \leq |A_{d,=}(\xi - i\tau)| \leq c_2'(1+|\hat{\zeta}^2|)^{\frac{\alpha-æ}{2}},$$

and constants c_1, c_1', c_2, c_2' non-depending on h, where

$$\hat{\zeta}^2 \equiv \hbar^2\left((e^{-ih(\xi_1+i\tau_1)} - 1)^2 + (e^{-ih(\xi_2+i\tau_2)} - 1)^2\right),$$

$$\xi = (\xi_1, \xi_2) \in \hbar\mathbf{T}^2, \quad \tau = (\tau_1, \tau_2) \in K.$$

The number $æ \in \mathbf{R}$ is called an index of periodic wave factorization.

2.3.2 *Solvability Conditions*

Using methods of [17, 23] one can obtain the following results on a solvability of the Eq. (2.2).

Theorem 1 *Let* $|æ - s| < 1/2$. *Then the Eq.* (2.3) *has zero solution only.*

Theorem 2 *Let* $æ - s = n + \delta, n \in \mathbf{N}, |\delta| < 1/2$. *Then a general solution of the Eq.* (2.3) *has the following form*

$$\tilde{u}_d(\xi) = A_{d,\neq}^{-1}(\xi)\left(\sum_{k=0}^{n-1} \tilde{c}_k(\xi_1)\zeta_2^k + \tilde{d}_k(\xi_2)\zeta_1^k\right), \quad \zeta_j = \hbar(e^{-i\xi_j h} - 1), \quad j = 1, 2, \tag{$*$}$$

where $\tilde{c}_k(\xi_1), \tilde{d}_k(\xi_2), k = 0, 1, \ldots, n-1$, *are arbitrary functions from* $\widetilde{H}^{s_k}(\hbar\mathbf{T})$, $s_k = s - æ + k - 1/2$.

The a priori estimate

$$||u_d||_s \leq const \sum_{k=0}^{n-1}([c_k]_{s_k} + [d_k]_{s_k}),$$

holds, where $[\cdot]_{s_k}$ *denotes a norm in* $H^{s_k}(h\mathbf{Z})$ *and const does not depend on* h.

2.3.3 Boundary Conditions

As we see, Theorem 2 asserts that for a certain case, we have a lot of solutions. To obtain the unique solution, we need to determine uniquely all arbitrary functions in the formula (*). We consider here the case $æ - s = 1+\delta$, $|\delta| < 1/2$ for the Eq. (2.3) and two different types of conditions.

2.3.4 Classical Variant: The Dirichlet Discrete Boundary Condition

We consider here first simple case with discrete Dirichlet boundary conditions. It follows from Theorem 2 that we have the following general solution of the Eq. (2.4)

$$\tilde{u}_d(\xi) = A^{-1}_{d,\neq}(\xi)(\tilde{c}_0(\xi_1) + \tilde{d}_0(\xi_2)), \tag{2.4}$$

where $c_0, d_0 \in H^{s-æ-1/2}(\hbar\mathbf{Z})$ are arbitrary functions. To determine uniquely these functions, we add the discrete Dirichlet conditions on angle sides

$$u_d|_{\tilde{x}_1=0} = f_d(\tilde{x}_2), \qquad u_d|_{\tilde{x}_2=0} = g_d(\tilde{x}_1). \tag{2.5}$$

Thus, we have the discrete Dirichlet problem (2.3), (2.5).

First, we apply the discrete Fourier transform to discrete conditions (2.5) and obtain the following form

$$\int\limits_{-\hbar\pi}^{\hbar\pi} \tilde{u}_d(\xi_1,\xi_2)d\xi_1 = \tilde{f}_d(\xi_2), \quad \int\limits_{-\hbar\pi}^{\hbar\pi} \tilde{u}_d(\xi_1,\xi_2)d\xi_2 = \tilde{g}_d(\xi_1).$$

Let us denote

$$\int\limits_{-\hbar\pi}^{\hbar\pi} A^{-1}_{d,\neq}(\xi)d\xi_1 \equiv \tilde{a}_0(\xi_2), \quad \int\limits_{-\hbar\pi}^{\hbar\pi} A^{-1}_{d,\neq}(\xi)d\xi_2 \equiv \tilde{b}_0(\xi_1)$$

and suppose that $\tilde{a}_0(\xi_2), \tilde{b}_0(\xi_1) \neq 0, \forall \xi_1 \neq 0, \xi_2 \neq 0$.

Therefore, we have the following system of two linear integral equations with respect to two unknown functions $\tilde{c}_0(\xi_1), \tilde{d}_0(\xi_2)$

$$\begin{cases} \int\limits_{-\hbar\pi}^{\hbar\pi} k_1(\xi)\tilde{c}_0(\xi_1)d\xi_1 + \tilde{d}_0(\xi_2) = \tilde{F}_d(\xi_2) \\ \tilde{c}_0(\xi_1) + \int\limits_{-\hbar\pi}^{\hbar\pi} k_2(\xi)\tilde{d}_0(\xi_2)d\xi_2 = \tilde{G}_d(\xi_1), \end{cases} \tag{2.6}$$

where we have used the following notations

$$\tilde{F}_d(\xi_2) = \tilde{f}_d(\xi_2)\tilde{a}_0^{-1}(\xi_2), \quad \tilde{G}_d(\xi_1) = \tilde{g}_d(\xi_1)\tilde{b}_0^{-1}(\xi_1),$$

$$k_1(\xi) = A_{d,\neq}^{-1}(\xi)\tilde{a}_0^{-1}(\xi_2), \quad k_2(\xi) = A_{d,\neq}^{-1}(\xi)\tilde{b}_0^{-1}(\xi_1).$$

Unique solvability conditions for the system (2.6) will be equivalent to unique solvability for the discrete Dirichlet problem (2.3), (2.5).

Thus, we obtain the following result.

Theorem 3 *Let* $f_d, g_d \in H^{s-1/2}(\mathbf{R}_+)$, $s > 1/2$, *and*

$$\inf|\tilde{a}_0(\xi_2)| > 0, \quad \inf|\tilde{b}_0(\xi_1)| > 0.$$

Then the discrete Dirichlet problem (2.3), (2.5) *is reduced to the equivalent system of linear integral equations* (2.6).

2.3.5 Nonlocal Discrete Boundary Condition

Another variant of a boundary condition is the following

$$\sum_{\tilde{x}_1 \in h\mathbf{Z}_+} u_d(\tilde{x}_1, \tilde{x}_2)h = f_d(\tilde{x}_2), \quad \sum_{\tilde{x}_2 \in h\mathbf{Z}_+} u_d(\tilde{x}_1, \tilde{x}_2)h = g_d(\tilde{x}_1),$$
$$\sum_{\tilde{x} \in h\mathbf{Z}_{++}} u_d(\tilde{x}_1, \tilde{x}_2)h^2 = 0. \tag{2.7}$$

These additional conditions will help us to determine uniquely the unknown functions c_0, d_0 in the solution (2.4).

Indeed, using the discrete Fourier transform, we rewrite the conditions (2.7) as follows

$$\tilde{u}_d(0, \xi_2) = \tilde{f}_d(\xi_2), \quad \tilde{u}_d(\xi_1, 0) = \tilde{g}_d(\xi_1), \quad \tilde{u}_d(0, 0) = 0. \tag{2.8}$$

Now we substitute the formulas (2.8) into (2.4). The first two equality are

$$\tilde{u}_d(0, \xi_2) = A_{d,\neq}^{-1}(0, \xi_2)(\tilde{c}_0(0) + \tilde{d}_0(\xi_2)) = \tilde{f}_d(\xi_2),$$

$$\tilde{u}_d(\xi_1, 0) = A_{d,\neq}^{-1}(\xi_1, 0)(\tilde{c}_0(\xi_1) + \tilde{d}_0(0)) = \tilde{g}_d(\xi_1).$$

It implies the following relations according to the third condition that $\tilde{f}_d(0) = \tilde{g}_d(0)$, and it gives

$$\tilde{c}_0(0) + \tilde{d}_0(0) = 0, \quad \tilde{c}_0(0) = \tilde{d}_0(0) = 0.$$

So, we have at least formally the following formula

$$\tilde{u}_d(\xi) = A^{-1}_{d,\neq}(\xi)\left(A_{d,\neq}(\xi_1, 0)\tilde{g}_d(\xi_1) + A_{d,\neq}(0, \xi_2)\tilde{f}_d(\xi_2)\right). \tag{2.9}$$

Theorem 4 *Let $f_d, g_d \in H^{s+1/2}(h\mathbf{Z})$. Then the discrete problem (2.3), (2.7) has unique solution which is given by the formula (2.9).*

The a priori estimate

$$||u_d||_s \leq const(||f_d||_{s+1/2} + ||g_d||_{s+1/2})$$

holds with a const non-depending on h.

2.4 Continuous Boundary Value Problems

Let A be a pseudo-differential operator with the symbol $A(\xi), \xi = (\xi_1, \xi_2)$ satisfying the condition

$$c_1(1 + |\xi|)^\alpha \leq |A(\xi)| \leq c_2(1 + |\xi|)^\alpha.$$

and admitting the wave factorization with respect to the quadrant K

$$A(\xi) = A_{\neq}(\xi)A_{=}(\xi)$$

with index æ such that $æ - s = 1 + \delta, |\delta| < 1/2$.

The continuous analogue of the discrete equation (2.3) is the following

$$(Au)(x) = 0, \quad x \in K. \tag{2.10}$$

2.4.1 *The Dirichlet Condition*

If we consider the Eq. (2.10), a general solution is written in the form [23]

$$\tilde{u}(\xi) = A^{-1}_{\neq}(\xi)(\tilde{C}_0(\xi_1) + \tilde{D}_0(\xi_2))$$

where arbitrary functions $\tilde{C}_0(\xi_1), \tilde{D}_0(\xi_2) \in \tilde{H}^{s-æ-1/2}(\mathbf{R})$ can be determined from the system of integral equations

$$\begin{cases} \int\limits_{-\infty}^{\infty} K_1(\xi)\tilde{C}_0(\xi_1)d\xi_1 + \tilde{D}_0(\xi_2) = \tilde{F}(\xi_2) \\ \tilde{C}_0(\xi_1) + \int\limits_{-\infty}^{\infty} K_2(\xi)\tilde{D}_0(\xi_2)d\xi_2 = \tilde{G}(\xi_1), \end{cases} \tag{2.11}$$

if we use the following boundary conditions

$$u_{|x_1=0} = f(x_2), \qquad u_{|x_2=0} = g(x_1) \tag{2.12}$$

and assume that the conditions $\inf|\tilde{A}_0(\xi_2)| \neq 0, \inf|\tilde{B}_0(\xi_1)| \neq 0$ hold. Here we have denoted

$$\int\limits_{-\infty}^{\infty} A_{\neq}^{-1}(\xi)d\xi_1 \equiv \tilde{A}_0(\xi_2), \quad \int\limits_{-\infty}^{\infty} A_{\neq}^{-1}(\xi)d\xi_2 \equiv \tilde{B}_0(\xi_1),$$

$$\tilde{F}(\xi_2) = \tilde{f}(\xi_2)\tilde{A}_0^{-1}(\xi_2), \quad \tilde{G}(\xi_1) = \tilde{g}(\xi_1)\tilde{B}_0^{-1}(\xi_1),$$

$$K_1(\xi) = A_{\neq}^{-1}(\xi)\tilde{A}_0^{-1}(\xi_2), \quad K_2(\xi) = A_{\neq}^{-1}(\xi)\tilde{B}_0^{-1}(\xi_1).$$

the following result is presented in the book [23].

Theorem 5 *If $s > 1/2$, conditions*

$$\inf|\tilde{A}_0(\xi_2)| \neq 0, \inf|\tilde{B}_0(\xi_1)| \neq 0$$

hold then the Dirichlet problem (2.10), (2.12) with data $f, g \in H^{s-1/2}(\mathbf{R}_+)$ is equivalent to the system of integral equations (2.11) with unknown functions $\tilde{C}_0, \tilde{D}_0 \in \tilde{H}^{s_0}(\mathbf{R})$ and right hand sides $\tilde{F}, \tilde{G} \in \tilde{H}^{s_0}(\mathbf{R})$.

2.4.2 Integral Condition

We consider the Eq. (2.10) with the following additional conditions

$$\int\limits_0^{+\infty} u(x_1, x_2)dx_1 = f(x_2), \quad \int\limits_0^{+\infty} u(x_1, x_2)dx_2 = g(x_1), \quad \int\limits_{-K} u(x)dx = 0. \tag{2.13}$$

A solution of the problem (2.10), (2.13) is sought in the space $H^s(K)$ [23], and boundary functions are taken from the space $H^{s+1/2}(\mathbf{R}_+)$. Such problem was considered in [27], and it has the solution

$$\tilde{u}(\xi) = A_{\neq}^{-1}(\xi)\left(A_{\neq}(\xi_1, 0)\tilde{g}(\xi_1) + A_{\neq}(0, \xi_2)\tilde{f}(\xi_2)\right) \tag{2.14}$$

under condition that the symbol $A(\xi)$ admits the wave factorization with respect to the quadrant K.

2.5 Error Estimates

To construct a discrete boundary value problem which is good approximation for (2.10), (2.12), and (2.10), (2.13), we need to choose $A_d(\xi)$ and f_d, g_d in a special way. First, we introduce the operator l_h which acts as follows. For a function u defined in $\mathbf{R}$, we take its Fourier transform $\tilde{u}$ then we take its restriction on $\hbar\mathbf{T}$ and periodically extend it to $\mathbf{R}$. Finally, we take its inverse discrete Fourier transform and obtain the function of discrete variable $(l_h u)(\tilde{x})$, $\tilde{x} \in h\mathbf{Z}$. Thus, we put

$$f_d = l_h f, \quad g_d = l_h g.$$

Second, the symbol of digital operator A_d we construct in the same way. If we have the wave factorization for the symbol $A(\xi)$, then we take restrictions of factors on $\hbar\mathbf{T}^2$, and the periodic symbol $A_d(\xi)$ is a product of these restrictions.

2.5.1 *The Discrete Dirichlet Problem*

We introduce the space $\widetilde{\mathbf{H}}^s(\mathbf{R})$ of vector-functions $f = (f_1, f_2)$, $f_j \in H^s(\widetilde{\mathbf{R}})$, $j = 1, 2$,

$$||f||_s \equiv ||f_1||_s + ||f_2||_s,$$

and matrix operators

$$K = \begin{pmatrix} K_1 & I \\ I & K_2 \end{pmatrix}, \qquad k = \begin{pmatrix} k_1 & I_h \\ I_h & k_2 \end{pmatrix},$$

acting in spaces $\widetilde{\mathbf{H}}^{s-æ-1/2}(\widetilde{\mathbf{R}})$ and $\widetilde{\mathbf{H}}^{s-æ-1/2}(\hbar\mathbf{T})$, respectively.

Let us remind that $s_0 = s - æ - 1/2$.

Lemma 2 *Under $s > 1$, $æ > 1$, the operator K is bounded in the space $\widetilde{\mathbf{H}}^{s_0}(\mathbf{R})$,*

$$K : \widetilde{\mathbf{H}}^{s_0}(\mathbf{R}) \to \widetilde{\mathbf{H}}^{s_0}(\mathbf{R}).$$

Proof It is enough to estimate $K_1 f$.

$$||K_1 f||_{s_0}^2 = \int\limits_{-\infty}^{+\infty} (1+|\xi_2|)^{2s_0} |(K_1 f)(\xi_2)|^2 d\xi_2 =$$

$$= \int\limits_{-\infty}^{+\infty} (1+|\xi_2|)^{2s_0} \left| \int\limits_{-\infty}^{+\infty} K_1(\xi_1,\xi_2) f(\xi_1) d\xi_1 \right|^2 d\xi_2 \leq$$

$$\leq \int\limits_{-\infty}^{+\infty} (1+|\xi_2|)^{2s_0} \left(\int\limits_{-\infty}^{+\infty} |K_1(\xi_1,\xi_2)||f(\xi_1)| d\xi_1 \right)^2 d\xi_2 \leq$$

$$\leq const \int\limits_{-\infty}^{+\infty} (1+|\xi_2|)^{2s_0} \left(\int\limits_{-\infty}^{+\infty} (1+|\xi_1|+|\xi_2|)^{-æ} ||f(\xi_1)| d\xi_1 \right)^2 d\xi_2$$

In the inner integral, we apply the Cauchy–Schwartz inequality using the factors $(1+|\xi_1|)^{-s_0}$ and $(1+|\xi_1|)^{s_0}$ for first and second term, respectively, and take into account that $(1+|\xi_1|)^{-s_0} \leq (1+|\xi_1|+|\xi_2|)^{-s_0}$. Thus,

$$||K_1 f||_{s_0}^2 \leq cons||f||_{s_0}^2 \int\limits_{-\infty}^{+\infty} (1+|\xi_2|)^{2s_0} \left(\int\limits_{-\infty}^{+\infty} (1+|\xi_1|+|\xi_2|)^{-2(s_0+æ)} d\xi_1 \right) d\xi_2 \leq$$

$$\leq const||f||_{s_0}^2 \int\limits_{0}^{+\infty} (1+|\xi_2|)^{2s_0} \left(\int\limits_{0}^{+\infty} (1+|\xi_1|+|\xi_2|)^{-2s+1} d\xi_1 \right) d\xi_2 \leq$$

$$\leq cons||f||_{s_0}^2 \int\limits_{0}^{+\infty} (1+|\xi_2|)^{2s_0-2s+2} d\xi_2 \leq const||f||_{s_0}^2$$

$$\times \int\limits_{0}^{+\infty} (1+|\xi_2|)^{-2æ+1} d\xi_2 \leq const||f||^2,$$

in view of the conditions $s > 1$, $æ > 1$. □

Let us denote by $\chi_h : H^s(\mathbf{R}) \to H^s(\hbar\mathbf{T})$ the restriction operator on the segment $\hbar\mathbf{T}$. The restriction operator on the segment $\hbar\mathbf{T}$ in the space $\mathbf{H}^s(\mathbf{R})$ will be denoted by Ξ_h so that for $f = (f_1, f_2) \in \mathbf{H}^s(\mathbf{R})$ we have

$$\Xi_h f = (\chi_h f_1, \chi_h f_2).$$

We remind that h is small enough, $0 < h < 1$.

Lemma 3 *Under* $s > 1$, $æ > 1$ *the operator* K *has the following property*

$$||\Xi_h K - K\Xi_h||_{\widetilde{\mathbf{H}}^{s_0}(\mathbf{R})\to\widetilde{\mathbf{H}}^{s_0}(\mathbf{R})} \leq const\ h^{s-1}.$$

Proof One can easily verify the following

$$\Xi_h K - K\Xi_h = \begin{pmatrix} \chi_h K_1 - K_1\chi_h & 0 \\ 0 & \chi_h K_2 - K_2\chi_h \end{pmatrix}$$

We conclude that

$$((\chi_h K_1 - K_1\chi_h)f)(\xi_2) = \begin{cases} \left(\int\limits_{-\infty}^{-\hbar\pi} + \int\limits_{\hbar\pi}^{+\infty}\right) K_1(\xi_1, \xi_2) f(\xi_1) d\xi_1, & \xi_2 \in \hbar\mathbf{T} \\ -\int\limits_{-\hbar\pi}^{+\hbar\pi} K_1(\xi_1, \xi_2) f(\xi_1) d\xi_1, & \xi_2 \notin \hbar\mathbf{T}. \end{cases}$$

Let us start from the first case and estimate one of integrals.

$$\left|\int\limits_{\hbar\pi}^{+\infty} K_1(\xi) f(\xi_1) d\xi_1\right| \leq \int\limits_{\hbar\pi}^{+\infty} |K_1(\xi)||f(\xi_1)| d\xi_1 \leq const \int\limits_{\hbar\pi}^{+\infty} (1+|\xi|)^{-æ} |f(\xi_1)| d\xi_1 eq$$

(we use Cauchy–Schwartz inequality)

$$\leq const \left(\int\limits_{\hbar\pi}^{+\infty} (1+|\xi|)^{-2æ}(1+|\xi_1|)^{-2s_0} d\xi_1\right)^{1/2}$$

$$\times \left(\int\limits_{\hbar\pi}^{+\infty} |f(\xi_1)|^2 (1+|\xi_1|)^{2s_0} d\xi_1\right)^{1/2} \leq$$

$$\leq const \left(\int\limits_{\hbar\pi}^{+\infty} (1+|\xi|)^{-2(æ+s_0)} d\xi_1\right)^{1/2} ||f||_{s_0},$$

Further,

$$\int_{\hbar\pi}^{+\infty}(1+|\xi|)^{-2(æ+s_0)}d\xi_1 \sim (1+|\xi_2|+\hbar\pi)^{-2(æ+s_0)+1},$$

since $-2(æ+s_0)+1=-2(s-1/2)+1=-2s+2<0$.

Thus, the following inequality is obtained

$$\left|\int_{\hbar\pi}^{+\infty}K_1(\xi)f(\xi_1)d\xi_1\right| \leq const||f||_{s_0}(1+|\xi_2|+\hbar\pi)^{-(æ+s_0)+1/2}.$$

Squaring the latter inequality, multiplying by $(1+|\xi_2|)^{2s_0}$ and integrating over $\hbar\mathbf{T}$ we obtain

$$\int_{\hbar\mathbf{T}}(1+|\xi_2|)^{2s_0}\left|\int_{\hbar\pi}^{+\infty}K_1(\xi)f(\xi_1)d\xi_1\right|^2 d\xi_2 \leq$$

$$\leq const||f||^2_{s_0}\int_{\hbar\mathbf{T}}(1+|\xi_2|+\hbar\pi)^{-2(æ+s_0)+1}(1+|\xi_2|)^{2s_0}d\xi_2 \leq$$

$$\leq const||f||^2_{s_0}(1+\hbar\pi)^{-2s+2}\int_{\hbar\mathbf{T}}(1+|\xi_2|)^{2s_0}d\xi_2,$$

since $1+|\xi_2|+\hbar\pi \geq 1+|\hbar\pi|$, $-2(æ+s_0)+1=-2s+2<0$. Let us note $2s_0<-1$. So, we have

$$\int_{\hbar\mathbf{T}}(1+|\xi_2|)^{2s_0}\left|\int_{\hbar\pi}^{+\infty}K_1(\xi)f(\xi_1)d\xi_1\right|^2 d\xi_2 \leq const||f||^2_{s_0}h^{2(s-1)}.$$

For the second case ($|\xi_2|>\hbar\pi$)

$$\left|\int_{-\hbar\pi}^{+\hbar\pi}K_1(\xi_1,\xi_2)f(\xi_1)d\xi_1\right| \leq const\int_{-\hbar\pi}^{+\hbar\pi}(1+|\xi|)^{-æ}|f(\xi_1)|d\xi_1 \leq$$

$$\leq const \left(\int_{-\hbar\pi}^{\hbar\pi} (1+|\xi|)^{-2æ}(1+|\xi_1|)^{-2s_0} d\xi_1 \right)^{1/2}$$

$$\times \left(\int_{-\hbar\pi}^{\hbar\pi} |f(\xi_1)|^2 (1+|\xi_1|)^{2s_0} d\xi_1 \right)^{1/2}.$$

Here we have used the Cauchy–Schwartz inequality once again. Taking into account the inequality $1+|\xi| \geq 1+|\xi_1|$ we obtain the estimate

$$\int_{-\hbar\pi}^{\hbar\pi} (1+|\xi|)^{-2æ}(1+|\xi_1|)^{-2s_0} d\xi_1 \leq const \int_{0}^{\hbar\pi} (1+\xi_1+|\xi_2|)^{-2(s_0+æ)} d\xi_1 \leq$$

$$\leq const(1+|\xi_2|)^{-2s+2} \leq const(1+\hbar\pi)^{-2s+2},$$

in view of $-2(s_0+æ) = -2(s-1/2) = -2s+1$.

Therefore, we obtain the inequality

$$\left| \int_{-\hbar\pi}^{+\hbar\pi} K_1(\xi_1, \xi_2) f(\xi_1) d\xi_1 \right| \leq const||f||_{s_0} h^{s-1}.$$

Multiplying by $(1+|\xi_2|)^{s_0}$ the latter inequality, squaring and integrating over $\mathbf{R} \setminus \hbar\mathbf{T}$ we find

$$\int_{\mathbf{R}\setminus\mathbf{T}} (1+|\xi_2|)^{2s_0} \left| \int_{\hbar\pi}^{+\infty} K_1(\xi) f(\xi_1) d\xi_1 \right|^2 d\xi_2 \leq const||f||_{s_0}^2 h^{2(s-1)} \int_{\hbar\pi}^{+\infty} (1+\xi_2)^{2s_0} d\xi_2.$$

The latter integral converges since $2s_0 < -1$.

The same estimates are valid for K_2. □

Corollary 1 *If $s > 1$, $æ > 1$ and the operator K is invertible then for the operator K^{-1} the same estimate holds*

$$||\Xi_h K^{-1} - K^{-1}\Xi_h||_{\widetilde{\mathbf{H}}^{s_0}(\mathbf{R}) \to \widetilde{\mathbf{H}}^{s_0}(\mathbf{R})} \leq const\ h^{s-1}.$$

Proof Indeed, we have

$$\Xi_h K^{-1} - K^{-1}\Xi_h = K^{-1} K \Xi_h K^{-1} - K^{-1}\Xi_h K K^{-1} = K^{-1}(\Xi_h K - K\Xi_h)K^{-1},$$

and therefore

$$||\Xi_h K^{-1} - K^{-1}\Xi_h||_{\widetilde{\mathbf{H}}^{s_0}(\mathbf{R})\to\widetilde{\mathbf{H}}^{s_0}(\mathbf{R})} \le ||K^{-1}||\cdot||\Xi_h K - K\Xi_h||_{\widetilde{\mathbf{H}}^{s_0}(\mathbf{R})\to\widetilde{\mathbf{H}}^{s_0}(\mathbf{R})}\cdot||K^{-1}||,$$

□

Lemma 4 *For* $æ > 1$ *the following estimate*

$$|K_1(\xi) - k_1(\xi)| \le const\ (1 + |\xi|)^{-æ} h^{æ-1}, \quad \xi \in \hbar\mathbf{T}^2.$$

holds.

Proof Indeed, according to our choice for $A^{-1}_{d,\neq}(\xi)$

$$|K_1(\xi) - k_1(\xi)| = |A^{-1}_{\neq}(\xi)\tilde{A}^{-1}_0(\xi_2) - A^{-1}_{d,\neq}(\xi)\tilde{a}^{-1}_0(\xi_2)|$$
$$\le const(1 + |\xi|)^{-æ}|\tilde{A}_0(\xi_2) - \tilde{a}_0(\xi_2)|.$$

Let us consider $|\tilde{A}_0(\xi_2) - \tilde{a}_0(\xi_2)|$. Then

$$|\tilde{A}_0(\xi_2) - \tilde{a}_0(\xi_2)| = \left| \int_{-\infty}^{\infty} A^{-1}_{\neq}(\xi)d\xi_1 - \int_{-\hbar\pi}^{\hbar\pi} A^{-1}_{d,\neq}(\xi)d\xi_1 \right| \le$$

$$\le const \int_{\hbar\pi}^{+\infty} (1 + |\xi|)^{-æ} d\xi_1 \le const\ (1 + |\xi_2| + \hbar)^{-æ+1} \le const\ h^{æ-1}$$

for enough small h. It implies the following implication $\inf|\tilde{A}_0(\xi_2)| \neq 0 \Longrightarrow \inf|\tilde{a}_0(\xi_2)| \neq 0$ for enough small h.

Collecting the obtained estimates we complete the proof. □

Let us introduce the operator $\Xi_h K \Xi_h$. Lemma 3 implies that for enough small h an invertibility of the operator $\Xi_h K \Xi_h$ in the space $\widetilde{\mathbf{H}}^{s-æ-1/2}(\hbar\mathbf{T})$ follows from an invertibility of the operator K in the space $\widetilde{\mathbf{H}}^{s-æ-1/2}(\mathbf{R})$ [24]. Moreover,

$$||(\Xi_h K \Xi_h)^{-1}||_{\widetilde{\mathbf{H}}^{s_0}(\hbar\mathbf{T})\to\widetilde{\mathbf{H}}^{s_0}(\hbar\mathbf{T})} \le const$$

for enough small h.

Lemma 5 *If* $æ > 1$ *then a comparison for norms of operators* $\Xi_h K \Xi_h$ *and* k *is given by the estimate*

$$||\Xi_h K \Xi_h - k||_{\widetilde{\mathbf{H}}^{s_0}(\hbar\mathbf{T})\to\widetilde{\mathbf{H}}^{s_0}(\hbar\mathbf{T})} \le const\ h^{æ-1}.$$

Proof The difference of the operators looks as follows

$$\Xi_h K \Xi_h - k = \begin{pmatrix} \chi_h K_1 \chi_h - k_1 & 0 \\ 0 & \chi_h K_2 \chi_h - k_2 \end{pmatrix},$$

and we need to estimate the norm $\chi_h K_j \chi_h - k_j$, $j = 1, 2$. Let us estimate K_1 using Lemma 3. So, we obtain

$$||\chi_h K_1 \chi_h f - k_1 f||_{s_0}^2 = \int_{\hbar\mathbf{T}} (1 + |\xi_2|)^{2s_0} \left| \int_{\hbar\mathbf{T}} [K_1(\xi) - k_1(\xi)] f(\xi_1) d\xi_1 \right|^2 d\xi_2 \leq$$

$$\leq \int_{\hbar\mathbf{T}} (1 + |\xi_2|)^{2s_0} \left(\int_{\hbar\mathbf{T}} |K_1(\xi) - k_1(\xi)||f(\xi_1)| d\xi_1 \right)^2 d\xi_2 \leq$$

$$\leq \; const \; h^{2æ-2} \int_{\hbar\mathbf{T}} (1 + |\xi_2|)^{2s_0} \left(\int_{\hbar\mathbf{T}} (1 + |\xi|)^{-æ} |f(\xi_1)| d\xi_1 \right)^2 d\xi_2.$$

In the inner integral, we apply the Cauchy–Schwartz inequality with the factor $(1 + |\xi_1|)^{s_0}$

$$\int_{\hbar\mathbf{T}} (1 + |\xi|)^{-æ} |f(\xi_1)| d\xi_1 \leq ||f||_{s_0} \left(\int_{\hbar\mathbf{T}} (1 + |\xi|)^{-2æ} (1 + |\xi_1|)^{-2s_0} d\xi_1 \right)^{1/2} \leq$$

$$\leq ||f||_{s_0} \left(\int_0^{+\infty} (1 + |\xi|)^{-2(æ+s_0)} d\xi_1 \right)^{1/2} \leq ||f||_{s_0} (1 + |\xi_2|)^{-(æ+s_0)+1/2}$$

$$= ||f||_{s_0} (1 + |\xi_2|)^{-s+1},$$

since $s_0 = s - æ - 1/2$. We have

$$||\chi_h K_1 \chi_h f - k_1 f||_{s_0}^2 \leq \; const \; h^{2æ-2} ||f||_{s_0}^2 \int_{-\hbar\pi}^{\hbar\pi} (1 + |\xi_2|)^{2s_0 - 2s + 2} d\xi_2 \leq$$

$$\leq const\ h^{2æ-2}||f||_{s_0}^2 \int\limits_0^{+\infty} (1+|\xi_2|)^{-2æ+1} d\xi_2 \leq const\ h^{2æ-2}||f||^2,$$

in view of $s_0 + 1 - s = -æ + 1/2$. Taking a square root, we obtain the required assertion. □

At this time, we are able to compare discrete and continuous solutions.

Theorem 6 *Let the conditions of Theorem 5 hold and $s > 1$, $æ > 1$. A comparison for solutions of problems (2.10), (2.12) (2.3), (2.5) for enough small h is given by the estimate*

$$||u - u_d||_{\tilde{H}^s(\hbar\mathbf{T}^2)} \leq const\ h^{s-1}(||f||_{s-1/2} + ||g||_{s-1/2})$$

where const does not depend on h.

Proof We start from a comparison of solutions of systems (2.6) and (2.11).

We have the continuous solution

$$\tilde{u}(\xi) = A_{\neq}^{-1}(\xi)(\tilde{C}_0(\xi_1) + \tilde{D}_0(\xi_2))$$

and the discrete one

$$\tilde{u}_d(\xi) = A_{d,\neq}^{-1}(\xi)(\tilde{c}_0(\xi_1) + \tilde{d}_0(\xi_2)).$$

Taking into account that $\xi \in \hbar\pi$, we obtain the conclusions below.

Let us denote by $\tilde{\Phi}_d$ and $\tilde{\Phi}$ vectors with components $(\tilde{F}_d, \tilde{G}_d)^T$ and $(\tilde{F}, \tilde{G})^T$, $\tilde{C}$ and $\tilde{c}$ are vectors with components– $(\tilde{C}_0, \tilde{D}_0)^T$ $(\tilde{c}_0, \tilde{d}_0)^T$, respectively. Then we write

$$\tilde{C} = K^{-1}\tilde{\Phi}, \qquad \tilde{c} = k^{-1}\tilde{\Phi}_d,$$

where C_1, C_2 and c_1, c_2, j-th coordinates of vectors $\tilde{C}$, $\tilde{c}$, $j = 1, 2$. Therefore,

$$(\chi_h\tilde{u})(\xi) - \tilde{u}_d(\xi) = \chi_h A_{\neq}^{-1}(\xi)\Big((\tilde{C}_0(\xi_1) - \tilde{c}_0(\xi_1)) + (\tilde{D}_0(\xi_2) - \tilde{d}_0(\xi_2))\Big) =$$

$$= \chi_h A_{\neq}^{-1}(\xi)\Big((K^{-1}\tilde{\Phi})_1(\xi_1) - (k^{-1}\tilde{\Phi}_d)_1(\xi_1) + (K^{-1}\tilde{\Phi})_2(\xi_2) - (k^{-1}\tilde{\Phi}_d)_2(\xi_2)\Big).$$

It implies that it is enough to estimate the norm $||\varXi_h K^{-1}\Phi - k^{-1}\Phi_d||_{\mathbf{H}^{s_0}(\hbar\mathbf{T})}$. We write

$$\varXi_h K^{-1}\Phi - k^{-1}\Phi_d = (\varXi_h K^{-1}\Phi - K^{-1}\varXi_h\Phi) + (K^{-1}\varXi_h\Phi - k^{-1}\Phi_d).$$

We use Corollary 1 for an estimate of first summand. We obtain

$$||\Xi_h K^{-1}\Phi - K^{-1}\Xi_h\Phi||_{s_0} \leq const\ h^{s-1}||\Phi||_{s_0}$$

and then the second summand, we represent as the sum

$$K^{-1}\Xi_h\Phi - k^{-1}\Phi_d = (K^{-1}\Xi_h\Phi - k^{-1}\Xi_h\Phi) + (k^{-1}\Xi_h\Phi - k^{-1}\Phi_d),$$

each summand we will estimate separately.

Let us consider $k^{-1}\Xi_h\Phi - k^{-1}\Phi_d$. Since norm of the operator k^{-1} is bounded by a constant non-depending on h, we obtain

$$\begin{aligned}||k^{-1}\Xi_h\Phi - k^{-1}\Phi_d||_{s_0} &\leq const||\Xi_h\Phi - \Phi_d||_{s_0}\\ &\leq const(||\chi_h F - F_d||_{s_0} + ||\chi_h G - G_d||_{s_0}).\end{aligned}$$

Last step is to estimate, for example, $||\chi_h F - F_d||_{s_0}$. We have

$$||\chi_h F - F_d||^2_{s_0} = \int\limits_{-\hbar\pi}^{\hbar\pi} |\tilde{f}(\xi_2)A_0^{-1}(\xi_2) - \tilde{f}_d(\xi_2)a_0^{-1}(\xi_2)|^2(1+|\xi_2|)^{2s_0}d\xi_2 \leq$$

$$\leq const\ h^{2æ-2}\int\limits_{-\hbar\pi}^{\hbar\pi} |\tilde{f}(\xi_2)|^2(1+|\xi_2|)^{2s_0}d\xi_2 \leq const\ h^{2æ-2}||f||^2_{s_0}$$

according to coincidence for f_d and f on $\hbar\mathbf{T}$ and the estimate of Lemma 4.

Left summands can be estimated by the following operator identity

$$K^{-1} - k^{-1} = K^{-1}(k-K)k^{-1}.$$

(Let us remind that an invertibility of the operator k follows from an invertibility of the operator K.) Therefore, comparing over $\hbar\mathbf{T}$

$$K^{-1}\Xi_h\Phi - k^{-1}\Xi_h\Phi = \Xi_h(K^{-1} - k^{-1})\Xi_h\Phi = \Xi_h K^{-1}(k-K)k^{-1}\Xi_h\Phi,$$

and taking into account Lemma 5, we have the estimate

$$||K^{-1}\Xi_h\Phi - k^{-1}\Xi_h\Phi||_{s_0} \leq const\ h^{æ-1}||\Phi||_{s_0} \leq const\ h^{æ-1}(||f||_{s_0} + ||g||_{s_0}).$$

Summing obtained estimates, we complete the proof, taking into account mapping properties of operators which admit to obtain H^s-norm. □

2.5.2 Nonlocal Discrete Boundary Value Problem

For such f_d, g_d and the symbol $A_d(\xi)$, we obtain the following result.

Theorem 7 *Let $f, g \in S(\mathbf{R})$, æ > 1. Then we have the following estimate for solutions u and u_d of the continuous problem (2.10), (2.13) and the discrete one (2.3), (2.7)*

$$|u(\tilde{x}) - u_d(\tilde{x})| \le C(f, g)h^{\beta},$$

where the const $C(f, g)$ depends on functions f, g, $\beta > 0$ can be an arbitrary number.

Proof First, let us note that solvability conditions for the problem (2.10), (2.13) guarantee satisfying solvability conditions of the problem (2.3), (2.7) for enough small h.

Further, we need to compare two functions (2.9) and (2.14), more exactly their inverse discrete Fourier transform and inverse Fourier transform at points $\tilde{x} \in K_d$. We have

$$u_d(\tilde{x}) - u(\tilde{x}) = \frac{1}{4\pi^2}\left(\int\limits_{\hbar\mathbf{T}^2} e^{i\tilde{x}\cdot\xi}\tilde{u}_d(\xi)d\xi - \int\limits_{\mathbf{R}^2} e^{i\tilde{x}\cdot\xi}\tilde{u}(\xi)d\xi\right) =$$

$$= \frac{1}{4\pi^2}\int\limits_{\mathbf{R}^2\setminus\hbar\mathbf{T}^2} e^{i\tilde{x}\cdot\xi}A_{\neq}^{-1}(\xi)\left(A_{\neq}(\xi_1, 0)\tilde{g}(\xi_1) + A_{\neq}(0, \xi_2)\tilde{f}(\xi_2)\right)d\xi,$$

since according to our choice for A_d, f_d, g_d the functions $\tilde{u}_d$ and $\tilde{u}$ coincide in points $\xi \in \hbar\mathbf{T}^2$.

We will estimate one summand.

$$\left|\frac{1}{4\pi^2}\int\limits_{\mathbf{R}^2\setminus\hbar\mathbf{T}^2} e^{i\tilde{x}\cdot\xi}A_{\neq}^{-1}(\xi)A_{\neq}(\xi_1, 0)\tilde{g}(\xi_1)d\xi\right|$$

$$\le C\int\limits_{\hbar\pi}^{+\infty}\frac{d\xi_2}{(1 + |\xi_1| + |\xi_2|)^{æ}}\int\limits_{\hbar\pi}^{+\infty}|\xi_1|^{-\gamma}d\xi_1,$$

since $\tilde{g} \in S(\mathbf{R})$. It implies the required estimate. □

2.6 Conclusion

We have considered very simple variants of discrete boundary value problems for digital operators. Particularly, our considerations are based on Theorem 2 which gives a general solution of our discrete equation. There are a lot of different situations in this studying, for example, the case $n > 1$ which permits to use more complicated boundary conditions, or the case $n \in \mathbf{N}, n < 0$ which admits to introduce more unknowns in Eq. (2.2) and potential like discrete operators similar considered ones in [18]. We work in this direction and present these studies in forthcoming publications.

References

1. C. Kuttler, J. Müller, *Methods and Models in Mathematical Biology* (Springer, New York, 2015)
2. A.V. Oppenheim, R.W. Schafer, *Discrete-Time Signal Processing* (Prentice Hall, Hoboken, 1989)
3. S.G. Mikhlin, S. Prößdorf, *Singular Integral Operators* (Springer, Berlin, 1986)
4. M. Taylor, *Pseudodifferential Operators* (Princeton University Press, Princeton, 1981)
5. F. Treves, *Introduction to Pseudodifferential Operators and Fourier Integral Operators* (Springer, New York, 1980)
6. G. Eskin, *Boundary Value Problems for Elliptic Pseudodifferential Equations* (AMS, Providence, 1981)
7. A. Samarskii, *The Theory of Difference Schemes* (CRC Press, Boca Raton, 2001)
8. V. Ryaben'kii, *Method of Difference Potentials and its Applications* (Springer, Berlin, 2002)
9. S.S. Cheng, *Partial Difference Equations* (Taylir & Francis, New York, 2003)
10. I.C. Gohberg, I.A. Feldman, *Convolution Equations and Projection Methods for Their Solution* (AMS, Providence, 1974)
11. R. Hagen, S. Roch, B. Silbermann, *C^*-Algebras and Numerical Analysis* (Dekker, New York, 2001)
12. S. Prößdorf, B. Silbermann, *Numerical Analysis for Integral and Related Operator Equations* (Birkhäuser, Basel, 1991)
13. A. Vasilyev, V. Vasilyev, Discrete singular operators and equations in a half-space. Azerb. J. Math. **3**, 84–93 (2013)
14. A. Vasil'ev, V. Vasil'ev, On the solvability of certain discrete equations and related estimates of discrete operators. Doklady Math. **92**, 585–589 (2015). https://doi.org/10.1134/S1064562415050312
15. V. Vasilyev, The periodic Cauchy kernel, the periodic Bochner kernel, discrete pseudo-differential operators, in ed. by T. Simos, C. Tsitouras, *Proceedings of the International Conference on Numerical Analysis and Applications (ICNAAM-2016), AIP Conference Proceedings*, vol. 1863 (AIP Publishing, Melville, 2017), pp. 140014-1–140014-4 .https://doi.org/10.1063/1.4992321
16. V. Vasilyev, Discreteness, periodicity, holomorphy, and factorization, in ed. by C. Constanda, M. Riva Dalla, P.D. Lamberti, P. Musolino, *Integral Methods in Science and Engineering, Theoretical Technique*, vol. 1 (Birkhäuser, New York, 2017), pp.315–324. https://doi.org/10.1007/978-3-319-59384-5_28
17. A.V. Vasilyev, V.B. Vasilyev, Pseudo-differential operators and equations in a discrete half-space. Math. Model. Anal. **23**, 492–506 (2018). https://doi.org/10.3846/mma.2018.029

18. A.V. Vasilyev, V.B. Vasilyev, On some discrete potential like operators. Tatra Mt. Math. Publ. **71**, 195–212 (2018). https://doi.org/10.2478/tmmp-2018-0016
19. O.A. Tarasova, V.B. Vasilyev, To the theory of discrete boundary value problems. 4Open **2**, 1–7 (2019). https://doi.org/10.1051/fopen/2019009
20. F.D. Gakhov, *Boundary Value Problems* (Dover Publications, Mineola, 1981)
21. N.I. Muskhelishvili, *Singular Integral Equations* (North Holland, Amsterdam, 1976)
22. V.S. Vladimirov, *Methods of the Theory of Functions of Many Complex Variables* (Dover Publications, Mineola, 2007)
23. V.B. Vasil'ev, *Wave Factorization of Elliptic Symbols: Theory and Applications* (Kluwer Academic Publishers, Dordrecht, 2000)
24. A.V. Kozak, On a projection method for solving operator equations in Banach space. Sov. Math. Dokl. **14**, 1159–1162 (1973)
25. V.S. Vladimirov, *Generalized Functions in Mathematical Physics* (Mir, Moscow, 1979)
26. L. Frank, Spaces of network functions. Math. USSR Sb. **15**, 182–226 (1971). https://doi.org/10.1070/SM1971v015n02ABEH001541
27. V. Vasil'ev, On some new boundary-value problems in nonsmooth domains. J. Math. Sci. **173**, 225–230 (2011)

Chapter 3
Lipschitz Quasistability of Impulsive Cohen–Grossberg Neural Network Models with Delays and Reaction-Diffusion Terms

Ivanka Stamova, Trayan Stamov, and Gani Stamov

3.1 Introduction

Due to their numerous applications, the literature on the class of neural networks with reaction-diffusion terms is very abundant [1, 2], including delayed models [3–6]. In fact the diffusion and delay effects are very common in the neural networks' modeling applications. Among other applied areas, neural networks with delays and reaction-diffusion terms are very useful tools for designing of epidemic reaction–diffusion models studied in medicine, biology, virology, and epidemiology [7–9].

There is also a very complete literature on the qualitative dynamical properties of the Cohen–Grossberg type [10–12] delayed reaction-diffusion neural networks [13–17]. Such type of neural network systems is very useful in different engineering design tasks [18].

In addition, the classes of reaction-diffusion neural networks and Cohen–Grossberg type neural networks with reaction-diffusion terms under impulsive perturbations are well studied by the researchers. We will refer to [19–22] for some results on the theory and applications of impulsive reaction-diffusion neural networks and to [23–32] for important qualitative results on impulsive Cohen–Grossberg type neural networks with and without reaction-diffusion terms.

The consideration of impulsive perturbations in delayed reaction-diffusion neural networks, including these of Cohen–Grossberg type, is motivated by the fact that different types of impulses exist in the real-world applications, and they can affect

I. Stamova (✉) · G. Stamov
University of Texas at San Antonio, San Antonio, TX, USA
e-mail: ivanka.stamova@utsa.edu; gani.stamov@utsa.edu

T. Stamov
Technical University of Sofia, Sofia, Bulgaria
e-mail: tstamov@tu-sofia.bg

C. M. A. Pinto, C. M. Ionescu (eds.), *Computational and Mathematical Models in Biology*, Nonlinear Systems and Complexity 38,
https://doi.org/10.1007/978-3-031-42689-6_3

the dynamical behavior of the models. Hence, it is essential to take into account the existence and to study the effects of some impulsive disruptions. The neural network models under impulsive perturbations are mainly modeled by impulsive differential equations [33–36]. Impulses are also very often considered as an impulsive control strategy. Such control has proven advantages since it can be applied only at some discrete times [37, 38]. As such, impulsive control can reduce control cost and the amount of transmitted information drastically [38]. In addition, in many cases when continuous control inputs are not appropriate, the impulsive control modeling approach can give an efficient way to study the model. The impulsive control strategies open up the possibility of creating impulsive control architectures for neuronal network models and successfully pursue long-term goals despite short-term attacks and shocks. Hence, understanding the control power of some impulsive effects on the dynamical behavior of neural network models is very important.

Note that, there are two main types of impulsive models—with fixed moments of impulsive effects and with variable impulsive perturbations. The second class is more general, and its investigations are more complicated due to the presence of phenomena such as "beating" of the solutions, loss of the property of autonomy, bifurcation, "merging" of solutions, and some others [39–41].

The above findings clarify the reason why most of the results on the impulsive Cohen-Grossberg-type neural networks with delays are devoted to models with fixed moments of impulsive controls. The general class of impulsive delayed Cohen–Grossberg-type neural networks with variable impulsive perturbations is a subject of consideration of a very few recent studies [42–44].

It is well known that the stability of the states is crucial for the design of a neural network model and its functioning. That is why numerous stability criteria are proposed for delayed Cohen–Grossberg-type neural networks with delays and reaction-diffusion terms under impulsive perturbations at fixed times [26, 27, 29, 31, 32]. Note that the corresponding stability theory for such models with variable impulsive perturbations is not completely developed. Also, most of the existing stability results are devoted to asymptotic and exponential stability of the states.

However, there are numerous real-world phenomena, where other stability concepts are more appropriate. The extended notion of Lipschitz stability introduced in [45] for ordinary differential systems lies between uniform stability and the asymptotic stability in variation concepts [46]. It coincides with uniform stability for linear systems and has the advantage that unlike uniform stability, the Lipschitz stability behavior of the considered nonlinear system is inherited by the linearized system [45, 46]. Some important Lipschitz stability results for systems under impulsive perturbations are reported in [47, 48]. Also, very recently, there have been many indications that the notion will be very relevant in the study of dynamical properties of neural networks given the Lipschitz continuity of their activation functions and Lipschitz measures used in such networks [49, 50]. Recently, the Lipschitz stability concept is used by numerous researchers in the study of important inverse problems [51–53]. For impulsive delayed Cohen–Grossberg-type neural networks with reaction-diffusion terms, the idea of Lipschitz stability is applied only in [54].

However, when the impulsive perturbations are not at fixed times, the impulsive instants for different states are not the same, and we cannot apply the Lipschitz stability notion for an arbitrary solution in the classical sense [45]. For such systems, the notion of Lipschitz quasistability is defined in [55, 56] and has been applied by several researchers [57, 58] including impulsive systems with delays [59].

Since this important concept is not yet applied to neural network systems, we will adopt it to delayed Cohen–Grossberg-type neural networks with reaction-diffusion terms under variable impulsive perturbations. We expect that our analysis will be helpful to the development of the Lipschitz stability theory of different classes of applied neural network models.

The main contributions of our research are:

(1) The concept of Lipschitz quasistability is introduced to an impulsive Cohen-Grossberg-type neural networks with delays and reaction-diffusion terms.
(2) An impulsive control methodology is designed by variable impulsive perturbations.
(3) By the use of comparison lemmas and the Lyapunov function methodology, uniform Lipschitz quasistability criteria based on the impulsive control law are provided.
(5) Examples are presented to demonstrate the strength of the derived criteria.

The details in our presentation are as follows. Section 3.2 is focused on some preliminary results related to the delayed Cohen–Grossberg-type neural networks with reaction-diffusion terms with unfixed moments of impulses. The notion of Lipschitz quasistability is defined, and some auxiliary lemmas are provided. In Sect. 3.3, the comparison approach and Lyapunov technique are applied to establish the main uniform Lipschitz quasistability results. Examples are discussed in Sect. 3.4. Some concluding notes are included in Sect. 3.5.

3.2 Model Description and Preliminaries

We will begin with the definition of the main notations. The set of all $n-$dimensional points will be defined by $\mathbb{R}^N$. The norm of a $z = (z_1, z_2, \ldots, z_N)^T \in \mathbb{R}^N$ will be defined as $||z|| = \sum_{p=1}^{N} |z_p|$. We will also consider an open and bounded set $\mathcal{B} \subset \mathbb{R}^N$ with a smooth boundary $\partial\mathcal{B}$ and a measure defined as mes $\mathcal{B} > 0$. Let $\mathbf{0} = (0, 0, \ldots, 0)^T \in \mathcal{B}$, and let $\mathbb{R}_+ = [0, \infty)$.

In this study, we will propose the Lipschitz quasistability concept for the following Cohen–Grossberg neural network model with reaction-diffusion terms, time-varying delays, and variable impulsive perturbations

$$\begin{cases} \dfrac{\partial y_o(t,z)}{\partial t} = \displaystyle\sum_{p=1}^{N} \frac{\partial}{\partial z_p}\Big(D_{op}\frac{\partial y_o(t,z)}{\partial z_p}\Big) - a_o(y_o(t,z))\Big[b_o(y_o(t,z)) \\ \qquad - \displaystyle\sum_{\iota=1}^{M} c_{o\iota}(t) f_\iota\big(y_\iota(t,z)\big) \\ \qquad - \displaystyle\sum_{\iota=1}^{M} \omega_{o\iota}(t) g_\iota\big(y_\iota(t-h_\iota(t),z)\big)\Big],\ t \neq \tau_i(y_o(t,z)), \\ \Delta y_o(t,z) = Y_{oi}(y_o(t,z)),\ t = \tau_i(y_o(t,z)), \end{cases} \tag{3.1}$$

where $t > 0$, $o = 1, 2, \ldots, M$, $M \geq 2$ determines the number of notes in the model, $y_o(t,z)$ is the state variable of the o-th neuron in space $z \in \mathcal{B}$, and at time t, $D_{op} = D_{op}(t,z) \geq 0,\ p = 1, 2, \ldots, N,\ o = 1, 2, \ldots, M$ is the diffusion coefficient, $c_{o\iota}(t)$ and $\omega_{o\iota}(t)$ represent the connection weights, $a_0(\cdot)$ express the amplification functions, $b_o(\cdot)$ stand for appropriately behaved functions, $o = 1, 2, \ldots, M$, $f_\iota(\cdot)$ and $g_\iota(\cdot)$, $\iota = 1, \ldots, M$ are the activation functions, $h_\iota(t)$ are the time-varying delays and $0 \leq h_\iota(t) \leq h$, $\iota = 1, \ldots, M$, $\tau_i(y)$ are continuous functions that determine the impulsive instants, the functions Y_{oi} express the abrupt variations in the state variable $y_o(t,z)$ for $t = \tau_i(y)$, $\Delta y_o(t,z) = y_o(t^+,z) - y_o(t^-,z)$, $y_o(t^-,z) = y_o(t,z)$ is the state variable of the o-th neuron before a jump effect at $t = \tau_i(y_o(t,z))$, $y_o(t^+,z)$ is the impulsively controlled output of the o-th node at $t = \tau_i(y_o(t,z))$, $o = 1, 2, \ldots, M$, $i = 1, 2, \ldots$.

The initial and boundary conditions related to the introduced model (3.1) are

$$y_o(s,z) = \vartheta_o(s,z),\ s \in [-h, 0],\ z \in \mathcal{B},\ o = 1, 2, \ldots, M, \tag{3.2}$$

$$y_o(t,z) = 0,\ t \in [-h, \infty),\ z \in \partial\mathcal{B},\ o = 1, 2, \ldots, M, \tag{3.3}$$

where $\vartheta = (\vartheta_1, \vartheta_2, \ldots, \vartheta_M)^T$ is defined on $[-h, 0] \times \mathcal{B}$, $\vartheta_o(s^+, z)$, $\vartheta_o(s^-, z)$ exist, $\vartheta_o(s^-, x) = \vartheta_o(s,z)$, $o = 1, 2 \ldots, M$, for all points $(s,z) \in [-h, 0] \times \mathcal{B}$ which must be finite number. The class of all such initial functions will be denoted by $\mathcal{PC}$ and PCB will be the notation of the class of all functions $\vartheta \in \mathcal{PC}$ that are bounded on $[h, 0] \times \mathcal{B}$.

The solution of the initial value boundary problem (3.1), (3.2), (3.3) will be denoted by $y(t,z) = y(t,z;\vartheta)$,

$$y(t,z) = (y_1(t,z), y_2(t,z), \ldots, y_M(t,z))^T,\ t > 0,\ z \in \mathcal{B}.$$

Introduce the notation

$$\sigma_i = \big\{(t,y) \in [0,\infty) \times \mathbb{R}^M :\ t = \tau_i(y(t,z))\big\},$$

i.e., σ_i, $i = 1, 2, \ldots$, are hypersurfaces with equations $t = \tau_i(y)$.

Denote by t_{l_1}, $t_{l_2}, \ldots$ the points at which the integral surface of $y(t, z)$ meets the hypersurfaces σ_i, i.e., each of the points t_{l_1} $t_{l_2}, \ldots$ is a solution of one of the equations $t = \tau_i(y(t, z))$, $i = 1, 2, \ldots,$

Following [39, 40, 42, 43, 57, 58], for the solution $y(t, z)$ of the model (1), we have

$$y_o(t_{l_i}^-, x) = y_o(t_{l_i}, z),$$

$$y_o(t_{l_i}^+, z) = y_o(t_{l_i}, x) + Y_{ol_i}(y_o(t_{l_i}, z)),$$

$z \in \mathcal{B}$, $o = 1, 2, \ldots, M$, $i = 1, 2, \ldots$.

Note that, in general, $i \neq l_i$. Also, different solutions may have different impulsive moments [57, 58]. Let $\vartheta^* \in PCB$, and consider the model:

$$\begin{cases} \dfrac{\partial y_o^*(t, z)}{\partial t} = \displaystyle\sum_{p=1}^{N} \frac{\partial}{\partial z_p}\left(D_{op}\frac{\partial y_o^*(t, z)}{\partial z_p}\right) - a_o(y_o^*(t, z))\Big[b_o(y_o^*(t, z)) \\ \qquad - \displaystyle\sum_{\iota=1}^{M} c_{o\iota}(t) f_\iota\big(y_\iota^*(t, z)\big) \\ \qquad - \displaystyle\sum_{\iota=1}^{M} \omega_{o\iota}(t) g_\iota\big(y_\iota^*(t - h_\iota(t), z)\big)\Big],\ t \neq \tau_i(y_o^*(t, z)), \\ \Delta y_o^*(t, z) = Y_{oi}(y_o^*(t, z)),\ t = \tau_i(y_o^*(t, z)), \end{cases} \tag{3.4}$$

where $y^*(t, z) = y^*(t, z; \vartheta^*)$,

$$y^*(t, z) = (y_1^*(t, z), y_2^*(t, z), \ldots, y_M^*(t, z))^T,\ t > 0,\ z \in \mathcal{B}$$

is the solution of (3.4) that satisfies

$$y_o^*(s, z) = \vartheta_o^*(s, z),\ s \in [-h, 0],\ z \in \mathcal{B},\ o = 1, 2, \ldots, M, \tag{3.5}$$

$$y_o^*(t, z) = 0,\ t \in [-h, \infty),\ z \in \partial\mathcal{B},\ o = 1, 2, \ldots, M, \tag{3.6}$$

with impulsive instants $t_{l_1}^*$, $t_{l_2}^*, \ldots$.

For a $y(t, z) = (y_1(t, z), y_2(t, z), \ldots, y_M(t, z))^T \in \mathbb{R}^M$, $z \in \mathcal{B}$, $t \in \mathbb{R}_+$, we consider the following norm

$$||y(t,\cdot)||_1 = \left[\int_{\mathcal{B}} \sum_{o=1}^{M} y_o^2(t,z)dz\right]^{1/2}$$

Also, for initial functions $\vartheta \in \mathcal{PC}$, we will use the norm $||\cdot||_{h_1}$ defined as

$$||\vartheta||_{h_1} = \sup_{-h\leq s\leq 0} ||\vartheta(s,\cdot)||_1.$$

We will study the behavior of the state variables of the models (3.1) and (3.4) under the following assumptions:

A1. The diffusion coefficients $D_{op} = D_{op}(t,z)$ are positive continuous functions, and there exist constants $k_{op} \geq 0$ such that $D_{op}(t,z) \geq k_{op}$ for $(t,z) \in \mathbb{R}_+ \times \mathcal{B}$, $o = 1, 2, \dots, M$, $p = 1, 2, \dots, N$.
A2. The connection weight s $c_{o\iota}(t)$ and $\omega_{o\iota}(t)$ are continuous real-valued functions, and $\overline{c}_{o\iota} = \sup_{t\in\mathbb{R}_+} |c_{o\iota}(t)|$, $\overline{\omega}_{o\iota} = \sup_{t\in\mathbb{R}_+} |\omega_{o\iota}(t)|$, $o, \iota = 1, 2, \dots, M$.
A3. The amplification functions $a_0(\cdot)$ are nonnegative and continuous, and there exist positive constants $\underline{a}_o$ and $\overline{a}_o$ such that $\underline{a}_o \leq a_o(\xi) \leq \overline{a}_o$ for $\xi \in \mathbb{R}$, $o = 1, 2, \dots, M$.
A4. The functions b_o, $o = 1, 2, \dots, M$, are continuous on $\mathbb{R}$, there exist positive constants B_0 with such that

$$\frac{b_o(\xi_1) - b_o(\xi_2)}{\xi_1 - \xi_2} \geq B_o > 0$$

for $\xi_1, \xi_2 \in \mathbb{R}$, $\xi_1 \neq \xi_2$, $o = 1, 2, \dots, M$.
A5. The activation functions $f_\iota(\cdot)$ and $g_\iota(\cdot)$, $\iota = 1, \dots, M$ are Lipschitz continuous with real values, i.e., and there exist positive constants f_ι^L, g_ι^L, with

$$|f_\iota(\xi_1) - f_\iota(\xi_2)| \leq f_\iota^L|\xi_1 - \xi_2|, \; |g_\iota(\xi_1) - g_j(\xi_2)| \leq g_\iota^L|\xi_1 - \xi_2|$$

for all $\xi_1, \xi_2 \in \mathbb{R}$, $\xi_1 \neq \xi_2$, and $f_\iota(0) = g_\iota(0) = 0$, $\iota = 1, 2, \dots, M$.
A6. The impulsive functions Y_{oi} are continuous on $\mathbb{R}$ for $o = 1, 2, \dots, M$, $i = 1, 2, \dots$.
A7. The functions $\tau_i(y_o)$, $o = 1, 2, \dots, M$, $i = 1, 2, \dots$ are continuous and

$$0 = \tau_0(y_o) < \tau_1(y_o) < \tau_2(y_o) < \dots, \; \tau_i(y_o) \to \infty \text{ as } i \to \infty$$

uniformly on $y_o \in \mathbb{R}$.
A8. The integral surface of each solution $y(t,z)$ of the initial value boundary problem (3.1), (3.2), (3.3) meets for $t > 0$ successively each one of the hypersurfaces $\sigma_1, \sigma_2, \dots$ exactly once.
A9. The integral surface of each solution $y^*(t,z)$ of the initial value boundary problem (3.4), (3.5), (3.6) meets for $t > 0$ successively each one of the hypersurfaces $\sigma_1, \sigma_2, \dots$ exactly once.

Remark 1 For impulsive models with variable impulsive perturbations, the phenomenon where the integral surface $y(t, z)$ meets several or infinitely many times one and the same hypersurface may occur. This phenomenon is called "beating" of the solutions and can lead to complications in the study of the qualitative properties of the solutions, including not existence of a solution after a certain moment of time. Assumptions A8 and A9 guarantee the absence of this phenomenon, i.e., both assumptions guarantee that

$$0 < t_{l_1} < t_{l_2} < \dots,$$

$$0 < t^*_{l_1} < t^*_{l_2} < \dots.$$

The concept of uniform Lipschitz quasistability [57–59] will be defined for the considered Cohen–Grossberg neural network model with reaction-diffusion terms, time-varying delays, and variable impulsive perturbations by the following definition.

Definition 1 The solution $y(t, z; \vartheta)$ of the impulsive Cohen–Grossberg neural network reaction-diffusion model (1) is said to be uniformly Lipschitz quasistable, if there exists a positive constant $M > 0$ such that for any $\eta > 0$ there exists $\delta = \delta(\eta) > 0$ such that for any initial function $\vartheta^* \in \mathcal{PC} : \ ||\vartheta - \vartheta^*||_{h_1} < \delta$ and any $t > 0 : \ |t - t_{l_i}| > \eta, l_i = 1, 2, \dots$ we have

$$||y(t, z; \vartheta) - y^*(t, z; \vartheta^*)||_1 \leq M||\vartheta - \vartheta^*||_{h_1}, \ z \in \mathcal{B}.$$

Since the moments of impulse effect for the solutions $y(t, z; \vartheta)$ and $y^*(t, z; \vartheta^*)$ of problems (3.1)–(3.3) and (3.4)–(3.6) are different, then we have to overcome a number of obstacles in the estimation of the difference of these solutions. To this end, in order to overcome these obstacles, we shall use a suitable comparison lemma [55, 59].

Consider the following scalar comparison impulsive equation

$$\begin{cases} \dot{\theta}(t) = m(t, \theta), \ t \notin (\underline{t}_i, \bar{t}_i], \\ \theta(\bar{t}_i^+) = \theta(\bar{t}_i + 0) = \phi_i(\theta(\underline{t}_i)), \ i = 1, 2, \dots, \end{cases} \tag{3.7}$$

where $m : \mathbb{R}_+ \times \mathbb{R}_+ \to \mathbb{R}, \phi_i : \mathbb{R}_+ \to \mathbb{R}_+, i = 1, 2, \dots, 0 < \underline{t}_1 \leq \bar{t}_1 < \underline{t}_2 \leq \bar{t}_2 < \dots < \underline{t}_i \leq \bar{t}_i < \dots$ and $\underline{t}_i \to \infty$ as $i \to \infty$.

Let $\theta_0 \in \mathbb{R}_+$ and denote the maximal solution of Eq. (3.7), which satisfies the initial condition

$$\theta^+(0^+; 0, \theta_0) = \theta_0 \tag{3.8}$$

by $\theta^+(t) = \theta^+(t; 0, \theta_0)$.

From a practical point of view, we will consider only positive solutions $\theta^+(t;0,\theta_0)$ of the comparison equation (3.7) that correspond to initial values $\theta_0 \in \mathbb{R}_+$.

The Lyapunov function strategy is applied very often in the study of the stability properties of different classes of models. However, in the investigation of the stability properties of nonzero solutions of impulsive systems with not fixed moments of impulsive effects, the application of the method requires some modifications. That is why, in our Lipschitz quasistability analysis, we will apply the Lyapunov function technique using piecewise continuous Lyapunov-type functions $L: \mathbb{R}_+ \times \mathbb{R}^M \to \mathbb{R}_+$ with points of discontinuity of the first kind $t = \underline{t}_i$ and $t = \overline{t}_i$ at which it is continuous from the left.

For such functions, we define the following derivative

$$D^+L(t,\vartheta(0,\cdot)) = \lim_{\chi\to 0^+} \sup \frac{1}{\chi}\big[L(t+\chi, y(t+\chi,\cdot;\vartheta(0,\cdot)) - L(t,\vartheta(0,\cdot))\big]$$

for $\vartheta \in \mathcal{PC}$, $t \notin (\underline{t}_i, \overline{t}_i)$, $i = 1, 2, \ldots$.

The following comparison lemma can be proved identically as the corresponding comparison lemma in [56]. It is also applied in [59] for impulsive delayed systems with variable impulsive perturbations.

Lemma 1 *Assume that:*

1. *The function* $m \in C[\mathbb{R}_+ \times \mathbb{R}_+, \mathbb{R}]$.
2. *The functions* $\phi_i \in C[\mathbb{R}_+, \mathbb{R}_+]$ *are continuous and nondecreasing with respect to* θ, $i = 1, 2, \ldots$.
3. *The maximal solution of Eq. (3.7)* $\theta^+(t;0,\theta_0)$ *is defined on* $\mathbb{R}_+ \setminus \bigcup_{i=1}^{\infty}(\underline{t}_i, \overline{t}_i]$.
4. *There exists a Lyapunov-type function* $L: \mathbb{R}_+ \times \mathbb{R}^M \to \mathbb{R}_+$ *such that for* $t \in \mathbb{R}_+$ *and* $\vartheta \in \mathcal{PC}$

$$L(\overline{t}_i^+, \vartheta(0,\cdot) + \Delta(\vartheta,\cdot)) \le \phi_i(L(\underline{t}_i, \vartheta(0,\cdot))), \ i = 1, 2, \ldots$$

and the inequality

$$D^+L(t,\vartheta(0,\cdot)) \le m(t, L(t,\vartheta(0,\cdot))), \ t \notin (\underline{t}_i, \overline{t}_i],$$

is valid whenever

$$L(t+s, \vartheta(s,\cdot)) \le L(t,\vartheta(0,\cdot)), \ -h \le s \le 0.$$

Then $\sup_{-h\le s\le 0} L(0,\vartheta(s,\cdot)) \le \theta_0$ *implies*

$$L(t, y(t,\cdot)) \le \theta^+(t;0,\theta_0), \ t \in \mathbb{R}_+ \setminus \bigcup_{i=1}^{\infty}(\underline{t}_i, \overline{t}_i].$$

Finally, a Poincarè-type integral inequality [61] for the set $\mathcal{B} = \prod_{p=1}^{N}[\alpha_p, \beta_p]$, $\alpha_p = const \in \mathbb{R}$, $\beta_p = const \in \mathbb{R}$, $p = 1, 2, \ldots, N$ will be applied in the proofs of our main results.

Lemma 2 ([60, 61]) *For any real-valued function $v(z)$ that belongs to $C^1(\mathcal{B})$ the following relation is valid*

$$\int_{\mathcal{B}} v^2(z)dz \le \frac{\Theta^2}{4N}\int_{\mathcal{B}} |\nabla v(z)|^2 dz$$

whenever $\mathcal{B} = \prod_{p=1}^{N}[\alpha_p, \beta_p]$, $\alpha_p, \beta_p \in \mathbb{R}$, $\mathbf{0} = (0, 0, \ldots, 0)^T \in \mathcal{B}$ *and* $\Theta = \max\{\beta_p - \alpha_p,\ p = 1, 2, \ldots, N\}$.

For some extensions of Lemma 2, we also refer to [13].

3.3 Uniform Lipschitz Quasistability Results

In this section, we will apply the comparison lemma, Lyapunov-function approach, and the Poincarè-type integral inequality to establish our main uniform Lipschitz quasistability results for the impulsive Cohen–Grossberg neural network model (1) with reaction-diffusion terms and time-varying delays.

Let $\alpha_p, \beta_p \in \mathbb{R}$, $p = 1, 2, \ldots, N$. We will consider the set $\mathcal{B} \subset \mathbb{R}^N$ of points z, $z = (x_1, x_2, \ldots, x_N)^T$ defined as $\alpha_p \le z_p \le \beta_p$, i.e, $\mathcal{B} = \prod_{p=1}^{N}[\alpha_p, \beta_p]$. Let $\mathbf{0} = (0, 0, \ldots, 0)^T \in \mathcal{B}$ and $\Theta = \max\{\beta_p - \alpha_p\}$, $p = 1, 2, \ldots, N$.

Theorem 1 *Assume that Assumptions A1–A9 are fulfilled, and*

1. *There exists a continuous for $t \ne t_i$, $t \ne t_i^*$ function $\bar{m}_1(t)$, $i = 1, 2, \ldots$, that satisfies*

$$\min_{1\le o\le M}\left[2\left(\frac{4N\underline{K}}{\Theta^2} + \underline{a}_o B_o\right) - \bar{a}_o\sum_{\iota=1}^{M}\left(f_\iota^L \overline{c}_{o\iota} + g_\iota^L \overline{\omega}_{o\iota} + f_o^L \overline{c}_{\iota o}\right)\right]$$
$$- \max_{1\le o\le M}\left(g_o^L \sum_{\iota=1}^{M} \overline{a}_\iota \overline{\omega}_{\iota o}\right) > \bar{m}_1(t) \ge 0,$$

where $\underline{K} = \min\{k_{op}\}$, $o = 1, 2, \ldots, M$, $p = 1, 2, \ldots, N$,

2. *The impulsive functions Y_{oi} are such that for $z \in \mathcal{B}$, $t \in \sigma_i$, $i = 1, 2, \ldots$*

$$Y_{oi}(y_o(t, z)) = -\gamma_{oi}\, y_o(t, z),\ 0 < \gamma_{oi} < 2,\ o = 1, 2, \ldots, M,\ i = 1, 2, \ldots.$$

3. *For $(t, z) \in \mathbb{R}_+ \times \mathcal{B}$, the functions $\tau_i(y(t, z))$ are nonincreasing with respect to t, $i = 1, 2, \ldots$.*

4. *The functions* τ_i *satisfy*

$$|\tau_i(y) - \tau_i(y^*)| \le \mu ||y - y^*||_1^2,$$

where $\mu > 0$ *is a constant,* $i = 1, 2, \dots$.

5. *There exist constants* $M > 0$ *and* $\delta_1 > 0$ *such that the solution of Eq. (3.7) with* $m(t,\theta) = -\bar{m}_1(t)\theta$ *and* $\phi_i(\theta) = \theta$, $i = 1, 2, \dots$, *satisfies* $\theta^+(t; 0, \theta_0) \le M\theta_0$ *for* $0 \le \theta_0 < \delta_1$ *and* $t \in \mathbb{R}_+ \setminus \bigcup_{i=1}^{\infty}(\underline{t}_i, \bar{t}_i]$.

Then, the solution $y(t, z; \vartheta)$ *of the impulsive Cohen–Grossberg neural network reaction-diffusion model (1) is uniformly Lipschitz quasistable.*

Proof Let $z \in \mathcal{B}$, $\vartheta, \vartheta^* \in \mathcal{PC}$. Let $\eta > 0$. Choose $\delta = \delta(\eta) = \min\{\sqrt{2\delta_1}, \sqrt{\frac{\eta}{2\mu M+1}}\}$. Let $||\vartheta - \vartheta^*||_h < \delta$ and $y^*(t, z) = y^*(t, z; \vartheta^*)$,

$$y^*(t, z) = (y_1^*(t, z), y_2^*(t, z), \dots, y_M^*(t, z))^T, \ t > 0, \ z \in \mathcal{B}$$

is the solution of (3.4) that satisfies (3.5) and (3.6).

Consider the points t_{l_i} $(0 < t_{l_1} < t_{l_2} < \dots)$ at which the integral surface of $y(t, z)$ meets the hypersurfaces σ_i, and the points $t_{l_i}^*$ $(0 < t_{l_1}^* < t_{l_2}^* < \dots)$ at which the integral surface of $y^*(t, z)$ meets the hypersurfaces σ_i, $i = 1, 2, \dots$.

Denote $\underline{t}_i = \min(t_{l_i}, t_{l_i}^*)$ and $\bar{t}_i = \max(t_{l_i}, t_{l_i}^*)$.

We define a Lyapunov-type function as follows

$$L_1(t, y) = \frac{1}{2}||y(t, .) - y^*(t, .)||_1^2, \ t \in \mathbb{R}_+ \tag{3.9}$$

or

$$L_1(t, y) = \frac{1}{2}\int_{\mathcal{B}}\sum_{o=1}^{M}\big(y_o(t, z) - y_o^*(t, z)\big)^2 dz, \ t \in \mathbb{R}_+.$$

For any $t \notin (\underline{t}_i, \bar{t}_i]$, $i = 1, 2, \dots$, for the derivative of the function $L_1(t, y(t, \cdot))$, we have

$$\frac{dL_1(t, y(t, \cdot))}{dt} = \sum_{o=1}^{M}\int_{\mathcal{B}}\big(y_o(t, z) - y_o^*(t, z)\big)\frac{\partial\big(y_o(t, z) - y_o^*(t, z)\big)}{\partial t}dz. \tag{3.10}$$

From systems (3.1) and (3.4), applying A2 and A3, we obtain

$$\frac{\partial(y_o(t, z) - y_o^*(t, z))}{\partial t} \le \sum_{p=1}^{N}\frac{\partial}{\partial z_p}\left(D_{op}\frac{\partial(y_o(t, z) - y_o^*(t, z))}{\partial z_p}\right)$$

$$-\underline{a}_o[b_o(y_o(t, z)) - b_o(y_o^*(t, z))]$$

$$+\overline{a}_o \sum_{\iota=1}^{M} \overline{c}_{o\iota} \left| f_\iota\big(y_\iota(t,z)\big) - f_\iota(y_\iota^*(t,z)) \right|$$
$$+\overline{a}_o \sum_{\iota=1}^{M} \overline{\omega}_{o\iota} \left| g_\iota\big(y_\iota(t-h_\iota(t),z)\big) - g_\iota(y_\iota^*(t-h_\iota(t),z)) \right|. \tag{3.11}$$

Taking into account (3.11), we get

$$\begin{aligned}
&\int_{\mathcal{B}} (y_o(t,z) - y_o^*(t,z)) \frac{\partial (y_o(t,z) - y_o^*(t,z))}{\partial t} \\
&\leq \int_{\mathcal{B}} \sum_{p=1}^{N} \frac{\partial}{\partial z_p} \left(D_{op} \frac{\partial (y_o(t,z) - y_o^*(t,z))}{\partial z_p} \right) (y_o(t,z) - y_o^*(t,z)) dz \\
&- \int_{\mathcal{B}} \underline{a}_o (y_o(t,z) - y_o^*(t,z)) [b_o(y_o(t,z)) - b_o(y_o^*(t,z))] dz \\
&+ \overline{a}_o \int_{\mathcal{B}} (y_o(t,z) - y_o^*(t,z)) \sum_{\iota=1}^{M} \overline{c}_{o\iota} \left| f_\iota\big(y_\iota(t,z)\big) - f_\iota(y_\iota^*(t,z)) \right| dz \\
&+ \overline{a}_o \int_{\mathcal{B}} (y_o(t,z) - y_o^*(t,z)) \sum_{\iota=1}^{M} \overline{\omega}_{o\iota} \big| g_\iota(y_\iota(t-h_\iota(t),z)) \\
&- g_\iota(y_\iota^*(t-h_\iota(t),z)) \big| dz.
\end{aligned} \tag{3.12}$$

In order to estimate the first term on the right-hand side of (3.12), we apply the Green's theorem, the boundary conditions, A1 and Lemma 2 to get

$$\begin{aligned}
&\int_{\mathcal{B}} \sum_{p=1}^{N} \frac{\partial}{\partial z_p} \left(D_{op} \frac{\partial (y_o(t,z) - y_o^*(t,z))}{\partial z_p} \right) (y_o(t,z) - y_o^*(t,z)) dz \\
&\qquad = - \sum_{p=1}^{N} \int_{\mathcal{B}} D_{op} \left(\frac{\partial (y_o(t,z) - y_o^*(t,z))}{\partial z_p} \right)^2 dz \\
&\qquad \leq - \sum_{p=1}^{N} \int_{\mathcal{B}} k_{op} \left(\frac{\partial (y_o(t,z) - y_o^*(t,z))}{\partial z_p} \right)^2 dz \\
&\qquad \leq - \frac{4N\underline{K}}{\Theta^2} \int_{\mathcal{B}} (y_o(t,z) - y_o^*(t,z))^2 dz.
\end{aligned} \tag{3.13}$$

Consequently, from A4 and A5, we obtain

$$\int_{\mathcal{B}} \underline{a}_o(y_o(t,z) - y_o^*(t,z))[b_o(y_o(t,z)) - b_o(y_o^*(t,z))]dz$$
$$\geq \underline{a}_o B_o \int_{\mathcal{B}} (y_o(t,z) - y_o^*(t,z))^2 dz, \tag{3.14}$$

$$\overline{a}_o \int_{\mathcal{B}} (y_o(t,z) - y_o^*(t,z)) \sum_{\iota=1}^{M} \overline{c}_{o\iota} \left| f_\iota\big(y_\iota(t,z)\big) - f_\iota(y_\iota^*(t,z))\right| dz$$
$$\leq \overline{a}_o \int_{\mathcal{B}} \sum_{\iota=1}^{M} \overline{c}_{o\iota} f_\iota^L |y_o(t,z) - y_o^*(t,z)||y_\iota(t,z) - y_\iota^*(t,z)|dx \tag{3.15}$$
$$\leq \frac{1}{2}\overline{a}_o \sum_{\iota=1}^{M} \int_{\mathcal{B}} \overline{c}_{o\iota} f_\iota^L \big[(y_o(t,z) - y_i^*(t,z))^2 + (y_\iota(t,z) - y_\iota(t,z))^2\big]dz,$$

and

$$\overline{a}_o \int_{\mathcal{B}} (y_o(t,z) - y_o^*(t,z)) \sum_{\iota=1}^{M} \overline{\omega}_{o\iota} \left| g_\iota(y_\iota(t - h_\iota(t), z)) - g_\iota(y_\iota^*(t - h_\iota(t), z))\right| dz$$
$$\leq \overline{a}_o \sum_{\iota=1}^{M} \int_{\mathcal{B}} \overline{\omega}_{o\iota} g_\iota^L |y_o(t,z) - y_o^*(t,z)| \left| y_\iota(t - h_\iota(t), z) - y_\iota^*(t - h_\iota(t), z)\right| dz$$
$$\leq \frac{1}{2}\overline{a}_o \sum_{\iota=1}^{M} \int_{\mathcal{B}} \overline{\omega}_{o\iota} g_\iota^L \big[(y_o(t,z) - y_o^*(t,z))^2$$
$$+ (y_\iota(t - h_\iota(t), z) - y_\iota^*(t - h_\iota(t), z))^2\big]dz. \tag{3.16}$$

Finally, from (3.10)–(3.16), we obtain

$$\frac{dL_1(t, y(t,\cdot))}{dt} \leq \sum_{o=1}^{M} \Big[-\left(\frac{4N\underline{K}}{\Theta^2} + \underline{a}_o B_o\right) \int_{\mathcal{B}} (y_o(t,z) - y_o^*(t,z))^2 dz$$
$$+ \frac{1}{2}\overline{a}_o \sum_{\iota=1}^{M} \int_{\mathcal{B}} \overline{c}_{o\iota} f_\iota^L \big[(y_o(t,z) - y_o^*(t,z))^2 + (y_\iota(t,z) - y_\iota^*(t,z))^2\big]dz$$
$$+ \frac{1}{2}\overline{a}_o \sum_{\iota=1}^{M} \int_{\mathcal{B}} \overline{\omega}_{o\iota} g_\iota^L \big[(y_o(t,z) - y_o^*(t,z))^2$$
$$+ (y_\iota(t - h_\iota(t), z) - y_\iota^*(t - h_\iota(t), z))^2\big]dz \Big]$$

$$\leq -\frac{1}{2}\sum_{o=1}^{M}\Big[2\Big(\frac{4N\underline{K}}{\Theta^2}+\underline{a}_o B_o\Big)$$

$$-\overline{a}_o \sum_{\iota=1}^{M}\big(f_\iota^L \overline{c}_{o\iota} + g_\iota^L \overline{\omega}_{o\iota} + f_o^L \overline{c}_{\iota o}\big)\Big]\int_{\mathcal{B}}(y_o(t,z)-y_i^*(t,z))^2 dz$$

$$+\frac{1}{2}\sum_{o=1}^{M}\sum_{\iota=1}^{M}\overline{a}_\iota \overline{\omega}_{\iota o} g_o^L \int_{\mathcal{B}} \sup_{-h\leq s\leq 0}(y_\iota(s,z)-y_\iota^*(s,z))^2 dz$$

$$\leq -\lambda_1 \frac{1}{2}\sum_{\iota=1}^{M}\int_{\mathcal{B}}(y_\iota(t,z)-y_\iota^*(t,z))^2 dz$$

$$+\lambda_2 \frac{1}{2}\sum_{\iota=1}^{M}\int_{\mathcal{B}} \sup_{-h\leq s\leq 0}(y_\iota(s,z)-y_\iota^*(s,z))^2 dz, \tag{3.17}$$

where

$$\lambda_1 = \min_{1\leq o\leq M}\Big[2\Big(\frac{4N\underline{K}}{\Theta^2}+\underline{a}_o B_o\Big) - \overline{a}_o \sum_{\iota=1}^{M}\big(f_\iota^L \overline{c}_{o\iota} + g_\iota^L \overline{\omega}_{o\iota} + f_o^L \overline{c}_{\iota o}\big)\Big] > 0,$$

$$\lambda_2 = \max_{1\leq o\leq M}\Big(M_i \sum_{\iota=1}^{M}\overline{a}_\iota \overline{\omega}_{\iota o}\Big) > 0.$$

Condition 1 of Theorem 1 implies that

$$D^+ L_1(t,\vartheta(0,\cdot)) \leq -\bar{m}_1(t) L_1(t,\vartheta(0,\cdot)),\ t \notin (\underline{t}_i, \overline{t}_i],\ i=1,2,\ldots \tag{3.18}$$

when $L_1(t+s,\vartheta(s,\cdot)) \leq L_1(t,\vartheta(0,\cdot)),\ -h\leq s\leq 0$.

For $t \in \sigma_i$, $i=1,2,\ldots$, we apply condition 2 of Theorem 1 to obtain

$$L_1(\overline{t}_i^+, y(\overline{t}_i^+,\cdot)) = \frac{1}{2}\int_{\mathcal{B}}\sum_{o=1}^{M}\big(y_o(\overline{t}_i^+,z)-y_o^*(\overline{t}_i^+,z)\big)^2 dz$$

$$= \frac{1}{2}\int_{\mathcal{B}}\sum_{i=1}^{M}(1-\gamma_{oi})^2\big(y_o(\overline{t}_i,z)-y_o^*(\overline{t}_i,z)\big)^2 dz \tag{3.19}$$

$$< \frac{1}{2}\int_{\mathcal{B}}\sum_{i=1}^{M}\big(y_o(\overline{t}_i,z)-y_o^*(\overline{t}_i,z)\big)^2 dz = L_1(\overline{t}_i, y(\overline{t}_i,\cdot)).$$

Also, using (3.17), we get

$$L_1(\bar{t}_i, y(\bar{t}_i, \cdot)) = L_1(\underline{t}_i, y(\underline{t}_i, \cdot)) + \int_{\underline{t}_i}^{\bar{t}_i} \dot{L}_1(t, y(t, \cdot))dt \\ \leq L_1(\underline{t}_i, y(\underline{t}_i, \cdot)), \ i = 1, 2, \ldots. \tag{3.20}$$

Then, for any $\vartheta \in \mathcal{PC}$ we have

$$L_1(\bar{t}_i^+, \vartheta(0, \cdot) + \Delta(\vartheta, \cdot)) \leq L_1(\underline{t}_i, \vartheta(0, \cdot)), \ i = 1, 2, \ldots. \tag{3.21}$$

Then, from Lemma 1, we obtain

$$L_1(t, y(t, \cdot)) \leq \theta^+(t; 0, \theta_0), \ t \in \mathbb{R}_+ \setminus \bigcup_{i=1}^{\infty} (\underline{t}_i, \bar{t}_i], \tag{3.22}$$

where $\theta^+(t; 0, \theta_0)$ is the solution of Eq. (3.7) with $\theta_0 = \frac{1}{2}||\vartheta^* - \vartheta||_{h_1}^2$.

From (3.22) and condition 5 of Theorem 1, we have

$$L_1(t, y(t, \cdot) \leq \frac{M}{2}||\vartheta^* - \vartheta||_{h_1}^2$$

for $0 \leq ||\vartheta^* - \vartheta||_{h_1} < \sqrt{2\delta_1}$ and $t \in \mathbb{R}_+ \setminus \bigcup_{i=1}^{\infty} (\underline{t}_i, \bar{t}_i]$.

We estimate the difference $\bar{t}_i - \underline{t}_i$, $i = 1, 2, \ldots$. Consider the following cases:

Case 1: $\bar{t}_i = t_{l_i}^*$, $\underline{t}_i = t_{l_i}$. In this case, from condition 3 of Theorem 1, it follows that

$$\tau_i(y^*(\bar{t}_i, \cdot)) \leq \tau_i(y^*(\underline{t}_i, \cdot)).$$

Then, from condition 4 of Theorem 1, we have

$$0 \leq \bar{t}_i - \underline{t}_i = \tau_i(y_i^*(\bar{t}_i, \cdot)) - \tau_i(y(\underline{t}_i, \cdot)) \\ \leq \tau_i(y_i^*(\underline{t}_i, \cdot)) - \tau_i(y(\underline{t}_i, \cdot)) \leq 2\mu L_1(\underline{t}_i, y(\underline{t}_i, \cdot)), \ i = 1, 2, \ldots. \tag{3.23}$$

Case 2: $\bar{t}_i = t_{l_i}$, $\underline{t}_i = t_{l_i}^*$. In this case, we again apply conditions 3 and 4 of Theorem 1, to get

$$0 \leq \bar{t}_i - \underline{t}_i = \tau_i(y_i(\bar{t}_i, \cdot)) - \tau_i(y^*(\underline{t}_i, \cdot)) \\ \leq \tau_i(y_i(\underline{t}_i, \cdot)) - \tau_i(y^*(\underline{t}_i, \cdot)) \leq 2\mu L_1(\underline{t}_i, y(\underline{t}_i, \cdot)), \ i = 1, 2, \ldots. \tag{3.24}$$

Hence, from (3.23), (3.24) and the choice of δ, we have

$$0 \leq \bar{t}_i - \underline{t}_i \leq 2\mu L_1(\underline{t}_i, y(\underline{t}_i, \cdot)) \\ \leq \mu M||\vartheta^* - \vartheta||_{h_1}^2 < \mu M \delta^2 < \frac{1}{2}\eta, \ i = 1, 2, \ldots. \tag{3.25}$$

It follows from (3.25) that

$$\{t \in \mathbb{R}_+, |t - t_{l_i}| > \eta\} \subset \mathbb{R}_+ \setminus \bigcup_{i=1}^{\infty} (\underline{t}_i, \overline{t}_i].$$

Therefore,

$$L_1(t, y(t, \cdot) \leq \frac{M}{2} ||\vartheta^* - \vartheta||^2_{h_1}$$

for $||\vartheta^* - \vartheta||_{h_1} < \delta$ and $t \in \mathbb{R}_+, |t - t_{l_i}| > \eta$, or

$$||y(t, z; \vartheta) - y^*(t, z; \vartheta^*)||_1 \leq \sqrt{M}||\vartheta - \vartheta^*||_{h_1} \tag{3.26}$$

for $z \in \mathcal{B}$, $||\vartheta^* - \vartheta||_{h_1} < \delta$ and $t \in \mathbb{R}_+, |t - t_{l_i}| > \eta$, which proves Theorem 1. □

Using the norm

$$||y(t, \cdot)||_2 = \int_{\mathcal{B}} \sum_{o=1}^{M} |y_o(t, z)| \, dz$$

of an $y(t, z) = (y_1(t, z), y_2(t, z), \ldots, y_M(t, z))^T \in \mathbb{R}^M$, $z \in \mathcal{B}$, $t \in \mathbb{R}_+$, and a different Lyapunov-type function, we will establish next result. The corresponding norm for the initial functions $\vartheta \in \mathcal{PC}$ is defined as

$$||\vartheta||_{h_2} = \sup_{-h \leq s \leq 0} ||\vartheta(s, \cdot)||_2.$$

Theorem 2 *Assume that Assumptions A1–A9 and conditions 2, 3 of Theorem 1 hold, and:*

1. *There exists a continuous for $t \neq t_i$, $t \neq t_i^*$ function $\bar{m}_2(t)$, $i = 1, 2, \ldots$, that satisfies*

$$\min_{1 \leq o \leq M} \left(\frac{4N\underline{K}}{\underline{a}\Theta^2} + \underline{B}_o - \frac{\overline{a}}{\underline{a}} f_o^L \sum_{\iota=1}^{M} \overline{c}_{\iota o} \right) - \left(\frac{\overline{a}}{\underline{a}} \right)^2 \max_{1 \leq o \leq M} \left(g_o^L \sum_{\iota=1}^{M} \overline{\omega}_{\iota o} \right) > \bar{m}_2(t) \geq 0,$$

   where $\underline{a} = \min_{1 \leq o \leq M} \underline{a}_o$, $\overline{a} = \max_{1 \leq o \leq M} \overline{a}_o$.
2. *For o=1,2,...,M, i=1,2,..., we have*

$$|1 - \gamma_{oi}| \leq \frac{\underline{a}}{\overline{a}}.$$

3. *The functions* τ_i *satisfy*

$$|\tau_i(y) - \tau_i(y^*)| \leq \mu||y - y^*||_2,$$

where $\mu > 0$ *is a constant,* $i = 1, 2, \ldots$.

4. *There exist constants* $M > 0$ *and* $\delta_1 > 0$ *such that the solution of Eq. (3.7) with* $m(t, \theta) = -\bar{m}_2(t)\underline{a}\theta$ *and* $\phi_i(\theta) = \theta$, $i = 1, 2, \ldots$, *satisfies* $\theta^+(t; 0, \theta_0) \leq M\theta_0$ *for* $0 \leq \theta_0 < \delta_1$ *and* $t \in \mathbb{R}_+ \setminus \bigcup_{i=1}^{\infty}(\underline{t}_i, \bar{t}_i]$.

Then, the solution $y(t, z; \vartheta)$ *of the impulsive Cohen–Grossberg neural network reaction-diffusion model (1) is uniformly Lipschitz quasistable.*

Proof Let $\eta > 0$. Choose $\delta = \delta(\eta) = \min\{\delta_1, \sqrt{\frac{\eta}{2\bar{a}\mu M+1}}\}$. Let $||\vartheta - \vartheta^*||_{h_2} < \delta$ and $y^*(t, z) = y^*(t, z; \vartheta^*)$,

$$y^*(t, z) = (y_1^*(t, z), y_2^*(t, z), \ldots, y_M^*(t, z))^T, \ t > 0, \ z \in \mathcal{B}$$

is the solution of (3.4) that satisfies (3.5) and (3.6).

Consider again the impulsive instants t_{l_i} $(0 < t_{l_1} < t_{l_2} < \ldots)$ and $t_{l_i}^*$ $(0 < t_{l_1}^* < t_{l_2}^* < \ldots)$ for the solutions $y(t, z)$ and $y^*(t, z)$, respectively, $i = 1, 2, \ldots$.

Denote $\underline{t}_i = \min(t_{l_i}, t_{l_i}^*)$ and $\bar{t}_i = \max(t_{l_i}, t_{l_i}^*)$.

We define a Lyapunov function

$$L_2(t, y) = \int_{\mathcal{B}} \left(\sum_{o=1}^{M} |w_o(y_o(t, z), y_o^*(t, z))| \right) dz,$$

where

$$w_o(\xi_1, \xi_2) = sgn(\xi_1 - \xi_2) \int_0^{\xi_1 - \xi_2} \frac{ds}{a_o(s)}, \ o = 1, 2, \ldots, M. \tag{3.27}$$

From (3.27) for any $o = 1, 2, \ldots, M$, we have that

$$\frac{|\xi_1 - \xi_2|}{\bar{a}_o} \leq w_o(\xi_1, \xi_2) \leq \frac{|\xi_1 - \xi_2|}{\underline{a}_o}. \tag{3.28}$$

Using Assumptions A2–A5, for any $t \notin (\underline{t}_i, \bar{t}_i]$, $i = 1, 2, \ldots$, for the derivative of the function $L_2(t, y(t, \cdot), y^*(t, \cdot))$, we have

$$
\begin{aligned}
&\frac{dL_2(t, y(t,\cdot))}{dt} \\
&= \int_{\mathcal{B}} \left(\sum_{o=1}^{M} \frac{sgn(y_o(t,z) - y_o^*(t,z))}{a_o(y_o(t,z) - y_o^*(t,z))} . \frac{\partial [y_o(t,z) - y_o^*(t,z)]}{\partial t} \right) dz \\
&\leq \sum_{o=1}^{M} \int_{\mathcal{B}} \left(\frac{1}{\underline{a}} \sum_{p=1}^{N} \frac{\partial}{\partial z_p} \left(D_{op} \frac{\partial |y_o(t,z) - y_o^*(t,z)|}{\partial z_p} \right) \right. \\
&\quad - \underline{B}_o |y_o(t,z) - y_o^*(t,z)| + \frac{\overline{a}}{\underline{a}} \sum_{\iota=1}^{M} f_\iota^L |\overline{c}_{o\iota}| |y_\iota(t,z) - y_\iota^*(t,z)| \\
&\quad \left. + \frac{\overline{a}}{\underline{a}} \sum_{\iota=1}^{M} g_\iota^L j |\overline{\omega}_{o\iota}| | \sup_{-h \leq s \leq 0} (y_\iota(s,z) - y_\iota^*(s,z))| \right) dz.
\end{aligned} \tag{3.29}
$$

For the term

$$
\int_{\mathcal{B}} \frac{1}{\underline{a}} \sum_{p=1}^{N} \frac{\partial}{\partial z_p} \left(D_{op} \frac{\partial |y_o(t,z) - y_o^*(t,z)|}{\partial z_p} \right)
$$

analogously to (3.13) by A1 and Lemma 2, we have

$$
\begin{aligned}
&\int_{\mathcal{B}} \frac{1}{\underline{a}} \sum_{p=1}^{N} \frac{\partial}{\partial z_p} \left(D_{op} \frac{\partial |y_o(t,z) - y_o^*(t,z)|}{\partial z_p} \right) \\
&= \int_{\mathcal{B}} \frac{1}{\underline{a}} \left(\frac{1}{|y_o(t,z) - y_o^*(t,z)|} \sum_{p=1}^{N} |y_o(t,z) - y_o^*(t,z)| \frac{\partial}{\partial z_p} \right. \\
&\quad \times \left. \left(D_{op} \frac{\partial (y_o(t,z) - y_o^*(t,z))}{\partial z_p} \right) \right). \\
&\leq -\frac{4N\underline{K}}{\underline{a}\Theta^2} \int_{\mathcal{B}} |y_o(t,z) - y_o^*(t,z)| dz.
\end{aligned} \tag{3.30}
$$

From (3.28), we obtain

$$
\begin{aligned}
&\frac{1}{\overline{a}} \int_{\mathcal{B}} \left(\sum_{o=1}^{M} |y_o(s,z) - y_o^*(s,z)| \right) dz \leq L_2(t, y(s,\cdot)) \\
&\leq L_2(t, y(t,\cdot)) \leq \frac{1}{\underline{a}} \int_{\mathcal{B}} \left(\sum_{o=1}^{M} |y_o(t,z) - y_o^*(t,z)| \right) dz
\end{aligned}
$$

or

$$\int_{\mathcal{B}} \left(\sum_{i=o}^{M} |y_o(s, z) - y_o^*(s, z)| \right) dz \leq \frac{\overline{a}}{\underline{a}} \int_{\mathcal{B}} \left(\sum_{i=1}^{m} |y_o(t, z) - y_o^*(t, z)| \right) dz, \tag{3.31}$$

for $L_2(t, y(s, \cdot), y^*(s, \cdot)) \leq L_2(t, y(t, \cdot), y^*(t, \cdot)),\ t - h \leq s \leq t, t \geq 0.$

From (3.29), (3.30) and (3.31), we get

$$\frac{dL_2(t, y(t, \cdot))}{dt} \leq -(\mu_1 - \mu_2) \int_{\mathcal{B}} \left(\sum_{i=1}^{m} |y_o(t, z) - y_o^*(t, z)| \right) dz,$$

where

$$\mu_1 = \min_{1 \leq o \leq M} \left(\frac{4N\underline{K}}{\underline{a}\Theta^2} + \underline{B}_o - \frac{\overline{a}}{\underline{a}} f_o^L \sum_{\iota=1}^{M} \overline{c}_{\iota o} \right),$$

$$\mu_2 = \left(\frac{\overline{a}}{\underline{a}} \right)^2 \max_{1 \leq o \leq M} \left(g_o^L \sum_{\iota=1}^{M} \overline{\omega}_{\iota o} \right).$$

Condition 1 of Theorem 2 and (3.28) imply, that

$$D^+ L_2(t, \vartheta(0, \cdot)) \leq -\bar{m}_2(t) \underline{a} L_2(t, \vartheta(0, \cdot)),\ t \notin (\underline{t}_i, \overline{t}_i],\ i = 1, 2, \ldots \tag{3.32}$$

when $L_2(t + s, \vartheta(s, \cdot)) \leq L_2(t, \vartheta(0, \cdot)),\ -h \leq s \leq 0.$

For $t \in \sigma_i$, $i = 1, 2, \ldots$, we apply condition 2 of Theorem 2 to obtain

$$\begin{aligned} L_2(\overline{t}_i^+, y(\overline{t}_i^+, \cdot)) &\leq \frac{1}{\underline{a}} \int_{\mathcal{B}} \sum_{o=1}^{M} |y_o(\overline{t}_i^+, z) - y_o^*(\overline{t}_i^+, z)| dz \\ &= \frac{1}{\underline{a}} \int_{\mathcal{B}} \sum_{i=1}^{M} |1 - \gamma_{oi}| |y_o(\overline{t}_i, z) - y_o^*(\overline{t}_i, z)| dz \\ &\leq \frac{1}{\overline{a}} \int_{\mathcal{B}} \sum_{i=1}^{M} |y_o(\overline{t}_i, z) - y_o^*(\overline{t}_i, z)| dz \leq L_2(\overline{t}_i, y(\overline{t}_i, \cdot)). \end{aligned} \tag{3.33}$$

Also, using (3.32), we get

$$\begin{aligned} L_2(\overline{t}_i, y(\overline{t}_i, \cdot)) &= L_2(\underline{t}_i, y(\underline{t}_i, \cdot)) + \int_{\underline{t}_i}^{\overline{t}_i} \dot{L}_2(t, y(t, \cdot)) dt \\ &\leq L_2(\underline{t}_i, y(\underline{t}_i, \cdot)),\ i = 1, 2, \ldots. \end{aligned} \tag{3.34}$$

Then, for any $\vartheta \in \mathcal{PC}$ we have

$$L_2(\overline{t}_i^+, \vartheta(0,\cdot) + \Delta(\vartheta,\cdot)) \le L_2(\underline{t}_i, \vartheta(0,\cdot)),\ i = 1, 2, \dots. \tag{3.35}$$

We again apply Lemma 1 to obtain

$$L_2(t, y(t,\cdot)) \le \theta^+(t; 0, \theta_0),\ t \in \mathbb{R}_+ \setminus \bigcup_{i=1}^{\infty} (\underline{t}_i, \overline{t}_i], \tag{3.36}$$

where $\theta^+(t; 0, \theta_0)$ is the solution of Eq. (3.7) with $\theta_0 = ||\vartheta^* - \vartheta||_{h_2}$.

Now, (3.36) and condition 4 of Theorem 2 imply

$$L_2(t, y(t,\cdot) \le M||\vartheta^* - \vartheta||_{h_2}$$

for $0 \le ||\vartheta^* - \vartheta||_{h_2} < \delta_1$ and $t \in \mathbb{R}_+ \setminus \bigcup_{i=1}^{\infty} (\underline{t}_i, \overline{t}_i]$.

In order to estimate the difference $\overline{t}_i - \underline{t}_i$, $i = 1, 2, \dots$, we apply condition 3 of Theorem 1 and condition 3 of Theorem 2 considering again both cases $\overline{t}_i = t^*_{l_i}$, $\underline{t}_i = t_{l_i}$ and $\overline{t}_i = t_{l_i}$, $\underline{t}_i = t^*_{l_i}$. Repeating the same steps as in the proof of Theorem 1, we obtain

$$0 \le \overline{t}_i - \underline{t}_i \le \mu\overline{a} L_2(\underline{t}_i, y(\underline{t}_i,\cdot)),\ i = 1, 2, \dots. \tag{3.37}$$

Hence, from (3.37) and the choice of δ, we have

$$\begin{aligned} 0 \le \overline{t}_i - \underline{t}_i &\le \mu\overline{a} L_2(\underline{t}_i, y(\underline{t}_i,\cdot)) \\ &\le \mu\overline{a}M||\vartheta^* - \vartheta||_{h_2} < \mu\overline{a}M\delta < \frac{1}{2}\eta,\ i = 1, 2, \dots. \end{aligned} \tag{3.38}$$

It follows from (3.38) that

$$\{t \in \mathbb{R}_+, |t - t_{l_i}| > \eta\} \subset \mathbb{R}_+ \setminus \bigcup_{i=1}^{\infty} (\underline{t}_i, \overline{t}_i].$$

Therefore,

$$L_2(t, y(t,\cdot)) \le M||\vartheta^* - \vartheta||_{h_2}$$

for $||\vartheta^* - \vartheta||_{h_2} < \delta$ and $t \in \mathbb{R}_+$, $|t - t_{l_i}| > \eta$, or

$$||y(t, z; \vartheta) - y^*(t, z; \vartheta^*)||_2 \le M\overline{a}||\vartheta - \vartheta^*||_{h_2} \tag{3.39}$$

for $z \in \mathcal{B}$, $||\vartheta^* - \vartheta||_{h_2} < \delta$ and $t \in \mathbb{R}_+$, $|t - t_{l_i}| > \eta$, which proves Theorem 2.

Remark 2 The main challenge in the study of neural network models with variable impulsive perturbations is in the fact that different neuronal states may have different impulsive instances. This leads to a number of difficulties in the analysis of the effect of impulsive perturbations on the dynamic activity patterns in the systems and necessity of deriving various stability and control strategies. In Theorems 1 and 2, we derive efficient extended Lipschitz stability criteria for impulsive Cohen–Grossberg-type neural networks with delays and reaction-diffusion terms and variable impulsive perturbations, which generalize and extend many existing stability results for delayed reaction-diffusion neural networks of Cohen–Grossberg type [10–13, 23–32].

3.4 Examples

Example 1 Consider the case $N = M = 2$ when the set $\mathcal{B} \subset \mathbb{R}^2$ is defined by $\mathcal{B} = [0, 2] \times [0, 1]$ and the following impulsive Cohen–Grossberg-type neural network model with reaction-diffusion terms

$$\begin{cases} \dfrac{\partial y_o(t,z)}{\partial t} = \displaystyle\sum_{p=1}^{N} \frac{\partial}{\partial z_p}\left(D_{op}\frac{\partial y_o(t,z)}{\partial z_p}\right) - a_o(y_o(t,z))\Big[b_o(y_o(t,z)) \\ \qquad - \displaystyle\sum_{\iota=1}^{M} c_{o\iota}(t) f_\iota\big(y_\iota(t,z)\big) \\ \qquad - \displaystyle\sum_{\iota=1}^{M} \omega_{o\iota}(t) g_\iota\big(y_\iota(t-h_\iota(t),z)\big)\Big],\ t \neq \tau_i(y_o(t,z)), \\ \Delta y_o(t,z) = Y_{oi}(y_o(t,z)),\ t = \tau_i(y_o(t,z)), \end{cases} \tag{3.40}$$

where $t \geq 0$, $z = (z_1, z_2)^T \in \mathcal{B}$, $f_\iota(y_\iota) = g_\iota(y_\iota) = 0.5(|y_\iota + 1| - |y_\iota - 1|)$, $h_1(t) = h_2(t) = e^t/(1+e^t)$, $0 \leq h_\iota(t) \leq 1$, $a_o(\xi) = 1 + 0.4\sin(\xi)$, $\iota, o = 1, 2$, $b_1(\xi) = 2\xi$, $b_2(\xi) = \xi$,

$$(c_{o\iota})(t) = \begin{pmatrix} c_{11}(t) & c_{12}(t) \\ c_{21}(t) & c_{22}(t) \end{pmatrix} = \begin{pmatrix} 0.1 - 0.2\sin(t) & 0.1 - 0.3\cos(t) \\ 0.2 - 0.1\cos(t) & 0.4 - 0.2\sin(t) \end{pmatrix},$$

$$(\omega_{o\iota})(t) = \begin{pmatrix} \omega_{11}(t) & \omega_{12}(t) \\ \omega_{21}(t) & \omega_{22}(t) \end{pmatrix} = \begin{pmatrix} 0.1 + 0.1\sin(t) & 0.1 - 0.2\cos(t) \\ 0.2 - 0.3\cos(t) & 0.1 - 0.3\sin(t) \end{pmatrix},$$

$$(D_{op})_{2\times 2} = \begin{pmatrix} D_{11} & D_{12} \\ D_{21} & D_{22} \end{pmatrix} = \begin{pmatrix} 2+\sin t & 0 \\ 0 & 3+\cos t \end{pmatrix},$$

the impulsive functions Y_{oi} are

$$Y_{oi}(y_o(t,z)) = \left(\frac{1}{4i} - 1\right) y_o(t,z),\ o = 1,2,\ i = 1,2,\ldots,$$

and the functions $\tau_i(y)$ are

$$\tau_i(y) = -||y||_1 + i,\ i = 1,2,\ldots$$

and satisfy

$$|\tau_i(y) - \tau_i(x)| \leq ||y-x||_1^2,$$

$i = 1,2,\ldots.$

We can check that the assumptions A1–A6 are satisfied for $\underline{a}_o = 0.6$, $\overline{a}_o = 1.4$, $o = 1,2$, $B_1 = 2$, $B_2 = 1$, $f_1^L = f_2^L = g_1^l = g_2^L = 1$ and

$$(k_{op})_{2\times 2} = \begin{pmatrix} k_{11} & k_{12} \\ k_{21} & k_{22} \end{pmatrix} = \begin{pmatrix} 1 & 0 \\ 0 & 2 \end{pmatrix}.$$

Also,

$$\tau_1(y) < \tau_2(y) < \ldots,\ \tau_i(y) \to \infty \text{ as } i \to \infty$$

uniformly on $y \in \mathbb{R}^2$ and the assumptions A7, A8, and A9 are all also true.

In addition, we have that $\underline{K} = 1$, $\Theta = 2$,

$$\lambda_1 = \min_{1\leq o\leq M}\left[2\left(\frac{4N\underline{K}}{\Theta^2} + \underline{a}_o B_o\right) - \overline{a}_o \sum_{\iota=1}^{M}\left(f_\iota^L \overline{c}_{o\iota} + g_\iota^L \overline{\omega}_{o\iota} + f_o^L \overline{c}_{\iota o}\right)\right] = 3.88,$$

$$\lambda_2 = \max_{1\leq o\leq M}\left(g_o^L \sum_{\iota=1}^{M} \overline{a}_\iota \overline{\omega}_{\iota o}\right) = 0.98,$$

or condition 1 of Theorem 1 is satisfied for any function $\bar{m}_1(t)$ such that $0 < \bar{m}_1(t) < 2.9$.

It follows from [59] that for the Eq. (3.7) with $m(t,\theta) = -\bar{m}_1(t)\theta$ and $\phi_i(\theta) = \theta$, $i = 1,2,\ldots$, condition 5 of Theorem 1 holds.

Since all conditions of Theorem 1 hold, the solution $y(t,z)$ of the impulsive Cohen–Grossberg neural network reaction-diffusion model (3.40) is uniformly Lipschitz quasistable.

Example 2 In this example, consider the model

$$\begin{cases} \dfrac{\partial y_o(t,z)}{\partial t} = \sum_{p=1}^{N} \dfrac{\partial}{\partial z_p}\left(D_{op}\dfrac{\partial y_o(t,z)}{\partial z_p}\right) - a_o(y_o(t,z))\Big[b_o(y_o(t,z)) \\ \quad - \sum_{\iota=1}^{M} c_{o\iota}(t) f_\iota\big(y_\iota(t,z)\big) \\ \quad - \sum_{\iota=1}^{M} \omega_{o\iota}(t) g_\iota\big(y_\iota(t-h_\iota(t),z)\big)\Big],\ t \neq \tau_i(y_o(t,z)), \\ \Delta y_o(t,z) = Y_{oi}(y_o(t,z)),\ t = \tau_i(y_o(t,z)), \end{cases} \tag{3.41}$$

for $M = 3$, $N = 2$, $\mathcal{B} = \{(z_1, z_2)^T : 0 \le z_1, z_2 \le 2\} \subset \mathbb{R}^2$, $y(t,z) = (y_1(t,z), y_2(t,z), y_3(t,z))^T$, $t > 0$, $f_o(y_o) = g_o(y_o) = 0.5(|y_o+1| - |y_o-1|)$, $0 \le h_o(t) \le 2$ $a_o(\xi) = 1.3 + 0.3\cos(\xi)$, $o = 1, 2, 3$, $b_1(\xi) = 3.2\xi$, $b_2(\xi) = 3\xi$, $b_3(\xi) = 3.1\xi$,

$$(c_{o\iota})(t) = \begin{pmatrix} 0.2 - 0.5\sin(t) & 0.1 + 0.4\cos(t) & 0.3 + 0.1\sin(t) \\ 0.2 + 0.2\sin(t) & 0.1 - 0.2\sin(t) & 0.3 + 0.3\cos(t) \\ 0.2 - 0.3\cos(t) & 0.1 - 0.2\cos(t) & 0.3 + 0.4\sin(t) \end{pmatrix},$$

$$(\omega_{o\iota})(t) = \begin{pmatrix} 0.4 & 0.1 & -0.3 \\ -0.2 & 0.3 & 0 \\ -0.3 & -0.5 & 0.1 \end{pmatrix},\quad D_{op} = k_{op} = \begin{pmatrix} 2.8 & 2.85 \\ 3.125 & 2.9 \\ 2.9 & 3.05 \end{pmatrix}.$$

Let the constants γ_{oi} be such that

$$\gamma_{oi} = 1.4 - \frac{1}{i},\ o = 1, 2, 3,\ i = 1, 2, \ldots. \tag{3.42}$$

Then, both condition 2 of Theorem 1 and condition 2 of Theorem 2 are satisfied. If we consider again functions $\tau_i(y)$ for which

$$\tau_i(y) = -||y||_2 + i,\ i = 1, 2, \ldots$$

and

$$|\tau_i(y) - \tau_i(x)| \le ||y - x||_2,$$

$i = 1, 2, \ldots$, then, assumptions A7, A8, and A9 are true.

Also, from the choice of the system parameters, we have that all assumptions A1–A6 are satisfied for $\underline{a}_o = 1$, $\overline{a}_o = 1.6$, $o = 1, 2, 3$, $\underline{B}_1 = 3.2$, $\underline{B}_2 = 3$, $\underline{B}_3 = 3.1$, $f_o^L = g_o^L = 1$, $o = 1, 2, 3$. We can also check that $\Theta = 2$ and $\underline{K} = 2.8$. In addition,

$$\mu_1 = \min_{1\le o\le M}\left(\frac{4N\underline{K}}{\underline{a}\Theta^2} + \underline{B}_o - \frac{\overline{a}}{\underline{a}} f_o^L \sum_{\iota=1}^{M} \overline{c}_{\iota o}\right) = 5.98$$

$$\mu_2 = \left(\frac{\overline{a}}{\underline{a}}\right)^2 \max_{1\le o\le M}\left(g_o^L \sum_{\iota=1}^{M} \overline{\omega}_{\iota o}\right) = 2.304$$

or condition 1 of Theorem 2 is satisfied for any function $\bar{m}_2(t)$ such that $0 < \bar{m}_2(t) < 3.676$.

It follows again from [59] that for the Eq. (3.7) with $m(t, \theta) = -\bar{m}_2(t)\underline{a}\theta$ and $\phi_i(\theta) = \theta$, $i = 1, 2, \ldots$, condition 4 of Theorem 2 holds.

Since all conditions of Theorem 2 hold, the solution $y(t, z)$ of the impulsive Cohen–Grossberg neural network reaction-diffusion model (3.41) is uniformly Lipschitz quasistable.

Remark 3 We can also check that for the model (3.40), we have

$$\mu_1 = \min_{1\le o\le M}\left(\frac{4N\underline{K}}{\underline{a}\Theta^2} + \underline{B}_o - \frac{\overline{a}}{\underline{a}} f_o^L \sum_{\iota=1}^{M} \overline{c}_{\iota o}\right) = 2$$

$$\mu_2 = \left(\frac{\overline{a}}{\underline{a}}\right)^2 \max_{1\le o\le M}\left(g_o^L \sum_{\iota=1}^{M} \overline{\omega}_{\iota o}\right) = 3.81$$

or condition 1 of Theorem 2 is not satisfied. Hence, the availability of different criteria provides a flexibility in the investigation of the qualitative behavior of the models.

Remark 4 In Example 2, we have

$$\lambda_1 = \min_{1\le o\le M}\left[2\left(\frac{4N\underline{K}}{\Theta^2} + \underline{a}_o B_o\right) - \overline{a}_o \sum_{\iota=1}^{M}\left(f_\iota^L \overline{c}_{o\iota} + g_\iota^L \overline{\omega}_{o\iota} + f_o^L \overline{c}_{\iota o}\right)\right] = 9.88,$$

$$\lambda_2 = \max_{1\le o\le M}\left(g_o^L \sum_{\iota=1}^{M} \overline{a}_\iota \overline{\omega}_{\iota o}\right) = 1.44,$$

or condition 1 of Theorem 1 is satisfied for any function $\bar{m}_1(t)$ such that $0 < \bar{m}_1(t) < 8.44$. In this case, we can make conclusions on the Lipschitz quasistability properties of the model (3.41) by both theorems.

3.5 Concluding Notes

In this paper, the extended concept of Lipschitz quasistability is introduced to impulsive Cohen–Grossberg-type neural networks with delays and reaction-diffusion terms. Such neural network models are widely applied as models of phenomena studied in biology and medicine. We design an impulsive control method by using variable impulsive perturbations. We establish efficient Lipschitz quasistability criteria based on the use of the impulsive control law, comparison lemmas, and the Lyapunov function method. Examples are also presented to demonstrate the efficiency of the derived criteria. The introduced concept and obtained results can be further applied to other important classes of neural network models. Considering the effect of some undetermined terms is one of the directions of our future research.

References

1. X. Lou, B. Cui, Boundedness and exponential stability for nonautonomous cellular neural networks with reaction-diffusion terms. Chaos Soliton Fract. **33**, 653–662 (2007)
2. H. Zhao, Z. Mao, Boundedness and stability of nonautonomous cellular neural networks with reaction-diffusion terms. Math. Comput. Simul. **79**, 1603–1617 (2009)
3. W.H. Chen, L. Liu, X. Lu, Intermittent synchronization of reaction-diffusion neural networks with mixed delays via Razumikhin technique. Nonlinear Dyn. **87**, 535–551 (2017)
4. W.-H. Chen, S. Luo, W.X. Zheng, Impulsive synchronization of reaction-diffusion neural networks with mixed delays and its application to image encryption. IEEE Trans. Neural Netw. Learn. Syst. **27**, 2696–2710 (2016)
5. J.G. Lu, Global exponential stability and periodicity of reaction-diffusion delayed recurrent neural networks with Dirichlet boundary conditions. Chaos Soliton Fract. **35**, 116–125 (2008)
6. R. Rakkiyappan, S. Dharani, Q. Zhu, Synchronization of reaction-diffusion neural networks with time-varying delays via stochastic sampled-data controller. Nonlinear Dyn. **79**, 485–500 (2015)
7. Y. Luo, S. Tang, Z. Teng, L. Zhang, Global dynamics in a reaction-diffusion multi-group SIR epidemic model with nonlinear incidence. Nonlinear Anal. Real World Appl. **50**, 365–385 (2019)
8. Y. Tong, C. Lei, An SIS epidemic reaction-diffusion model with spontaneous infection in a spatially heterogeneous environment. Nonlinear Anal. Real World Appl. **41**, 443–460 (2018)
9. N. Wang, L. Zhang, Z. Teng, Dynamics in a reaction-diffusion epidemic model via environmental driven infection in heterogenous space. J. Biol. Dyn. **16**, 373–396 (2022)
10. M.A. Cohen, S. Grossberg, Absolute stability of global pattern formation and parallel memory storage by competitive neural networks. IEEE Trans. Syst. Man Cybern. **13**, 815–826 (1983)
11. M.M. Ketencigil, O. Faydasicok, S. Arik, Novel criteria for robust stability of Cohen-Grossberg neural networks with multiple time delays. Discrete Contin. Dyn. Syst.-S **15**, 3189–3203 (2022)
12. V. Kumar, J. Heiland, P. Benner, Exponential lag synchronization of Cohen–Grossberg neural networks with discrete and distributed delays on time scales (2022). arXiv:2209.00401
13. R. Li, J. Cao, A. Alsaedi, B. Ahmad, Passivity analysis of delayed reaction-diffusion Cohen-Grossberg neural networks via Hardy-Poincarè inequality. J. Franklin Inst. **354**, 3021–3038 (2017)
14. W. Chen, Y. Huang, S. Ren, Passivity and robust passivity of delayed Cohen-Grossberg neural networks with and without reaction-diffusion terms. Circuits Syst. Signal Process. **37**, 2772–2804 (2018)

15. Z. Wang, H. Zhang, Global asymptotic stability of reaction-diffusion Cohen-Grossberg neural networks with continuously distributed delays. IEEE Trans. Neral Netw. **21**, 39–49 (2010)
16. P. Yan, T. Lv, Periodicity of delayed reaction-diffusion high-order Cohen-Grossberg neural networks with Dirichlet boundary conditions. Rocky Mountain J. Math. **41**, 949–970 (2011)
17. H. Zhao, K. Wang, Dynamical behaviors of Cohen-Grossberg neural networks with delays and reaction-diffusion terms. Neurocomputing **70**, 536–543 (2006)
18. T. Stamov, Discrete bidirectional associative memory neural networks of the Cohen-Grossberg type for engineering design symmetry related problems: practical stability of sets analysis. Symmetry **14**, 216 (2022)
19. J. Cao, G. Stamov, I. Stamova, S. Simeonov, Almost periodicity in reaction-diffusion impulsive fractional neural networks. IEEE Trans. Cybern. **51**, 151–161 (2021)
20. C. Hu, H. Jiang, Z. Teng, Impulsive control and synchronization for delayed neural networks with reaction-diffusion terms. IEEE Trans. Neural Netw. **21**, 67–81 (2010)
21. I. Stamova, G. Stamov, Mittag-Leffler synchronization of fractional neural networks with time-varying delays and reaction-diffusion terms using impulsive and linear controllers. Neural Netw. **96**, 22–32 (2017)
22. X. Yang, J. Cao, Z. Yang, Synchronization of coupled reaction-diffusion neural networks with time-varying delays via pinning-impulsive control. SIAM J Control Optim. **51**, 3486–3510 (2013)
23. C. Aouiti, F. Dridi, New results on impulsive Cohen-Grossberg neural networks. Neural Proc. Lett. **49**, 1459–1483 (2019)
24. X. Li, Existence and global exponential stability of periodic solution for impulsive Cohen-Grossberg-type BAM neural networks with continuously distributed delays. Appl. Math. Comput. **215**, 292–307 (2009)
25. X. Li, Exponential stability of Cohen-Grossberg-type BAM neural networks with time-varying delays via impulsive control. Neurocomputing **73**, 525–530 (2009)
26. Z. Li, K. Li, Stability analysis of impulsive Cohen-Grossberg neural networks with distributed delays and reaction-diffusion terms. Appl. Math. Model. **33**, 1337–1348 (2009)
27. J. Pan, X. Liu, S. Zhong, Stability criteria for impulsive reaction-diffusion Cohen-Grossberg neural networks with time-varying delays. Math. Comput. Modell. **51**, 1037–1050 (2010)
28. G. Stamov, S. Tomasiello, I. Stamova, C. Spirova, Stability of sets criteria for impulsive Cohen-Grossberg delayed neural networks with reaction-diffusion terms. Math. **8**, 27 (2020)
29. J. Tan, C. Li, T. Huang, The stability of impulsive stochastic Cohen-Grossberg neural networks with mixed delays and reaction-diffusion terms. Cogn. Neurodyn. **9**, 213–220 (2015)
30. C. Wu, Existence of periodic solutions for Cohen-Grossberg neural networks with time-varying delays and impulses, in *Advances in Neural Networks*. ed. by D. Liu, H. Zhang, M. Polycarpou, C. Alippi, H. He (Springer, Berlin, 2011), pp. 521–528
31. J.L. Wang, H.N. Wu, L. Guo, Stability analysis of reaction-diffusion Cohen-Grossberg neural networks under impulsive control. Neurocomputing **106**, 21–30 (2013)
32. X. Zhang, S. Wu, K. Li, Delay-dependent exponential stability for impulsive Cohen-Grossberg neural networks with time-varying delays and reaction-diffusion terms. Commun. Nonlinear Sci. Numer. Simul. **16**, 1524–1532 (2011)
33. M. Benchohra, J. Henderson, J. Ntouyas, *Impulsive Differential Equations and Inclusions* (Hindawi, New York, 2006)
34. W.M. Haddad, V.S. Chellaboina, S.G. Nersesov, *Impulsive and Hybrid Dynamical Systems, Stability, Dissipativity, and Control* (Princeton University Press, Princeton, 2006)
35. X. Li, S. Song, *Impulsive Systems with Delays: Stability and Control* (Science Press & Springer, Singapore, 2022)
36. X. Liu, K. Zhang, *Impulsive Systems on Hybrid Time Domains* (Springer, Cham, 2019)
37. T. Yang, *Impulsive Control Theory* (Springer, Berlin, 2001)
38. X. Yang, D. Peng, X. Lv, X. Li, Recent progress in impulsive control systems. Math. Comput. Simul. **155**, 244–268 (2019)
39. M. Benchohra, J. Henderson, S.K. Ntouyas, A. Ouahab, Impulsive functional differential equations with variable times. Comput. Math. Appl. **47**, 1659–1665 (2004)

40. I.M. Stamova, G.T. Stamov, *Applied Impulsive Mathematical Models* (Springer, Cham, 2016)
41. E. Yilmaz, Almost periodic solutions of impulsive neural networks at non-prescribed moments of time. Neurocomputing **141**, 148–152 (2014)
42. M. Bohner, G.T. Stamov, I.M. Stamova, Almost periodic solutions of Cohen–Grossberg neural networks with time-varying delay and variable impulsive perturbations. Commun. Nonlinear Sci. Numer. Simul. **80**, 104952 (2020)
43. G. Stamov, I. Stamova, G. Venkov, T. Stamov, C. Spirova, Global stability of integral manifolds for reaction-diffusion delayed neural networks of Cohen–Grossberg-type under variable impulsive perturbations. Math. **8**, 1082 (2020)
44. I. Stamova, S. Sotirov, S. Simeonov, G. Stamov, Effects of variable impulsive perturbations on the stability of fractional-order Cohen-Grossberg neural networks with respect to functions, in *Contemporary Methods in Bioinformatics and Biomedicine and Their Applications, pp 185–194.* ed. by S.S. Sotirov, T. Pencheva, J. Kacprzyk, K.T. Atanassov, E. Sotirova, G. Staneva (Springer, Cham, 2022)
45. F.M. Dannan, S. Elaydi, Lipschitz stability of nonlinear systems of differential equations. J. Math. Anal. Appl. **113**, 562–577 (1986)
46. F.M. Dannan, S. Elaydi, Lipschitz stability of nonlinear systems of differential equations. II. Liapunov functions. J. Math. Anal. Appl. **143**, 517–529 (1989)
47. D. Bainov, I. Stamova, Lipschitz stability of impulsive functional differential equations. ANZIAM J. **42**, 504–514 (2001)
48. G.K. Kulev, D.D. Bainov, Lipschitz stability of impulsive systems of differential equations. Int. J. Theor. Phys. **30**, 737–756 (1991)
49. C. Aouiti, E.A. Assali, Nonlinear Lipschitz measure and adaptive control for stability and synchronization in delayed inertial Cohen-Grossberg-type neural networks. Int. J. Adapt. Control **33**, 1457–1477 (2019)
50. H. Gouk, E. Frank, B. Pfahringer, M.J. Cree, Regularisation of neural networks by enforcing Lipschitz continuity. Mach. Learn. **110**, 393–416 (2021)
51. B. Harrach, H. Meftahi, Global uniqueness and Lipschitz stability for the inverse Robin transmission problem. SIAM J. Appl. Math. **79**, 525–550 (2019)
52. A. Kawamoto, M. Machida, Global Lipschitz stability for a fractional inverse transport problem by Carleman estimates. Appl. Anal. **100**, 752–771 (2021)
53. A. Ruland, E. Sincich, Lipschitz stability for finite dimensional fractional Calderón problem with finite Cauchy data. Inverse Probl. Imag. **13**, 1023–1044 (2019)
54. I. Stamova, T. Stamov, G. Stamov, Lipschitz stability analysis of fractional-order impulsive delayed reaction-diffusion neural network models. Chaos Soliton Fract. **162**, 112474 (2022)
55. V. Lakshmikantham, D.D. Bainov, P.S. Simeonov, *Theory of Impulsive Differential Equations* (World Scientific, Singapore, 1989)
56. V. Lakshmikantham, X. Liu, On quasistability of impulsive differential systems. Nonlinear Anal. **13**, 819–828 (1989)
57. G.K. Kulev, D.D. Bainov, Lipschitz quasistability of impulsive systems of differential equations. Int. J. Theor. Phys. **30**, 1151–1162 (1991)
58. G.K. Kulev, D.D. Bainov, Lipschitz quasistability of impulsive differential equations. J. Math. Anal. Appl. **172**, 24–32 (1993)
59. D.D. Bainov, A.B. Dishliev, I.M. Stamova, Lipschitz quasistability of impulsive differential-difference equations with variable impulsive perturbations. J. Comput. Appl. Math. **70**, 267–277 (1996)
60. W.-S. Cheung, Some new Poincarè-type inequalities. Bull. Austral. Math. Soc. **63**, 321–327 (2001)
61. X. Lai, T. Yao, Exponential stability of impulsive delayed reaction-diffusion cellular neural networks via Poincarè integral inequality. Abstr. Appl. Anal. **2013**, Art. ID 131836, 10 (2013)

Chapter 4
Rate-Induced Tipping and Chaos in Models of Epidemics

Jochen Merker

4.1 Introduction

In this chapter, we analytically explore models of epidemics with time-dependent parameters. Mathematically, these models are nonlinear compartmental systems of ordinary differential equations (ODEs), whose parameters depend on time so that the systems are nonautonomous [1]. Such models of epidemics divide the population into compartments and assume a certain form of the time rates for the transfer from one compartment to another. The foundations of these models have been laid by ROSS [2–4], MCKENDRICK and KERMACK [5–7] prior to 1935; see references [8–10] for a recent general discussion of mathematical models of epidemiology. Of course, such simple ODE models can be criticised, because in contrast to models consisting of partial differential equations (PDEs)—or networks models considering social networks as graphs—they assume a homogeneous mixing of the population and do not allow spatial dependence. Further, in contrast to delay differential equations (DDEs), they do not allow delays to model a period of temporary immunity.

Yet, we concentrate in this chapter on compartmental ODE models as most simple class of models of epidemics, where interesting effects of time-dependent parameters can be studied, as such ODE models can be considered as finite-dimensional projections of infinite-dimensional dynamical systems generated by PDEs or DDEs. Particularly, we study rate-induced tipping leading to a die-out of an epidemic due to a sufficiently fast implementation of measures against a disease, and periodically forced chaos leading to an unpredictable occurence of epidemic waves. If these effects of time-dependence of parameters can be seen in the most simple

J. Merker (✉)
Leipzig University of Applied Sciences, Leipzig, Germany
e-mail: jochen.merker@htwk-leipzig.de

C. M. A. Pinto, C. M. Ionescu (eds.), *Computational and Mathematical Models in Biology*, Nonlinear Systems and Complexity 38,
https://doi.org/10.1007/978-3-031-42689-6_4

class of ODE models, then also in more advanced models by PDEs and DDEs via model reduction to ODEs.

A general discussion of rate-induced tipping is given in [11] for one-dimensional (scalar) ODEs and in [12] for ODE systems. From literature, it is known that rate-induced tipping occurs generically, e.g. in climate models [13], in two-dimensional models of ecosystems [14], in predator-prey systems [15] or in chaotic systems [16]. For an introduction to periodically forced chaos, particularly Melnikov's method, see [17], and for a relation between these both phenomena [18].

Models of epidemics often contain many parameters like, e.g. the transmission rate or the recovery rate, and these parameters may change in time, e.g. when measures against an epidemic are implemented so that the transmission rate decreases, or when a medication is found such that the recovery rate increases. Particularly, autonomous compartment models for epidemics are not suitable for describing the beginning of a disease outbreak or a change of the disease, e.g. due to virus mutation, while these situations may be mimicked to some degree by choosing appropriate time-dependent parameters.

In this chapter, we concentrate on two different types of parameter changes. On the one hand, we consider time-dependent parameter paths connecting two parameter values without any bifurcation on the path, but where nonetheless a drastic change of behaviour of the system occurs due to rate-induced tipping in the compartmental system modelling an epidemic [19, 20]. For an accessible mathematical framework, which provides testable criteria for rate-induced tipping in multidimensional non-autonomous dynamical systems, see [12]. On the other hand, we consider periodic or almost periodic time-dependent parameters leading to chaotic epidemic waves via a homoclinic bifurcation, in agreement with the conjecture "that homoclinic bifurcations must play a role in all the routes to chaos."[17, Chapter 4: Homoclinic Bifurcations].

Common to our study of these two situations is a certain kind of non-smoothness of the systems [21]. To observe rate-induced tipping in the most simple situation of a two-dimensional compartmental system of ODEs on a centre manifold, a certain degree of non-smoothness is necessary. In fact, for frozen parameters in a smooth system after a transcritical bifurcation of the globally asymptotic stable disease-free equilibrium (DFE), an endemic equilibrium (EE) attracts all interior states, and thus basin instability as well as rate-induced tipping cannot occur for non-frozen time-dependent parameters in the smooth case. Yet, if the system is non-smooth at the DFE, beneath the EE also the DFE can attract an open interior set of the state space; thus basin instability leading to rate-induced tipping can occur.

A certain kind of non-smoothness is necessary to observe periodically forced chaos in the most simple situation, too. In fact, as the DFE lies on the boundary of the state space, it can have a homoclinic orbit only if the system is non-smooth at the DFE. While homoclinic orbits have early been found in many mathematical models in epidemiology, see, e.g. [22], chaotic dynamics in an epidemiological model seemingly have been obtained for the first time in [23], resp., [24] for a SIR model with vital dynamics. However, not only the rigor of proof [25] but also the mechanism, how endemic chaotic waves occur in this model, may be criticised, because the chaotic waves merely connect two interior endemic equilibria but do not touch the DFE. Chaos near the DFE can only arise if the DFE has a homoclinic orbit, and due to its position on the boundary, this can only happen, if the system is non-smooth at the DFE. Our strategy to prove a chaotic behaviour is to use Melnikov's function in this non-smooth setting, where correspondingly the perturbed Hamiltonian equation is non-smooth.

4.1.1 *Outline*

In Sect. 4.2, we define the most basic notions of compartment models with continuous time and time-dependent parameters. In Sect. 4.3, we discuss occurrence of rate-induced tipping in compartment models of epidemics with time-dependent parameter paths connecting two parameter values without any bifurcation on the parameter path. In Sect. 4.4, we explore the occurrence of (almost) periodically forced chaos near the DFE. The final Sect. 4.5 concludes this chapter by a summary and open questions.

4.2 Preliminaries

We consider parameter-dependent compartment models in epidemiology given by (reduced) ODE systems

$$\dot{x}(t) = f(x(t), \lambda) \tag{4.1}$$

on the (reduced probability) simplex

$$\Sigma^n := \{x \in \mathbb{R}^n \mid x \geq 0\,,\ 1^T x \leq 1\}\,, \tag{4.2}$$

which is assumed to be positively invariant due to validity of

(A1) $f_i(x, \lambda) \geq 0$ holds for every $x \geq 0$ with $1^T x < 1$ and $x_i = 0$,

(A2) $\sum_{i=1}^{n} f_i(x, \lambda) \leq 0$ holds for every $x \geq 0$ with $1^T x = 1$,
for all parameters λ. We are interested in models with a persistent DFE at a corner $x^* \in \partial\Sigma^n$ and therefore additionally require

(A3) $f(x^*, \lambda) = 0$ for a corner $x^* \in \partial\Sigma^n$ and all λ.

Hereby, each component x_i of x gives the percentage by which the i-th compartment is occupied in the state x. The reduced ODE system (4.1) can be extended by $x \mapsto (x, 1 - 1^T x)$ to a system with one more dimension, where the components always sum up to a total of 1, and sometimes we consider the so extended compartmental system instead of the reduced system.

If a parameter $\lambda \in \Lambda$ in the finite-dimensional vector space Λ is fixed, then the compartmental system (4.1) is said to be autonomous, while in case of time-dependence $\lambda = \lambda(t)$ of parameters the system becomes non-autonomous. The image of the parameter curve $\lambda(t)$ is called the parameter path in Λ, and by substituting the parameter curve with $\lambda(rt)$ for some rate $r > 1$ (or with $\lambda(\sigma(t))$ for a strictly monotonic increasing function σ), the same parameter path in Λ can be traced out more fast. If a fixed value $\lambda \in \Lambda$ on a parameter path is chosen as parameter value, then the resulting autonomous system is said to be a system with frozen parameters.

4.3 Rate-Induced Tipping

In this section, we consider a time-dependent parameter curve $\lambda(t)$ connecting two parameter values λ_- as $t \to -\infty$ and λ_+ as $t \to +\infty$ without any bifurcation on the parameter path $P \subset \Lambda$. More precisely, in accordance with [12], we assume during the whole section that the parameter curve $\lambda(t)$ is exponentially bi-asymptotically constant, i.e. that $\lim\limits_{t\to\pm\infty} \frac{\lambda'(t)}{\exp(\mp\rho t)}$ exist for some $\rho > 0$. Although there does not occur any bifurcation on the parameter path P, nonetheless a drastic change of the behaviour of the non-autonomous system may occur if the parameter is changed sufficiently fast, e.g. if the parameter curve is replaced by $\lambda(rt)$ with a sufficiently large $r > 1$, and in this case, we informally say that rate-induced tipping occurs (for a precise definition, see [12, Definition 5.1]).

Note that if a bifurcation occurs at a parameter value λ^*, a drastic change of the behaviour of the system occurs, regardless how slow the parameter runs through the parameter value λ^*. In contrast, rate-induced tipping only occurs for sufficiently fast parameter changes, while for a too slow change of parameters, no drastic change of the behaviour of the system occurs.

For a locally asymptotically stable equilibrium $e(\lambda)$ of the frozen system varying smoothly with $\lambda \in P \subset \Lambda$, the main mechanism how rate-induced tipping occurs is due to basin instability. In [11], the notion of forward basin stability was used for one-dimensional (scalar) systems to give sufficient conditions for rate-induced tipping to occur or to be excluded. In [14], basin instability has been defined as follows.

Definition 1 ([14, Section 4.2]) Suppose $e(\lambda)$ is a locally asymptotically stable equilibrium of the autonomous ODE (4.1) for every λ on the parameter path P, and let $B(e(\lambda))$ denote the basin of attraction of $e(\lambda)$. Then, $e(\lambda)$ is said to be basin unstable on the parameter path, if there are two λ_1, λ_2 on the parameter path such that $e(\lambda_1)$ is outside the closure of the basin of attraction of $e(\lambda_2)$, i.e. $e(\lambda_1) \notin \overline{B(e(\lambda_2))}$.

In case of basin instability, a sufficiently fast time-dependent change $\lambda(t)$ from the parameter $\lambda_- := \lim_{t\searrow-\infty} \lambda(t)$ to the parameter $\lambda_+ := \lim_{t\searrow\infty} \lambda(t)$ leads to irreversible rate-induced tipping, i.e. the state starting at the locally asymptotically stable equilibrium $e(\lambda_-)$ leaves at some time t the basin of attraction of $e(\lambda(t))$ and eventually is attracted by a different local attractor, even if no bifurcation occurs on the whole parameter path P.

Theorem 1 ([14, Section 4.2]) *If a stable equilibrium $e(\lambda)$ of the autonomous frozen ODE (4.1) is basin unstable on a parameter path, then there is a time-varying external input $\lambda(t)$ of sufficiently fast rate that traces out the path and gives irreversible rate-induced tipping from $e(\lambda(t))$ in the non-autonomous system.*

In this chapter, we instead like to point to the newer reference [12] for a definition of basin instability in a rather general multidimensional case.

Definition 2 ([12, Definition 4.5]) Suppose the autonomous frozen system has a hyperbolic sink $e(\lambda)$. Consider a parameter path P such that $e(\lambda)$ varies C^1-smoothly with $\lambda \in P$.

(a) We say $e(\lambda)$ is threshold unstable on P if there exists a C^1-smooth family of regular thresholds $\theta(\lambda)$ and $(\lambda_a, \lambda_b) \in P \times P$ such that $e(\lambda_a) \in \theta(\lambda_b)$ and the signed distance $d_s(e(\lambda_1), \theta(\lambda_2))$ takes both signs in any neighbourhood of (λ_a, λ_b) in $P \times P$.
(b) We say $e(\lambda)$ is basin unstable on P if it is threshold unstable, and the threshold $\theta(\lambda_b)$ is contained in a multi-basin boundary (i.e. if $\theta(\lambda_b)$ is contained in the boundary of at least two different basins of attraction and thus separates at least two local attractors).

Hereby, a regular threshold in the frozen system is a codimension-one embedded orientable forward-invariant manifold θ of the state space that is normally repelling, and particularly due to orientation, one can define a signed distance from θ. Correspondingly, a regular edge state η of the regular threshold θ is defined as a minimal compact normally hyperbolic invariant set $\eta \subset \theta$ such that $\theta \subset W^s(\eta)$ is a subset of the stable manifold $W^s(\eta)$ of η, i.e. all states of the frozen system in θ tend to η as $t \to +\infty$.

If the locally asymptotically stable equilibrium $e(\lambda)$ is basin unstable, then for a sufficiently fast parameter change, every state of the non-autonomous system starting near $e(\lambda^-)$ as $t \to -\infty$ will fail to track $e(\lambda)$ by leaving the basin of attraction of $e(\lambda)$ through the regular threshold $\theta(\lambda)$ and tending to a different attractor.

Theorem 2 ([12, Theorem 7.3]) *Consider a nonautonomous system with a parameter path P. Suppose the autonomous frozen system has a hyperbolic sink $e(\lambda)$ that varies C^1-smoothly with $\lambda \in P$, and an equilibrium regular edge state $\eta(\lambda)$ with a regular threshold $\theta(\lambda)$.*

(a) If $e(\lambda)$ is threshold unstable on P due to $\theta(\lambda)$, then there is an exponentially bi-asymptotically constant input $\lambda(t)$ that traces out P and gives rate-induced tipping from $e(\lambda_-)$ in the non-autonomous system.

(b) *Consider a given exponentially bi-asymptotically constant input $\lambda(t)$ tracing out P such that $e(\lambda(t))$ is forward threshold unstable due to $\theta(\lambda(t))$, and $\eta(\lambda(t))$ limits to η_+ as $t \to +\infty$. Then, there is rate-induced tipping from $e(\lambda_-)$ in the non-autonomous system for parameter curves λ with suitably reparametrised time, i.e. for some $\lambda(\sigma(t))$ tracing out the same path P, where σ is a strictly monotonic increasing function.*

In Sect. 4.3.1, we will see that there can be no rate-induced tipping in the most simple smooth compartment models of epidemics, while in Sect. 4.3.2, we show that basin instability and hence rate-induced tipping can occur for modified systems, which are non-smooth at the disease-free equilibrium. Note that the systems studied here are different from the systems studied in [19, 20].

4.3.1 No Rate-Induced Tipping in a Smooth Unfolding of a Codimension Two Bifurcation of the DFE

The most simple smooth compartment models of epidemics are those, where the disease-free equilibrium (DFE) is asymptotically stable for a certain parameter and then loses stability due to a bifurcation. If f satisfying (A1),(A2),(A3) is smooth and a bifurcation of codimension two occurs at the formerly asymptotically stable equilibrium x^* in a corner of Σ^n for the parameter value $\lambda = \lambda^*$, then the linearization $A := \frac{\partial f}{\partial x}(x^*, \lambda^*)$ generically has a zero eigenvalue of algebraic multiplicity two, but geometric multiplicity one. In this case, the system can be reduced to a two-dimensional center manifold, on which the ODE generically has the unfolding

$$\begin{aligned} \dot{x} &= (\zeta - \gamma)x + \alpha y - \zeta xy - \zeta x^2 \\ \dot{y} &= (\beta - \alpha)y - \beta xy - \beta y^2 . \end{aligned} \tag{4.3}$$

with parameters $\alpha, \beta, \gamma, \zeta > 0$, and where the two-parameter bifurcation occurs at the corner $x^* = (0, 0)^T$ for parameter values $\zeta = \gamma$ and $\beta = \alpha$. By the transformation $x = R = 1 - S - I$, $y = I$, this normal form (4.3) reads (in non-reduced form) as

$$\begin{aligned} \dot{S} &= -(\beta I + \zeta R)S + \gamma R \\ \dot{I} &= (\beta S - \alpha)I \\ \dot{R} &= \alpha I + (\zeta S - \gamma)R . \end{aligned} \tag{4.4}$$

and occurs in epidemiology as an extension of the SIRS model, where susceptibles additionally become directly recovered with rate $\zeta > 0$ on contact with a recovered. The Jacobian of the right-hand side of (4.3) is

$$D\hat{f}(x,y) = \begin{pmatrix} (\zeta-\gamma)-\zeta(2x+y) & \alpha-\zeta x \\ -\beta y & (\beta-\alpha)-\beta(x+2y) \end{pmatrix}, \tag{4.5}$$

thus the DFE $x^* = (0,0)^T$ is asymptotically stable for $\beta < \alpha$, $\zeta < \gamma$. The DFE loses stability during the two-parameter bifurcation, and for $\zeta > \gamma$, the epidemic equilibrium (EPE) $(x,y) = (1-\frac{\gamma}{\zeta}, 0)$, resp., for $\beta > \alpha$ and $\frac{\beta}{\alpha} > \frac{\zeta}{\gamma}$ the endemic equilibrium (ENE) $(x,y) = (\frac{1-\frac{\alpha}{\beta}}{1+\frac{\gamma}{\alpha}-\frac{\zeta}{\beta}}, \frac{1-\frac{\alpha}{\beta}}{1+\frac{1}{\frac{\gamma}{\alpha}-\frac{\zeta}{\beta}}})$ arises in Σ^2 (note that if $\frac{\gamma}{\alpha}-\frac{\zeta}{\beta} < 0$, then either the denominator of x or the denominator of y is negative, and thus x and y have different signs, i.e. the ENE lies outside Σ^2).

The EPE is asymptotically stable if $\zeta > \gamma$ and $\frac{\zeta}{\gamma} > \frac{\beta}{\alpha}$ due to

$$Df\left(1-\frac{\gamma}{\zeta}, 0\right) = \begin{pmatrix} -(\zeta-\gamma) & \alpha-\zeta+\gamma \\ 0 & \frac{\beta\gamma}{\zeta}-\alpha \end{pmatrix} \tag{4.6}$$

while the ENE is asymptotically stable if $\beta > \alpha$ and $\frac{\beta}{\alpha} > \frac{\zeta}{\gamma}$, as

$$Df\left(\frac{1-\frac{\alpha}{\beta}}{1+\frac{\gamma}{\alpha}-\frac{\zeta}{\beta}}, \frac{1-\frac{\alpha}{\beta}}{1+\frac{1}{\frac{\gamma}{\alpha}-\frac{\zeta}{\beta}}}\right) = \begin{pmatrix} \zeta(-\frac{\gamma}{\zeta}-x+\frac{\alpha}{\beta}) & \alpha-\zeta x \\ -\beta y & -\beta y \end{pmatrix} \tag{4.7}$$

has determinant $-\beta\zeta y(\frac{\alpha}{\beta}-\frac{\gamma}{\zeta}-x)+\beta y(\alpha-\zeta x) = \beta\zeta(\frac{\alpha}{\zeta}-\frac{\alpha}{\beta}+\frac{\gamma}{\zeta})y > 0$, because $\beta > \alpha$ and $\frac{\beta}{\alpha} > \frac{\zeta}{\gamma}$ imply $-\frac{\alpha}{\beta}+\frac{\gamma}{\zeta} > 0$ and trace $\zeta(-\frac{\gamma}{\zeta}+\frac{\alpha}{\beta})-(\zeta x+\beta y) = \zeta(\frac{\alpha}{\beta}-\frac{\gamma}{\zeta})-\beta\frac{\gamma}{\alpha}x < 0$ due to $y+(\frac{\zeta}{\beta}-\frac{\gamma}{\alpha})x = 0$ and $\frac{\alpha}{\beta}-\frac{\gamma}{\zeta} < 0$, i.e. if both EPE and ENE are present, then the ENE attracts all interior points. Therefore, regardless how the parameters are chosen, in the smooth case, one equilibrium attracts all interior points, and there are never two disjoint basins of attraction with non-empty interior. Hence, in the non-autonomous case, rate-induced tipping cannot happen, regardless how the parameters depend on time.

4.3.2 *Rate-Induced Tipping in a Non-smooth Unfolding*

In contrast to the former section, rate-induced tipping can happen in non-smooth systems, where the DFE lying at a corner is unstable but still attracts points in the interior of the state space. Consider the non-smooth unfolding of the SIRS model

$$\begin{aligned} \dot{x} &= -\gamma x + \alpha y + g(x,y) \\ \dot{y} &= (\beta-\alpha)y - \beta xy - \beta y^2 - g(x,y), \end{aligned} \tag{4.8}$$

with a continuous $g : \Sigma^2 \to \mathbb{R}$, which is smooth at every $(x, y) \neq (0, 0)$ and satisfies $\lim_{x\searrow 0} \frac{g(x,0)}{x} = 0 = \lim_{y\searrow 0} \frac{g(0,y)}{y}$, but is not necessarily totally differentiable at $x^* = (0, 0)^T$, e.g.

$$g(x, y) := \begin{cases} \zeta \frac{xy}{x+y} & (x, y) \neq (0, 0) \\ 0 & (x, y) = (0, 0) \end{cases} \tag{4.9}$$

with $\zeta > 0$. Assume that g is such that the nullclines $\dot{x} = 0$, $\dot{y} = 0$, which for $y \neq 0$ are given by

$$\begin{aligned} -\gamma x + \alpha y + g(x, y) &= 0 \\ (\beta - \alpha) - \beta x - \beta y - \frac{g(x, y)}{y} &= 0 \, , \end{aligned} \tag{4.10}$$

intersect exactly at one interior point ENE in Σ^2 such that this equilibrium is asymptotically stable. For (4.9), this is the case, if for $\delta := \sqrt{\alpha(\alpha + 2\gamma + 2\zeta) + (\zeta - \gamma)^2}$ the inequalities $\gamma + \delta > \alpha + \zeta$ and $(\beta + \gamma)(\delta - \alpha) + \beta(\zeta - \gamma) > (\delta + \alpha)\zeta + (\zeta - \gamma)^2$ are valid. Using polar coordinates $x = r(1 - \phi)$, $y = r\phi$, for $r \geq 0$, $0 \leq \phi \leq 1$, w.r.t. the 1-norm on the simplex Σ^2, we obtain due to $r = x + y$, $\phi = \frac{y}{x+y}$ and $dr = dx + dy$, $d\phi = \frac{1}{r}((1 - \phi)\, dy - \phi\, dx)$, from (4.8) the equations

$$\begin{aligned} \dot{r} &= r\,(\beta(1 - r)\phi - \gamma(1 - \phi)) \\ \dot{\phi} &= \phi \Bigg((1 - \phi)(\beta - \alpha - \beta r - \frac{1}{r\phi} g(r(1 - \phi), r\phi) + \gamma) \\ &\quad -\alpha\phi - \frac{1}{r} g(r(1 - \phi), r\phi) \Bigg) \end{aligned} \tag{4.11}$$

Thus, if both r and ϕ are small, then in the first equation, the negative term $-\gamma(1 - \phi)$ dominates the bracket. For a similar result in the second equation, we need $\beta - \alpha - \frac{1}{r\phi} g(r(1-\phi), r\phi) + \gamma < 0$ for r and ϕ sufficiently small. For (4.9) this is the case, if $\beta - \alpha < \zeta - \gamma$, because of $\frac{1}{r\phi} g(r(1-\phi), r\phi)) = \frac{1}{y} g(x, y) = \zeta \frac{x}{x+y} = \zeta(1-\phi)$ and $-\frac{1}{r} g(r(1-\phi), r\phi) = -\zeta\phi(1-\phi)$. Therefore, the DFE attracts in the original system with coordinates (x, y) a subset with non-empty interior near the origin between the line $y = 0$ and the nullcline $(\beta - \alpha) - \beta x - \beta y - \frac{g(x,y)}{y} = 0$ near the origin $x^* = (0, 0)^T$. If now $\beta > \alpha$ and $\zeta > \gamma$ decrease, then the basin of attraction of the ENE shrinks so that the former ENE lies in the basin of attraction of the unstable DFE, i.e.

Proposition 1 *For the non-smooth system* (4.8)*, there exist parameter paths, for which the ENE is basin unstable and which do not cross a bifurcation parameter.*

Remark 1 By further modifying the vectorfield f, the curve separating the basins of attraction of DFE and ENE may even tend to the DFE as $t \to -\infty$ instead of leaving Σ^2, i.e. the separating curve becomes a homoclinic orbit of the DFE.

While we have seen in Sect. 4.3.1 that in the most simple smooth compartment models, there are never two disjoint basins of attraction with non-empty interior, Proposition 1 paves a way to irreversible rate-induced tipping induced by basin instability: Combining Theorem 2 with Proposition 1 leads for $\lambda = (\beta, \zeta)$ and the ENE as initial sink $e(\lambda_-)$ to the result that in the nearby non-smooth non-linear compartment model the state tends to the DFE under a sufficiently fast change of parameters from $\beta_- > \alpha$, $\zeta_- > \gamma$ to sufficiently smaller values $\beta_+ > \alpha$, $\zeta_+ > \gamma$, although no bifurcation parameter is crossed and in the autonomous system with frozen parameters β_+, ζ_+ the ENE is locally asymptotically stable. Particularly, for a disease governed by the non-smooth system (4.8) on a centre manifold, a sufficiently fast implementation of measures may lead to eradication of an epidemic, while for a too slow implementation, the state would have forever tracked the ENE, and the disease would have become endemic. Further, this irreversible rate-induced tipping from the ENE to the DFE in a non-smooth system also induces observable artefacts in nearby smooth systems, where for the same parameter change the state stays for a long time near the DFE, before finally tending to the ENE.

4.4 Periodically Forced Chaos

While in the former section we have focussed on proving rate-induced tipping for parameter paths connecting an initial parameter λ_- with a final parameter λ_+, here we consider periodic or almost-periodic parameter changes $\lambda = \lambda(t)$ and aim to prove chaotic dynamics. According to the conjecture that homoclinic bifurcations must play a role in all routes to chaos, we try to find a homoclinic orbit at the disease-free equilibrium (DFE) in a compartment model of epidemics. Hereby, the difficulty is that the DFE lies on the boundary, which due to positive invariance of the simplex Σ^2 prevents chaotic orbits surrounding the DFE. Again, the way out of this difficulty is to consider systems which are non-smooth near the DFE. But before we turn to such systems in the second subsection, we review in the first section a main result about periodically forced chaos in a smooth compartmental system.

4.4.1 Chaos Between Two Endemic Equilibria in a Smooth SIR Model with Vital Dynamics

Homoclinic orbits have early been found in many mathematical models in epidemiology; see, e.g. [22]. Yet, chaotic dynamics in an epidemiological model seemingly have been obtained for the first time in [23], resp., [24] for the SIR model with vital dynamics

$$\begin{aligned}\dot{S} &= \mu - \beta S I^p - \mu S \\ \dot{I} &= (\beta S I^{p-1} - \alpha - \mu) I \\ \dot{R} &= \alpha I - \mu R\end{aligned} \tag{4.12}$$

with exponent $p > 1$, a constant birth/death rate $\mu > 0$ and a constant recovery rate $\alpha > 0$ in case of a certain periodic, resp., almost periodic transmission rate $\beta = \beta(t)$. However, not only the rigor of proof [25] and the restriction $p > 1$ may be considered critically but also the mechanism, how endemic chaotic waves occur in this model, may be criticised as having not much to do with real epidemics, because the chaotic waves merely connect two interior endemic equlibria but do not touch the DFE. In fact, for a frozen transmission rate $\beta > 0$, the DFE $(S, I, R) = (1, 0, 0)$ in the SIR model with vital dynamics (4.12) is always locally asymptotically stable, as the linearization at the DFE

$$Df(1, 0) = \begin{pmatrix} -\mu & 0 \\ 0 & -\alpha - \mu \end{pmatrix} \tag{4.13}$$

has negative eigenvalues $\lambda = -\mu$ and $\lambda = -\alpha - \mu$. A saddle-node bifurcation occurs for the critical parameter $\beta = \beta_c := \frac{p^p(\mu+\alpha)^p}{\mu^{p-1}(p-1)^{p-1}}$ at the interior endemic equilibrium

$$(S, I, R) = \left(\frac{1}{p}, \frac{\mu(p-1)}{p(\mu+\alpha)}, \frac{\alpha(p-1)}{p(\mu+\alpha)}\right). \tag{4.14}$$

In fact, at this point for $\beta = \beta_c$, the equations $\mu - \beta S I^p - \mu S = \mu - \frac{\mu(p-1)}{p} - \mu \frac{1}{p} = 0$, $\beta S I^{p-1} - \alpha - \mu = (\mu + \alpha) - \alpha - \mu = 0$ and $\alpha I - \mu R = \alpha \frac{\mu(p-1)}{p(\mu+\alpha)} - \mu \frac{\alpha(p-1)}{p(\mu+\alpha)} = 0$ hold. Further, the linearization w.r.t. (S, I) at this equilibrium

$$\begin{aligned} Df\left(\frac{1}{p}, \frac{\mu(p-1)}{p(\mu+\alpha)}\right) &= \begin{pmatrix} -\beta I^p - \mu & -p\beta S I^{p-1} \\ \beta I^p & p\beta S I^{p-1} - \alpha - \mu \end{pmatrix} \Big|_{(S,I)=(\frac{1}{p}, \frac{\mu(p-1)}{p(\mu+\alpha)}), \beta=\beta_c} && (4.15) \\ &= \begin{pmatrix} -\mu p & -p(\mu+\alpha) \\ \mu(p-1) & (p-1)(\mu+\alpha) \end{pmatrix} && (4.16) \end{aligned}$$

has linear dependent rows and thus an eigenvalue $\lambda = 0$. More precisely, the characteristic polynomial is $(-\mu p - \lambda)((p-1)(\mu+\alpha) - \lambda) + p(\mu+\alpha)\mu(p-1) = \lambda^2 + (\mu - (p-1)\alpha)\lambda$ leading to two endemic equilibria for $\beta > \beta_c$. For the particular case $\mu = (p-1)\alpha$, the equilibrium (4.14) even undergoes a two-parameter Bogdanov-Takens bifurcation (as then the characteristic polynomial is λ^2 and thus has a double zero, while the linearization is not the zero matrix), whose normal form is given by

$$\begin{aligned}\dot{x} &= y \\ \dot{y} &= \beta_1 + \beta_2 x + x^2 - xy\,.\end{aligned} \tag{4.17}$$

Particularly, a homoclinic bifurcation occurs in (4.17) along a parameter curve $\beta_1 = -\frac{6}{25}\beta_2^2 + O(\beta_2^3)$ as $\beta_2 \nearrow 0$. For such parameters, in certain coordinates, the system is near to the smooth Hamiltonian system

$$\begin{aligned}\dot{q} &= p \\ \dot{p} &= 1 - q^2\end{aligned} \tag{4.18}$$

with Hamiltonian $H(q, p) = \frac{1}{2}p^2 + q - \frac{1}{3}q^3$, which has an equilibrium $(1, 0)^T$ with a homoclinic orbit, and Melnikov's method can be used to predict the occurrence of chaotic orbits under a non-autonomous perturbation. Therefore, also (4.12) has a homoclinic bifurcation for a parameter curve (α, β) originating at $(\frac{\mu}{p-1}, \beta_c)$, and a time-dependent periodic [23] or almost periodic [24] perturbation $\beta = \beta(t)$ of the transmission rate can lead to chaotic behaviour.

To prove the statements below (4.17), start with these equations, write $\beta_1 + \beta_2 x + x^2 - xy = \beta_1 - \frac{\beta_2^2}{4} + (x + \frac{\beta_2}{2})^2 + \frac{\beta_2}{2}y - (x + \frac{\beta_2}{2})y$ and substitute $x + \frac{\beta_2}{2} = \epsilon^2 q$, $y = \epsilon^3 p$, $\beta_1 = (\nu - 3)\epsilon^4$, $\beta_2 = -2\sqrt{\nu + 1}\epsilon^2$, $t = \tau/\epsilon$, to obtain the equivalent system

$$\begin{aligned}\frac{dq}{d\tau} &= p \\ \frac{dp}{d\tau} &= -4 + q^2 - \epsilon(\sqrt{\nu + 1} + q)p\,.\end{aligned} \tag{4.19}$$

For $\epsilon = 0$, this system (4.19) is Hamiltonian with $H(q, p) = \frac{1}{2}p^2 + 4q - \frac{1}{3}q^3$, and its equilibrium $(2, 0)$ has a homoclinic orbit in the level set $H(q, p) = \frac{16}{3}$ $(= H(2, 0))$ given by $q(\tau) = 2 - \frac{6}{\cosh^2(\tau)}$, $p(\tau) = 12\frac{\tanh(\tau)}{\cosh^2(\tau)}$, as

$$H(q(t), p(t)) = \frac{1}{2}\left(12\frac{\tanh(\tau)}{\cosh^2(\tau)}\right)^2 + \left(\frac{16}{3} - 72\frac{\tanh^2(\tau)}{\cosh^4(\tau)}\right) = \frac{16}{3}\,.$$

As the solution of $H(q, p) = \frac{16}{3}$ is given by the union of the two graphs $p(q) = \pm\sqrt{\frac{32}{3} - 8q + \frac{2}{3}q^3}$ over $[-4, 2]$, Melnikov's function for the perturbed system reduces to

$$
\begin{aligned}
&- \int_{-\infty}^{\infty} p(\tau)(\sqrt{\nu+1} + q(\tau))p(\tau)\,d\tau \\
&= -2 \int_{-4}^{2} (\sqrt{\nu+1} + q)\sqrt{\frac{32}{3} - 8q + \frac{2}{3}q^3}\,dq \\
&= -\frac{96}{35}(14\sqrt{\nu+1} - 20)
\end{aligned}
$$

due to $d\tau = dq/p$ and has a simple zero iff $\sqrt{\nu+1} = \frac{10}{7}$, i.e. $\nu = \frac{51}{49}$. Therefore, for this parameter ν satisfying $\beta_1/\beta_2^2 = \frac{\nu-3}{(-2\sqrt{\nu+1})^2} = -\frac{6}{25}$ (as $\nu - 3 = \frac{96}{49}$ and $-\frac{96}{49}/(-2^2\frac{100}{49}) = \frac{6}{25}$), we can deduce the existence of a homoclinic orbit in the two-dimensional autonomous system (4.17) for sufficiently small $\epsilon > 0$, and further the occurrence of chaotic waves between the two endemic equilibria of (4.12) for time-dependent periodic [23] or almost periodic [24] perturbations $\beta = \beta(t)$ of the transmission rate.

4.4.2 *Chaos Between DFE and ENE in a Non-smooth SIRS Model*

In contrast to the former subsection, where the homoclinic orbit at one endemic equilibrium surrounded another endemic equilibrium, and chaotic epidemic waves between these two interior equilibria occurred on a break-up of this homoclinic orbit, we aim to obtain in this subsection the occurrence of chaotic epidemic waves between the disease-free equilibrium (DFE) and a unique endemic equilibrium (ENE).

Let us start with the smooth normal form (4.3) in the case $\zeta = 0$, which reads as

$$
\begin{aligned}
\dot{x} &= -\gamma x + \alpha y \\
\dot{y} &= (\beta - \alpha)y - \beta xy - \beta y^2 \,.
\end{aligned}
\tag{4.20}
$$

Consider the scaling $x = \epsilon q$, $y = \epsilon^2 p$, $t = \tau/\epsilon$, $\gamma = \epsilon^2 \nu$, $\beta - \alpha = \epsilon\mu$, use $\beta = \alpha + (\beta - \alpha) = \alpha + \epsilon\mu$divide the second equation by p, substitute p for $\ln|p|$and use the approximation $\exp(p) \approx 1 + p$ to obtain

$$
\begin{aligned}
\dot{q} &= \alpha(1 + p) - \epsilon\nu|q| + \alpha(\exp(p) - 1 - p) \\
\dot{p} &= 1 - \alpha|q| - \epsilon(\mu|q| + \alpha p) - \epsilon^2\mu p \,.
\end{aligned}
\tag{4.21}
$$

where we use that $q = |q|$ for $q \geq 0$ (note $0 \leq x \leq 1$). Assuming that $\exp(p) - 1 - p$ is of order ϵ and putting $\epsilon = 0$ gives the non-smooth Hamiltonian system

$$\begin{aligned}\dot{q} &= \alpha(1+p)\\ \dot{p} &= 1-\alpha|q|\end{aligned} \tag{4.22}$$

with Hamiltonian $H(q, p) := \frac{\alpha}{2}(1+p)^2 - q + \frac{\alpha}{2}|q|q$.

Lemma 1 *In the non-smooth Hamiltonian system* (4.22), *the equilibrium* $(-\frac{1}{\alpha}, -1)$ *has a homoclinic orbit surrounding the equilibrium* $(\frac{1}{\alpha}, -1)$.

Proof The homoclinic orbit lies within the level set $H(q, p) = \frac{1}{2\alpha} = H(-\frac{1}{\alpha}, -1)$ of the equilibrium $(-\frac{1}{\alpha}, -1)$. Using linearity of the corresponding second-order equations, we obtain from $\ddot{q} = \alpha+\alpha^2 q$ for $q \le 0$ the solutions $q(t) = C_1 \exp(\alpha t) + C_2 \exp(-\alpha t) - \frac{1}{\alpha}$ and from $\ddot{q} = \alpha - \alpha^2 q$ for $q \ge 0$ the solutions $q(t) = C_1 \cos(\alpha t) + C_2 \sin(\alpha t) + \frac{1}{\alpha}$. Combining these solutions so that the homoclinic orbit starts and ends at the equilibrium $(\frac{1}{\alpha}, -1)$, the homoclinic orbit can be explicitly stated as

$$q_h(t) := \begin{cases} \frac{1}{\alpha}\exp(\alpha(t+\arccos(-\frac{1}{\alpha}))) - \frac{1}{\alpha}, & t \le -\arccos\left(-\frac{1}{\alpha}\right) \\ \cos(\alpha t) + \frac{1}{\alpha}, & -\arccos(-\frac{1}{\alpha}) \le t \le \arccos\left(-\frac{1}{\alpha}\right) \\ \frac{1}{\alpha}\exp(-\alpha(t-\arccos(-\frac{1}{\alpha}))) - \frac{1}{\alpha}, & \arccos\left(-\frac{1}{\alpha}\right) \le t \end{cases} \tag{4.23}$$

and $p_h(t) := \frac{1}{\alpha}\dot{q}_h(t) - 1$. □

To obtain a chaotic behaviour for the perturbed system (4.21), we use Melnikov's method and aim to prove that Melnikov's function has a simple zero for certain (almost) periodic parameter changes. The Melnikov function, which in general is given for a Hamiltonian system $\dot{x} = JDH(x) + \epsilon g(x, t, \epsilon)$ with $J = \begin{pmatrix} 0 & 1 \\ -1 & 0 \end{pmatrix}$ by

$$M := \int_{-\infty}^{\infty} DH(x_h(t))g(x_h(t), \omega(t-t_0), 0)\, dt\,,$$

reads in this case as

$$M = \int_{-\infty}^{\infty} (\alpha|q_h| - 1)\Big(\frac{\alpha}{\epsilon}(\exp(p_h) - 1 - p_h) - \nu|q_h|\Big) - \alpha(1+p_h)(\mu|q_h| + \alpha p_h)\, dt\,. \tag{4.24}$$

Let us calculate separately the different parts of M: For the first part

$$\begin{aligned} M_0(\alpha) &:= \int_{-\infty}^{\infty} (\alpha|q_h| - 1)\frac{\alpha}{\epsilon}(\exp(p_h) - 1 - p_h)\, dt \\ &= \frac{2\alpha}{e\epsilon}\left(\sinh(\sin(\alpha \sec^{-1}(-\alpha))) - \sinh\left(\frac{1}{\alpha}\right)\right) \end{aligned} \tag{4.25}$$

and for the second part

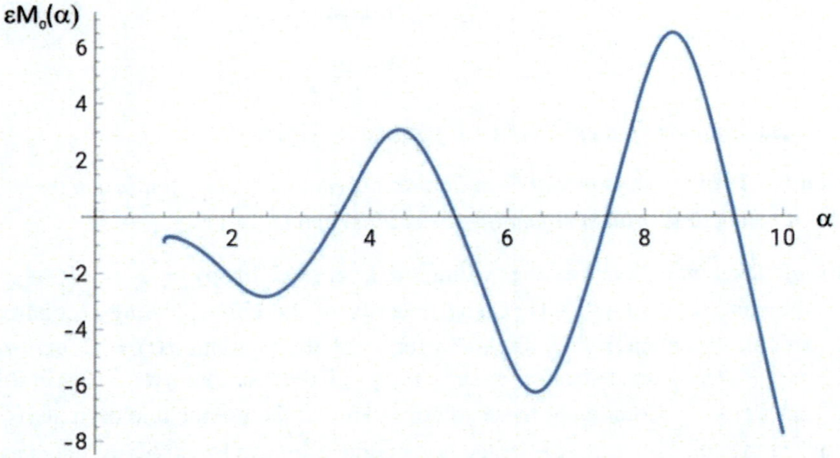

Fig. 4.1 In this figure, up to the factor $\frac{1}{\epsilon}$ the function $M_0(\alpha)$ from (4.25) is plotted over α, and the plot shows that there are zeros of M_0 w.r.t. α

$$
\begin{aligned}
M_1(\alpha) &:= \int_{-\infty}^{\infty} (\alpha|q_h| - 1)|q_h|\, dt \\
&= \alpha \cos^{-1}\left(-\frac{1}{\alpha}\right) + \frac{1}{\alpha}\left(2 + \alpha \cos\left(\alpha \cos^{-1}\left(-\frac{1}{\alpha}\right)\right)\right) \\
&\quad \times \sin\left(\alpha \cos^{-1}\left(-\frac{1}{\alpha}\right)\right) - \frac{1}{\alpha^2}
\end{aligned} \tag{4.26}
$$

Therefore, for a time-dependent (almost) periodic $\mu = \mu(t)$ and constant ν, we obtain

$$
M = M_0(\alpha) - \nu M_1(\alpha) - \int_{-\infty}^{\infty} \alpha(1 + p_h)(\mu(t)|q_h| + \alpha p_h)\, dt\,. \tag{4.27}
$$

As $M_0(\alpha)$ depends on ϵ only by the factor $\frac{1}{\epsilon}$, the plot of

$$
\epsilon M_0(\alpha) = \frac{2\alpha}{\mathrm{e}}\left(\sinh(\sin(\alpha \sec^{-1}(-\alpha))) - \sinh\left(\frac{1}{\alpha}\right)\right) \tag{4.28}
$$

in dependence on α in Fig. 4.1 shows that there are zeros of M_0. Thus, we can choose $\alpha > 0$ so that $M_0(\alpha) = 0$ for every $\epsilon > 0$. Further, Fig. 4.2 contains a plot of $M_1(\alpha)$ showing that $M_1(\alpha)$ is positive for $\alpha > 0$.

Now we aim to choose $\nu > 0$ constant such that the negative $-\nu M_1(\alpha)$ cancels out the last integral for a sufficiently large amplitude A in $\mu(t) = A\sin(\omega(t - t_0))$ and appropriate frequency $\omega \neq \alpha$ as well as phase ωt_0. Note that for constant μ, the integral $-\int_{-\infty}^{\infty} \alpha(1 + p_h)(\mu(t)|q_h| + \alpha p_h)\, dt$ does not depend on μ and is always

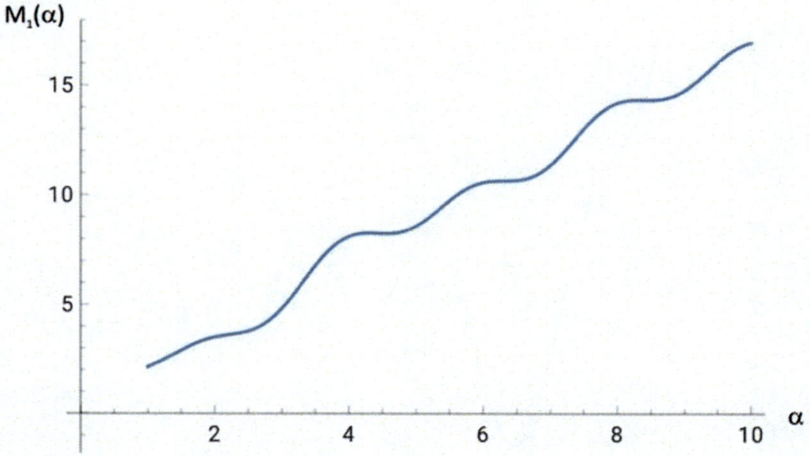

Fig. 4.2 In this figure, the function $M_1(\alpha)$ from (4.26) is plotted over α, and the plot shows that M_1 is positive and increasing

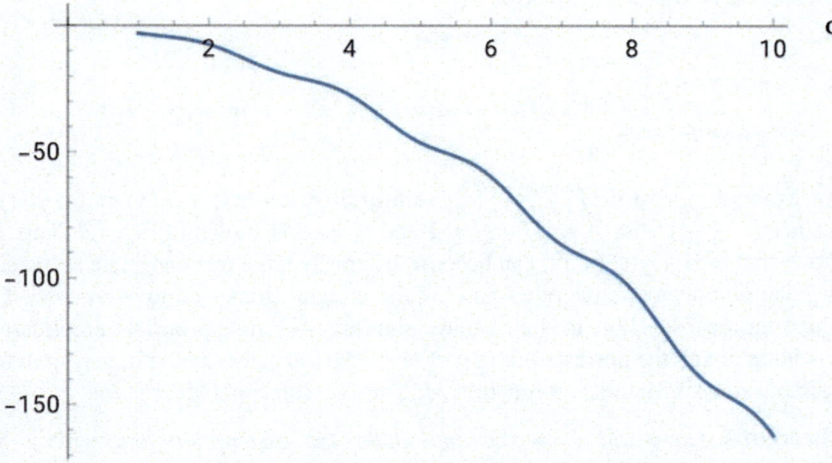

Fig. 4.3 In this figure, the integral $-\int_{-\arccos(-1/a)}^{\arccos(-1/a)} \alpha(1+p_h)\alpha p_h\,dt$ is plotted over α, and the plot shows that this integral is negative and decreasing

negative, as shown by Fig. 4.3. Thus, for constant μ, ν this integral and the negative term $-\nu M_1(\alpha)$ cannot cancel out. Consequently, as expected, there is no chaos for constant μ, ν.

Yet, for periodic $\mu = \mu(t)$ it may happen that these terms cancel out so that there is a simple zero of M. Note that the part of the integral $-\int_{-\infty}^{\infty} \alpha(1+p_h)(\mu(t)|q_h|+\alpha p_h)\,dt$ corresponding to the middle case in (4.23) reads as

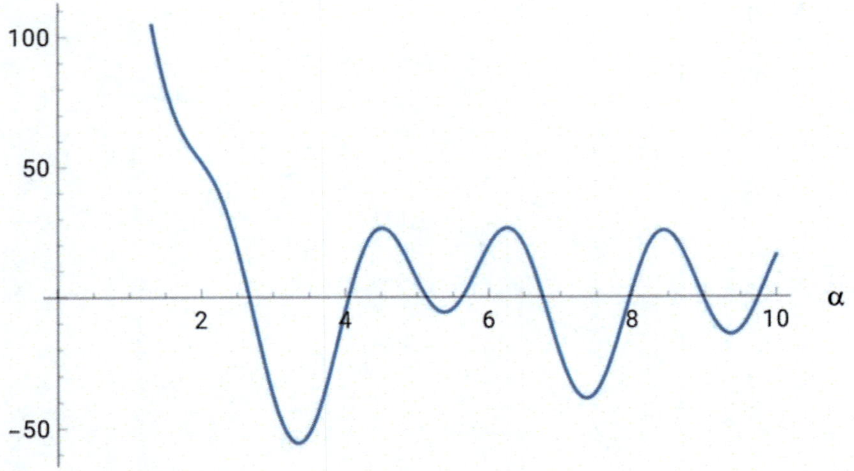

Fig. 4.4 Plot of the integral $-\int_{-\arccos(-1/a)}^{\arccos(-1/a)} \alpha(1+p_h)\mu(t)|q_h|\,dt$ for $\mu(t) = A\sin(t)$, $A = 50$, over α, showing that this integral for some α positive, and by choosing the amplitude A even higher its values can become arbitrarily large

$$-\int_{-\arccos(-1/a)}^{\arccos(-1/a)} a\sin(at)(\mu(t)(\cos(at)+1/a) - a\sin(at) - a)\,dt\,. \tag{4.29}$$

For example, a plot of $\int_{-\arccos(-1/a)}^{\arccos(-1/a)} a\sin(at)\mu(t)(\cos(at)+1/a)\,dt$ for $\mu(t) = A\sin(\omega(t-t_0))$ with $A = 50$, $\omega = 1$ and $t_0 = 0$ is shown in Fig. 4.4. This plot shows that the integral (4.29) can become arbitrarily large by choosing a sufficiently large amplitude A. As the other parts of the integral from $-\infty$ to $-\arccos(-1/a)$ and from $\arccos(-1/a)$ to $+\infty$ behave similarly, for an appropriate amplitude A, this integral and the corresponding part of $-\nu M_1(\alpha)$ can cancel out, giving in total a simple zero of Melnikov's function M. Thus we can conclude:

Theorem 3 *For small $\epsilon > 0$, and under the assumption that with ϵ also $\exp(p) - 1 - p$ is for the homoclinic solution of* (4.22) *sufficiently small, the perturbation* (4.21) *of the non-smooth Hamiltonian system* (4.22) *shows chaotic behaviour for $\alpha > 0$ chosen so that $M_0(\alpha) = 0$ and for $\nu =$ const and $\mu = \mu(t) = A\sin(\omega(t-t_0))$ with amplitude A, frequency ω and phase t_0 chosen so that Melnikov's function M in* (4.27) *has a simple zero.*

4.5 Conclusion

In comparison with [19, 20], in the first part of this chapter, we have have been able to show the occurence of rate-induced tipping in a two-dimensional non-smooth non-autonomous non-linear compartmental system obtained by a non-smooth unfolding of the SIRS model of epidemics at a two-parameter bifurcation of the disease-free equilibrium. Hereby, we argued that in the frozen system, not only the endemic equilibrium (EE) but also the disease-free equilibrium (DFE) has a basin of attraction intersecting the interior of the probability simplex, making possible basin instability and therefore rate-induced tipping for a sufficiently large rate of parameter change. Notably, we have not discussed how to determine the critical rate, but this can be done following the standard methods mentioned, e.g. in [12]. Thus, in models of epidemics which are non-smooth, it can happen that in two countries, the same measures against the diseases are decided at the same time, corresponding to the same initial and final values of a parameter path at the same times, but that due to a slow implementation of measures, the disease becomes endemic in one country, while in the other country, due to a fast implementation, the disease is eradicated.

In the second part, we have studied the question, whether a periodical forcing can lead to chaotic epidemic waves in a system near to the SIRS model, where the disease seems to have died out for some time, but then again strikes back and seems to be endemic for some time. Invoking again non-smoothness of a nearby Hamiltonian system, we have been able to apply Melnikov's method to obtain for certain constant or periodic parameters a simple zero of Melnikovs function and therefore chaotic behaviour near the DFE.

Yet, there are various open research questions:

1. Determine the critical rates for which rate-induced tipping happens in the model studied in Sect. 4.3.2.
2. Check whether the additional assumption on $\exp(p) - 1 - p$ in Theorem 3 is automatically satisfied.
3. Are there indications for non-smoothness of compartment models at the DFE in data observed for an epidemic?
4. Are other phenomena than the occurence of rate-induced tipping or (almost) periodically forced chaos near the DFE overseen by only considering systems which are smooth near the DFE?

Maybe an answer to these questions can help us to obtain better compartment models for epidemics showing interesting dynamical phenomena.

References

1. P. Kloeden, M. Rasmussen, *Nonautonomous Dynamical Systems*. AMS Mathematical Surveys and Monographs, vol. 176 (American Mathematical Society, Providence, 2011)
2. R. Ross, H.P. Hudson, An application of the theory of probabilities to the study of a priori pathometry II. Proc. R. Soc. A **92**, 204–230 (1916)
3. R. Ross, H.P. Hudson, An application of the theory of probabilities to the study of a priori pathometry II. Proc. R. Soc. A **93**, 212–225 (1917)
4. R. Ross, H.P. Hudson, An application of the theory of probabilities to the study of a priori pathometry III. Proc. R. Soc. A **93**, 225–240 (1917)

5. W.O. Kermack, A.G. McKendrick, Contributions to the mathematical theory of epidemics I. Proc. R. Soc. A **115**, 700–721 (1927)
6. W.O. Kermack, A.G. McKendrick, Contribution to the mathematical theory of epidemics II. Proc. R. Soc. A **138**, 55–83 (1932)
7. W.O. Kermack, A.G. McKendrick, Contributions to the mathematical theory of epidemics III. Proc. R. Soc. A **141**, 94–122 (1933)
8. H.W. Hethcote, The mathematics of infectious diseases. SIAM Rev. **42**, 599–653 (2000)
9. R. Smith, *Modeling Disease Ecology with Mathematics*. AIMS Series in Differential Equations & Dynamical Systems, vol. 2 (American Mathematical Society, Providence, 2008)
10. F. Brauer, C. Castillo-Chavez, *Mathematical Models in Population Biology and Epidemiology* (Springer, New York, 2012)
11. P. Ashwin, C. Perryman, S. Wieczorek, Parameter shifts for nonautonomous systems in low dimension: Bifurcation- and Rate-induced tipping. Nonlinearity **30**, 2185–2210 (2017)
12. S. Wieczorek, C. Xie, P. Ashwin, Rate-induced tipping: thresholds, edge states and connecting orbits. Nonlinearity **36**, 3238 (2023)
13. P. Ashwin, S. Wieczorek, R. Vitolo, P. Cox, Tipping points in open systems: bifurcation, noise-induced and rate-dependent examples in the climate system. Phil. Trans. R. Soc. A **370**, 1166–1184 (2012)
14. P.E. O'Keeffe, S. Wieczorek, Tipping phenomena and points of no return in ecosystems: beyond classical bifurcations. SIAM J. Appl. Dyn. Syst. **19**, 2371–2402 (2020)
15. A. Vanselow, S. Wieczorek, U. Feudel, When very slow is too fast - collapse of a predator-prey system. J. Theoret. Biol. **479**, 64–72 (2019)
16. B. Kaszás, U. Feudel, T. Tél, Tipping phenomena in typical dynamical systems subjected to parameter drift. Sci. Rep. (2019). https://doi.org/10.1038/s41598-019-44863-3
17. A. Fowler, M. McGuinness, *Chaos: An Introduction for Applied Mathematicians* (Springer, Cham, 2019)
18. C. Kuehn, I.P. Longo, Estimating rate-induced tipping via asymptotic series and a melnikov-like method (2020). Preprint arXiv:2011.04031
19. J. Merker, B. Kunsch, Rate-induced tipping phenomena in compartment models of epidemics, in *Analysis of Infectious Disease Problems (Covid-19) and Their Global Impact*. ed. by P. Agarwal, J.J. Nieto, M. Ruzhansky, D.F.M. Torres (Springer, Singapore, 2021), pp.307–328
20. J. Merker, B. Kunsch, G. Schuldt, Nonlinear compartment models with time-dependent parameters. Mathematics **9**, 1657 (2021)
21. T. Küpper, Concepts for non-smooth dynamical systems, in *Mathematics and the 21st Century*. ed. by A.A. Ashour, A.-S.F. Obada (World Scientific, Singapore, 2001), pp.123–140
22. W. Lih-Ing, Z. Feng, Homoclinic Bifurcation in an SIQR model for childhood diseases. J. Differ. Equ. **168**, 150–167 (2000)
23. O. Diallo, Y. Koné, Melnikov analysis of Chaos in a general epidemiological model. Nonlinear Analy. Real World Appl. **8**, 20–26 (2007)
24. O. Diallo, Y. Koné, A. Maiga, Melnikov analysis of Chaos in an epidemiological model with almost periodic incidence rates. Appl. Math. Sci. **2**, 1377–1386 (2008)
25. H. Li, X. Liao, L. Xiao, Comments on: Melnikov analysis of Chaos in a general epidemiological model [Nonlinear Anal. RWA 8 (2007) 20]. Nonlinear Analy. RWA **13**, 39–41 (2012)

Chapter 5
Study of the Nonelementary Singular Points and the Dynamics Near the Infinity in Predator-Prey Systems

Érika Diz-Pita, Jaume Llibre, and M. Victoria Otero-Espinar

5.1 Introduction

Dynamical systems that represent the interaction between coexisting species, as it is the particular case of predator-prey systems, have been widely studied in the literature. Starting from the most classical models, such as the one proposed by A. Lotka [1] and V. Volterra [2], researchers have made efforts to add new features to these models, making them more realistic and allowing them to be adapted to more real-world situations.

In [3], different proposals for modeling predator-prey systems have been studied comparatively. Special attention has been paid to the inclusion of characteristics such as the Allee effect, immigration, fear, and, in general, to the indirect effects that the presence of predators can produce on prey, apart from direct attack.

One thing that can be appreciated is that, while trying to make the models more realistic from a biological and ecological point of view, from a mathematical point of view appear more difficulties in the study of these differential systems.

For example, there are many population models focused on the analysis of finite singular points and their stability, but especially in hyperbolic cases. We will give a summary on a desingularization technique that can improve or complete the existing works in the sense that it allows one to study any type of singularities even if they are not elementary.

É. Diz-Pita (✉) · M. V. Otero-Espinar
Universidade de Santiago de Compostela, Santiago de Compostela, Spain
e-mail: erikadiz.pita@usc.es; mvictoria.otero@usc.es

J. Llibre
Universitat Autònoma de Barcelona, Barcelona, Spain
e-mail: jllibre@mat.uab.cat

C. M. A. Pinto, C. M. Ionescu (eds.), *Computational and Mathematical Models in Biology*, Nonlinear Systems and Complexity 38,
https://doi.org/10.1007/978-3-031-42689-6_5

On the other hand, we will also present the Poincaré compactification which allows one to accomplish a complete study of the global dynamics of the systems, making possible to know the behavior of the orbits in a neighborhood the infinity.

In addition to the description of these techniques from a theoretical point of view, we present two systems in the field of population dynamics to which we have applied them, and we present the results obtained.

The first model we have studied is obtained from the classical Rosenzweig and MacArthur model, introduced in [4]. Considering $x \geq 0$ as the prey density and $y \geq 0$ as the predator density, the original model has the form

$$\dot{x} = rx\left(1 - \frac{x}{K}\right) - y\frac{mx}{b+x},$$

$$\dot{y} = y\left(-\delta + c\frac{mx}{b+x}\right),$$

where the parameter $\delta > 0$ represents the death rate of the predator species and $c > 0$ is the rate of conversion of prey to predator. In this system, the functional response is given by the function $mx/(b+x)$, so it is a Holling type II functional response.

As we can see with detail in [5], the previous system can be reduced to a polynomial differential system. Summarizing it is necessary to do a rescaling $(\overline{x}, \overline{y}, \overline{b}, \overline{c}, \overline{\delta}) = (x/K, (m/rK)y, b/K, cm/r, \delta/r)$ and a time rescaling multiplying by $b + x$. Thus a polynomial system of degree three is obtained, being b, c and δ positive parameters:

$$\begin{aligned} \dot{x} &= x(-x^2 + (1-b)x - y + b), \\ \dot{y} &= y((c-\delta)x - \delta b). \end{aligned} \tag{5.1}$$

This is a particular case of Kolmogorov systems, which were proposed in [6] as an extension of the Lotka-Volterra systems to arbitrary dimension and degree. Kolmogorov systems are differential systems of the form

$$\dot{x_i} = x_i P_i(x_1, \ldots, x_n), \quad i = 1, \ldots, n,$$

where P_i are polynomials. The techniques we are describing in this chapter have been used to classify all the global dynamics of some families of general Kolmogorov systems in [7–10].

We study system (5.1) in the positive quadrant of the plane $\mathbb{R}^2$ where it has ecological meaning, more precisely, we wanted to complete the study of the dynamics of system (5.1) and classify all their phase portraits on the closed positive quadrant of the Poincaré disc, that we will introduce later on, so in this way we can control the dynamics of the system near the infinity. More details can be found in [11].

The second system we will consider in this chapter is the system

$$\begin{aligned} \dot{x} &= x\left(a_0 + c_1 x + c_2 z^2 + c_3 z\right), \\ \dot{z} &= z\left(c_0 + c_1 x + c_2 z^2 + c_3 z\right), \end{aligned} \tag{5.2}$$

This system, which is also a Kolmogorov system, is important as it has been obtained from a general Lotka-Volterra system on dimension three. Those kinds of Lotka-Volterra systems have been used for modeling different problems as in hydrodynamics [12], chemical reactions [13], economic, and social problems [14–16]. They have been also used in the field in which we are focusing our attention: population dynamics. For example, the the interaction between species has been modeled with these kinds of systems in [17–21].

Only very particular cases of Lotka-Volterra systems in dimension three had been studied, as, for example, the cases in [22], where the authors give the global phase portraits in the Poincaré ball of a system related with the study of black holes or in [23], where the authors study a family depending only on two parameters.

In view of the lack of general results, and although the complete study of all these systems does not seem directly approachable, it seemed interesting to address the study of larger classes of systems within the Lotka-Volterra in dimension three. From general 3-dimensional Lotka-Volterra systems depending on 12 parameters, we have obtained two big subfamilies in [7] by applying the Darboux theory of integrability.

System (5.2) corresponds with one of those subfamilies by setting the value of one of the parameters equal to -1. We present this particular case as it has a line of singular points at infinity, i.e., all the infinity is filled up with singular points. We consider this system interesting as there are few works that study these kinds of systems when they have a line filled up of singular points.

Next, in Sect. 5.3, we describe the desingularization technique using directional blow-ups and then apply it to the systems that we have just presented. With the same structure, we describe the Poincaré compactification in Sect. 5.2 and apply it to both systems, explaining the results obtained.

5.2 The Dynamics Near the Infinity

Many systems used in population dynamics are planar systems, for example, all those which deal with the evolution of a predator species and a prey species that coexist. When trying to study the behavior of those systems in the whole plane $\mathbb{R}^2$, an obvious difficulty arises if we try to define what happens globally, because we cannot fully control the orbits as they go to or come from infinity.

What we present here is a special kind of compactification, which can be used provided that the functions defining the vector field are polynomials. This technique, introduced by Poincaré [24], allows one to control the orbits which tend to or come from infinity and to draw the phase portrait in a finite region, the sphere $\mathbb{S}^2$, or, even simpler, in the planar disk $\mathbb{D}^2$.

In the following, we describe the construction of this compactification for a general polynomial system, and the we apply it to the predator-prey systems given in Sect. 5.1.

5.2.1 The Poincaré Compactification

Let us consider a polynomial vector field of degree d of the form:

$$X = P\frac{\partial}{\partial x_1} + Q\frac{\partial}{\partial x_2},$$

where P and Q are polynomials such that d is the maximum of the degrees of P and Q. Let us consider that the variables of the polynomial are x_1 and x_2, so the vector field defines a polynomial differential system of the form

$$\begin{aligned} \dot{x}_1 &= P(x_1, x_2), \\ \dot{x}_2 &= Q(x_1, x_2). \end{aligned} \tag{5.3}$$

For this polynomial system, we give the construction of the Poincaré compactification. The main idea is that we want to project our vector field from an infinite surface, which is $\mathbb{R}^2$ onto the bounded surface of the sphere $\mathbb{S}^2 = \{y \in \mathbb{R}^3 : y_1^2 + y_2^2 + y_3^2 = 1\}$. Thus, we can place $\mathbb{R}^2$ as a tangent plane to the sphere at the point $(0, 0, 1)$, i.e., we consider $\mathbb{R}^2$ as the plane in $\mathbb{R}^3$ defined by $(y_1, y_2, y_3) = (x_1, x_2, 1)$. We call the *Poincaré sphere* to the sphere $\mathbb{S}^2$, which is usually divided into three regions: the northern hemisfere,

$$H_+ = \left\{y \in \mathbb{S}^2 : y_3 > 0\right\};$$

the southern hemisphere,

$$H_- = \left\{y \in \mathbb{S}^2 : y_3 < 0\right\};$$

and the equator

$$\mathbb{S}^1 = \left\{y \in \mathbb{S}^2 : y_3 = 0\right\}.$$

Now we need a way to project what is on $\mathbb{R}^2$ onto the Poincaré sphere. For doing that, we can consider the central projections $f^+ : \mathbb{R}^2 \to \mathbb{S}^2$ and $f^- : \mathbb{R}^2 \to \mathbb{S}^2$.

The image $f^+(x)$ of a point x is defined as the intersection of the straight line passing through the point x and the origin with the northern hemisphere of $\mathbb{S}^2$, and, respectively, the image $f^-(x)$ of a point x is the intersection of the straight line passing through the point x and the origin with the southern hemisphere. The analytical expressions of these projections are

$$f^+(x) = \left(\frac{x_1}{\Delta(x)}, \frac{x_2}{\Delta(x)}, \frac{1}{\Delta(x)}\right), \quad f^-(x) = \left(\frac{-x_1}{\Delta(x)}, \frac{-x_2}{\Delta(x)}, \frac{-1}{\Delta(x)}\right),$$

where $\Delta(x) = \sqrt{x_1^2 + x_2^2 + 1}$.

With the differential of these projections, we obtain induced vector fields in the northern and southern hemispheres. The induced vector field on H_+ is

$$\overline{X}(y) = Df^+(x)X(x), \text{ where } y = f^+(x),$$

and the one in H_- is

$$\overline{X}(y) = Df^-(x)X(x), \text{ where } y = f^-(x).$$

Now we have a vector field on $\mathbb{S}^2 \backslash \mathbb{S}^1$, that we name by $\overline{X}$, that is everywhere tangent to $\mathbb{S}^2$.

But our motivation for doing this compactification was to study the dynamics of the orbits in a neighborhood of the infinity. Note that the points at the infinity of $\mathbb{R}^2$ are in bijective correspondence with the points of the equator of $\mathbb{S}^2$, and for the moment, our induced vector field $\overline{X}$ is not defined over the equator.

So the next step is to carry out this extension of the induced vector field $\overline{X}$ from $\mathbb{S}^2 \backslash \mathbb{S}^1$ to $\mathbb{S}^2$. In general, the field is not bounded as we get close to $\mathbb{S}^1$, so, to make possible the extension, we should multiply the vector field by the factor $\rho(x) = x_3^{d-1}$, and then the extension is feasible.

In short, we call the *Poincaré compactification* of the vector field X on $\mathbb{R}^2$ to this extended vector field, and we denote it by $\rho(X)$.

Once the idea of this method has been exposed, we need to specify what will be the expression of this compactification, in order to be able to work with it, as we will do later on with the predator-prey systems. As we are working on a surface $\mathbb{S}^2$, to make calculations, we need to use local charts in the surface.

We consider the six local charts of $\mathbb{S}^2$ given by

$$U_i = \left\{y \in \mathbb{S}^2 \mid y_k > 0\right\}, \quad V_i = \left\{y \in \mathbb{S}^2 \mid y_k < 0\right\},$$

and the local maps

$$\phi_k : U_k \longrightarrow \mathbb{R}^2 \quad \text{and} \quad \psi_k : V_k \longrightarrow \mathbb{R}^2,$$

for $i = 1, 2, 3$. The previous maps are defined by

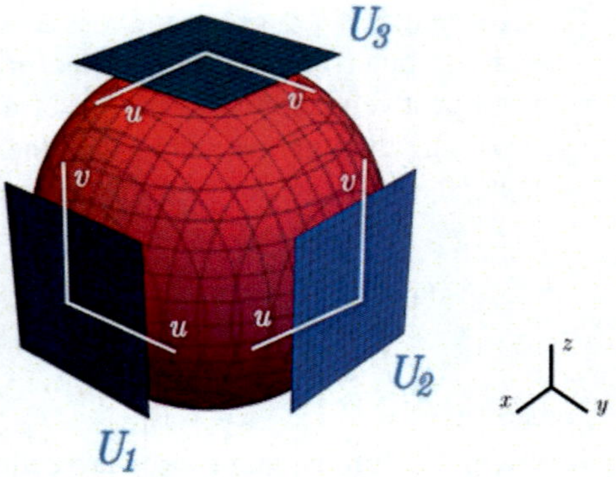

Fig. 5.1 The Poincaré sphere with the local charts (U_j, ϕ_j) for $j = 1, 2, 3$

$$\phi_i(y) = \psi_i(y) = \left(\frac{y_m}{y_i}, \frac{y_n}{y_i}\right),$$

with $m < n$ and $m, n \neq i$. We denote by $z = (u, v)$ the value of the maps $\phi_i(y)$ or $\psi_i(y)$, for any value of i. Then (u, v) has different roles depending on the selected local chart. Geometrically, the coordinates (u, v) can be expressed as in Fig. 5.1. In all the charts, the points which are over the equator have the coordinate $v = 0$.

Let us consider the chart U_1, and let us calculate the expression of $\rho(X)$ in this chart (see [25]). Our initial polynomial vector field was

$$X(x) = (P(x_1, x_2), Q(x_1, x_2)),$$

and with the central projection f^+, we get the vector field on the northern hemisphere

$$\overline{X}(y) = Df^+(x)X(x) \text{ with } y = f^+(x),$$

with the local map ϕ_1:

$$D\phi_1(y)\overline{X}(y) = D\phi_1(y) \circ Df^+(x)X(x) = D(\phi_1 \circ f^+)(x)X(x).$$

Then

$$(\phi_1 \circ f^+)(x) = \phi_1\left(\frac{x_1}{\Delta(x)}, \frac{x_2}{\Delta(x)}, \frac{1}{\Delta(x)}\right) = \left(\frac{x_2}{x_1}, \frac{1}{x_1}\right) = (u, v),$$

and

$$D(\phi_1 \circ f^+)(x) = \begin{pmatrix} -\dfrac{x_2}{x_1^2} & \dfrac{1}{x_1} \\ -\dfrac{1}{x_1^2} & 0 \end{pmatrix}.$$

If we denote by $\overline{X}\mid_{U_1}$, the system defined as $D\phi_1(y)\overline{X}(y)$, we have

$$\overline{X}\mid_{U_1} = \begin{pmatrix} -\dfrac{x_2}{x_1^2} & \dfrac{1}{x_1} \\ -\dfrac{1}{x_1^2} & 0 \end{pmatrix} \begin{pmatrix} P(x_1, x_2) \\ Q(x_1, x_2) \end{pmatrix},$$

and so the components of the field can be expressed as

$$\begin{aligned} \overline{X}\mid_{U_1} &= \left(-\frac{x_2}{x_1^2} P(x_1, x_2) + \frac{1}{x_1} Q(x_1, x_2), -\frac{1}{x_1^2} P(x_1, x_2) \right) \\ &= \frac{1}{x_1^2} \left(-x_2 P(x_1, x_2) + x_1 Q(x_1, x_2), -P(x_1, x_2) \right). \end{aligned}$$

Also

$$\rho(y) = y_3^{d-1} = \left(\frac{1}{\Delta(x)} \right)^{d-1} = \left(\frac{1}{x_1} \right)^{d-1} m(z),$$

where

$$m(z) = \left(1 + \left(\frac{x_2}{x_1} \right)^2 + \left(\frac{1}{x_1} \right)^2 \right)^{\frac{1-d}{2}}.$$

We can multiply the field $\overline{X}\mid_{U_1}$ by $\rho(y)$, which is equivalent to change the time variable t for a new variable s, so that $dt = \rho(y)ds$, and the only change on the phase portrait is the velocity at which orbits are traveled.

Now we have a compactification of the field on the local charts that has a well-defined polynomial expression

$$\rho(y)(\overline{X}\mid_{U_1}) = \frac{m(z)}{x_1^2} \left(-x_2 P(x_1, x_2) + x_1 Q(x_1, x_2), -P(x_1, x_2) \right),$$

and, although $\overline{X}\mid_{U_1}$ is not defined when at the points of the equator, $p(X)\mid_{U_1} = \rho\overline{X}\mid_{U_1}$ is well defined at the infinity, so the extension of $\rho\overline{X}$ to $p(X)$ is defined on the whole of $\mathbb{S}^1$. Moreover, in order to simplify the extended vector field, we also make a change in the time variable and remove the factor $m(z)$.

If the variables in the local charts are (u, v), the Poincaré compactification of the vector field X is given by

$$\begin{aligned} \dot{u} &= v^d \left[-u\, P\left(\frac{1}{v}, \frac{u}{v}\right) + Q\left(\frac{1}{v}, \frac{u}{v}\right) \right], \\ \dot{v} &= -v^{d+1}\, P\left(\frac{1}{v}, \frac{u}{v}\right), \end{aligned} \tag{5.4}$$

in local chart (U_1, ϕ_1). In (U_2, ϕ_2) we have

$$\begin{aligned} \dot{u} &= v^d \left[P\left(\frac{u}{v}, \frac{1}{v}\right) - uQ\left(\frac{u}{v}, \frac{1}{v}\right) \right], \\ \dot{v} &= -v^{d+1}\, Q\left(\frac{u}{v}, \frac{1}{v}\right), \end{aligned} \tag{5.5}$$

and finally, in (U_3, ϕ_3):

$$\begin{aligned} \dot{u} &= P(u, v), \\ \dot{v} &= Q(u, v). \end{aligned} \tag{5.6}$$

Even the complete geometrical construction is important to understand the idea of this method, once we want to apply it to some particular equations, we can obtain the expression in the following way.

To obtain (5.4), we start with (5.3) and introduce coordinates (u, v) by the formulas $(x_1, x_2) = (1/v, u/v)$. This leads to a vector field $\overline{X}^u$ which we must multiply by v^{d-1}.

To obtain (5.5), we start with (5.3) and introduce coordinates (u, v) by the formulas $(x_1, x_2) = (u/v, 1/v)$. We again multiply the obtained vector field $\overline{X}^v$ by v^{d-1}.

The expressions in (5.6) do not need any elaboration, and we just have to replace (x_1, x_2) by (u, v).

In the remaining charts (V_k, ψ_k), with $k = 1, 2, 3$, the expression for $\rho(X)$ is the same as for (U_k, ϕ_k) but multiplied by $(-1)^{d-1}$. Therefore, it is not necessary to study the system in these charts, as it is enough to determine the behavior of the orbits based on the behavior on the charts (U_i, ϕ_i), with $i = 1, 2, 3$.

We recall that the importance of this compactification relies on the fact that the points at the infinity of $\mathbb{R}^2$ are now the finite point on the equator of the sphere. All the singular points of $\rho(X)$ which lie in the equator are called the *infinite singular points* of X, and the following result holds:

Proposition 1 *If $y \in \mathbb{S}^1$ is an infinite singularity of the Poincaré compactification $\rho(X)$ of a field X, then the opposite point $-y$ is an infinite singularity of the compactification $\rho(X)$ and both singularities have the same stability if the degree of vector field is odd; they have opposite stability if the degree is even.*

As mentioned in the introduction, we can make a new projection that simplifies the representation of the phase portraits. The idea is now to project the northern

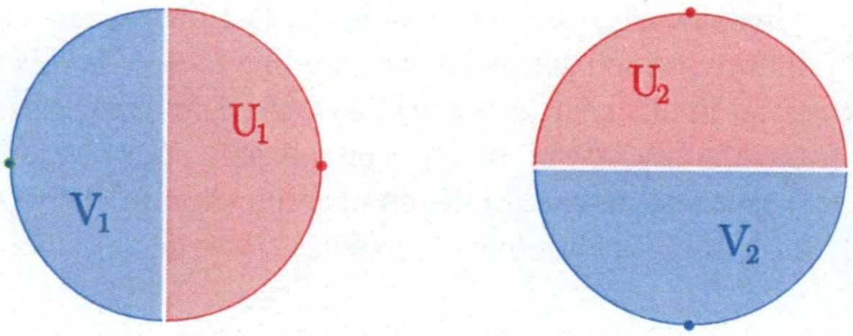

Fig. 5.2 The projection of the northern hemisphere on the Poincaré disk, with charts U_1, V_1, U_2 and V_2

hemisphere of $\mathbb{S}^2$ onto the plane $y_3 = 0$ with the orthogonal projection π. The image of the hemisphere is called the *Poincaré disk*, and we denote it by $\mathbb{D}^2$.

It will be enough to study the system in the Poincaré disk since the orbits of $\rho(X)$ on $\mathbb{S}^2$ are symmetric with respect to the origin of $\mathbb{R}^3$, so we only need to consider the flow of $\rho(X)$ in the closed northern hemisphere. In Fig. 5.2, we include the Poincaré disk with the charts U_1, U_2, v_1, and V_2, and in the examples that we will provide, we will also include the phase portraits in the Poincaré disk.

5.2.2 *Application of the Poincaré Compactification to Predator-Prey Systems*

In this subsection, we obtain the Poincaré compactifications of systems (5.1) and (5.2).

A Kolmogorov System Obtained from the Rosenzweig-MacArthur System

First, we consider (5.1), and we are going to obtain the expression of the compactification in the charts U_1 and U_2. In this case, as the system is proposed in the field on population dynamics, and the variables only have biological meaning if they are nonnegative, we are going to restrict the study of the orbits to the positive quadrant of the Poincaré disk. In any case, it is necessary to study the compactification in the charts U_1 and U_2, as the origin of U_2 is not included in the chart U_1. Following the structure in system (5.3), for system (5.1), we have

$$P(x_1, x_2) = x_1(-x_1^2+(1-b)x_1-x_2+b) \quad \text{and} \quad Q(x_1, x_2) = x_2((c-\delta)x_1-\delta b).$$

The degree of the equations is $d = 3$. According with the expression in (5.4), the compactification in chart U_1 is

$$\begin{aligned} \dot{u} &= uv^2 - b(\delta+1)uv^2 + (b+c-\delta-1)uv + u, \\ \dot{v} &= uv^2 - bv^3 + (b-1)v^2 + v. \end{aligned} \tag{5.7}$$

This new expression of system (5.1) allows one to study the singular points at the infinity, which are those over the line $v = 0$. In this case, the only infinite singular point in this chart is the origin of U_1. The linear part of system (5.7) at the origin is the identity matrix, so O_1 is an unstable node as there are two positive eigenvalues.

We deal now with the compactification in chart U_2. According with the expression in (5.5), system (5.1) in chart U_2 writes

$$\begin{aligned} \dot{u} &= -u^3 + (\delta + 1 - b - c)u^2 v + b(\delta + 1)uv^2 - uv, \\ \dot{v} &= (\delta - c)uv^2 + b\delta v^3. \end{aligned} \tag{5.8}$$

We recall that we are interested in the phase portrait in the positive quadrant of the Poicaré disk. The only infinite point in the disk which is not covered by the chart U_1 is the origin of chart U_2. Then we only need to determine if the origin of system (5.8) is a singular point. In this case, the origin is indeed a singular point, but in contrast to the origin of U_1, the linear part of system (5.8) at O_2 is identically zero, so we need a desingularization technique to study it. In the following section, we will introduce the blow-up technique, and then we will deal again with this system.

Now we consider system (5.2). Again we will obtain the compactification in the local charts U_1 and U_2, taking into account that for studying all the infinite singular points, it is enough to study the singular points over $v = 0$ in the chart U_1 and the origin of the chart U_2. In this case, we consider the complete Poincaré disk because, although in the case of the population dynamics, the variables only have biological meaning when they are positive, and these equations can be used in many other contexts, as explained in the Introduction, and in some of them, it may make sense to consider negative values of the variables.

A Kolmogorov System Obtained from the Spatial Lotka-Volterra Systems

System (5.2) is a polynomial system of degree 3 of the form (5.3) with

$$P(x_1, x_2) = x_1(a_0 + c_1 x_1 + c_2 x_2^2 + c_3 x_2) \text{ and } Q(x_1, x_2) = x_2(c_0 + c_1 x + c_2 x_2^2 + c_3 z).$$

We start in this case with the study of the chart U_2, which is simpler. According to Eqs. (5.5), the system (5.2) in this chart has the expression:

$$\begin{aligned} \dot{u} &= (a_0 - c_0)uv^2, \\ \dot{v} &= -c_1 uv^2 - c_0 v^3 - c_3 v^2 - c_2 v. \end{aligned} \tag{5.9}$$

As we are interested in the infinite singular points, we study the singular points appearing over the line $v = 0$. In this case, over this line, we get that

$$\dot{u} \mid_{v=0} = \dot{v} \mid_{v=0} = 0,$$

and then all points at infinity are singular points, including the origin which is the only we must study in this chart.

We see that the eigenvalues of the Jacobian matrix at the origin are zero and $-c_2$; therefore, it can be concluded that if $c_2 > 0$, there is exactly one orbit that goes from outside the infinity to the origin of the chart U_2 and if $c_2 < 0$ there is exactly one orbit that leaves the origin of U_2 and goes outside the infinity. This result is obtained with the results about normally hyperbolic surfaces. As it is not our objective here, we will not give more details, but interested readers can see [9].

Now we address the study of the infinite singular points in the local chart U_1, where according to Eqs. (5.4), the expression of the systems is

$$\begin{aligned} \dot{u} &= (c_0 - a_0)uv^2, \\ \dot{v} &= -c_2u^2v - c_3uv^2 - a_0v^3 - c_1v^2. \end{aligned} \tag{5.10}$$

Taking $v = 0$, we get again that all points at infinity in this chart are singular points. At any point $(u_0, 0)$ with $u_0 \neq 0$, the eigenvalues are one zero and the other $-c_2u_0^2$ so, with the same results mentioned for the origin of U_2, we can conclude that if $c_2 > 0$ exactly one orbit outside the infinity arrives at each infinite singular point on the chart U_1 distinct from the origin; and if $c_2 < 0$ from each inifinite singular point on the chart U_1 distinct from the origin leaves exactly one orbit from outside the infinity.

However at the origin of chart U_1 the eigenvalues of the Jacobian matrix are both zero, so it is a linearly zero singular point. Again we will return to this problem after Sect. 5.3, in which we will introduce the blow-ups.

5.3 A Desingularization Technique

The study of the dynamics around the singular points of a differential system is important as it can determine behaviors of practical interest, such as stabilization of populations at a certain level, or the extinction of some of the populations. However, given the difficulty involved in the study of some of these points, we find in the literature that the study of the singular points is often omitted when they are not elementary.

In addition to the well-known case of hyperbolic singularities, for the case of planar polynomial systems, there are very good results that give a complete classification in the semi-hyperbolic and nilpotent cases, i.e., when the eigenvalues of the Jacobian matrix at the singular points have one eigenvalue equal to zero or both equal to zero but the Jacobian matrix is not identically zero. These results can be found in Chapters 2 and 3 of [25].

The result for the nilpotent case in [25] is obtained by using the so-called *homogeneous blow up's*. This technique is based essentially in the use of polar coordinates to transform the singular points which are nilpotent. As the result is completed in that case, and it can be directly applied to classify the nilpotent singular point determining the local behavior of the orbits, we will not give details about the proof, which can be found in [25].

However for the linearly zero singular points, there are no results that allow us to classify them. Then we will give a short description on a technique that we have used in the context of population dynamics for the desingularization of these points: the *directional blow up's.*

5.3.1 Theoretical Introduction of the Directional Blow-Ups

The basic idea behind this method is to convert the singular point that we want to study into a line. For doing that, we do a change in the variables that "explodes" the singularity. The new singular points that appear on the line in which we have converted the singular point can be easier to study. If this is not the case, the process is repeated until we get to a system which does not have linearly zero singular points. Then we can study these singularities which are simpler and go back to our original system undoing the variable changes. This method is always valid since Dumortier demonstrated the finiteness of this iterative desingularization procedure (see [26]).

We introduce the directional blow-ups for polynomial equations. To this end, we consider a differential system of the form

$$\begin{aligned} \dot{x} &= P(x, y) = P_m(x, y) + P_h(x, y), \\ \dot{y} &= Q(x, y) = Q_m(x, y) + G_h(x, y), \end{aligned} \tag{5.11}$$

where P and Q are coprime polynomials, P_m and Q_m are homogeneous polynomials of degree $m \in \mathbb{N}$ and P_h and Q_h are higher order terms in x and y. Our objective is to study the origin of this system, as it is a singular point since $m > 0$.

For a system of this form, we call the polynomial

$$\mathcal{F}(x, y) := x Q_m(x, y) - y P_m(x, y),$$

the *characteristic polynomial* of system (5.11). With the directional blow-ups, we transform the origin of this system into a line, and we call that line the *exceptional divisor.*

Homogeneous Vertical Blow-Up

The *homogeneous directional blow-up in the vertical direction* is the correspondence

$$(x, y) \to (x, z) = (x, y/x),$$

where z is a new variable. Then the exceptional divisor in this case is the line $x = 0$, as we transform the origin of system (5.11) into this line. After doing the variable change, the expression of the original system (5.11) is

$$\begin{aligned} \dot{x} &= P(x, xz), \\ \dot{z} &= \frac{Q(x, xz) - zP(x, xz)}{x}, \end{aligned} \tag{5.12}$$

that is always well-defined since we are assuming that the origin is a singular point. In the obtained system, all the points at the exceptional divisor are singular points. The next step in this process is to cancel some common factors, more precisely:

- We cancel a common factor x^{m-1} if $\mathcal{F} \not\equiv 0$.
- We cancel a common factor x^m if $\mathcal{F} \equiv 0$.

We have to take into account, while studying the transformation of the orbits, that in this kind of blow-up, the variable change swaps the second and third quadrants with respect to the original system.

Homogeneous Horizontal Blow-Up

The *homogeneous directional blow-up in the horizontal direction* is the correspondence

$$(x, y) \to (z, y) = (x/y, y),$$

where z is a new variable. The exceptional divisor in this case is the line $y = 0$, as we transform the origin of system (5.11) into this line. After doing the variable change, the expression of the original system (5.11) becomes

$$\begin{aligned} \dot{z} &= \frac{P(yz, y) - zQ(yz, y)}{y}, \\ \dot{x} &= P(yz, y), \end{aligned}$$

that is always well-defined since we are assuming that the origin is a singular point. As in the case of the vertical blow-up, in the obtained system, all the points at the exceptional divisor are singular points, so we have to cancel some common factors, more precisely:

- We cancel a common factor x^{m-1} if $\mathcal{F} \not\equiv 0$.
- We cancel a common factor x^m if $\mathcal{F} \equiv 0$.

Now, the quadrants that are swapped are the third and fourth quadrants, so we have to take this into account to studying the configuration around the original singular point.

Depending on the expression of the characteristic polynomial, the origin can be either a *nondicritical* singular point if $\mathcal{F} \not\equiv 0$ or a *dicritical* singular point if $\mathcal{F} \equiv 0$.

In the dicritical case, we have

$$P_m = xW_{m-1} \quad \text{and} \quad Q_m = yW_{m-1},$$

with $W_{m-1} \not\equiv 0$ a homogeneous polynomial of degree $m-1$. If this polynomial W_{m-1} has a factor of the form $y - vx$ where $v = \tan\theta^*$, $\theta^* \in [0, 2\pi)$, then we say that θ^* is a *singular direction*.

The advantage of this method is that it allows one to study and determine the behavior of the solutions around the origin of our initial system (5.11) by studying the singular points of system (5.12) on the exceptional divisor, which can be of a simpler nature. If this does not occur, then some of the singular points on the exceptional divisor are linearly zero, and we can repeat the process until all the points obtained are nonelementary. This method always works because it has been proven in [26] that this chain of blow-ups necessary to get only elementary singular points is finite. We recall that singular points on the exceptional divisor we have to study correspond to either characteristic directions in the nondicritical case or singular directions in the dicritical case.

Once we have studied the dynamics around the exceptional divisor, to obtain the dynamics around the origin of the original system, we must undo the variable changes, taking into account the changes in the orbits and their orientations.

For example, if we have performed a blow-up in the vertical direction, by undoing it, the third and fourth quadrants will swap their position and all the orbits arriving or leaving from a point that is on the exceptional divisor with the second coordinate equal to p will become orbits arriving or leaving from the origin, respectively, with a slope p.

We must also study the flow over the axes in the original system and, combining the information, determine the sectors that appear in the phase portrait. The blow-ups always determine the behavior of the orbits except, at most, around the exceptional divisor. For this reason, if when undoing the blow-up we find some indeterminacy, it will be necessary to carry out a directional blow-up in another direction. For example, if we have performed a blow-up in the vertical direction, and we find an indeterminacy around the vertical axis of the original system, to solve it, it can be enough to perform a blow-up in the horizontal direction.

The theoretical results that provide the relationship between the original singular point of system (5.11) and the new singularities of system (5.12) are the following: (for more details see [27]).

Proposition 2 *Let* $\varphi_t = (x(t), y(t))$ *be a solution of system* (5.11) *which tends to the origin when* t *goes to* $\pm\infty$. *Suppose that the origin of the system is a nondicritical singular point. Assume that* φ_t *is tangent to one of the two angle directions* $\tan\theta = v$, $v \neq \infty$. *Then the following statements hold.*

1. *The two angle directions* $\theta = \arctan v$ *(in* $[0, 2\pi)$*) are characteristic directions.*
2. *The point* $(0, v)$ *on the* (x, z)*-plane is an isolated singular point of system* (5.12).

Table 5.1 The finite singular points in the closed positive quadrant for system (5.1)

Case	Conditions	Finite singular points
1	$b\delta > c - \delta$	P_0 saddle, P_1 stable node
2	$b\delta = c - \delta$	P_0 saddle, P_1 saddle-node
3	$0 < b\delta < c - \delta$, $B \geq 0$, $A > 0$	P_0 saddle, P_1 saddle, P_2 unstable node
4	$0 < b\delta < c - \delta$, $B \geq 0$, $A < 0$	P_0 saddle, P_1 saddle, P_2 stable node
5	$0 < b\delta < c - \delta$, $B < 0$, $A > 0$	P_0 saddle, P_1 saddle, P_2 unstable focus
6	$0 < b\delta < c - \delta$, $B < 0$, $A < 0$	P_0 saddle, P_1 saddle, P_2 stable focus
7	$0 < b\delta < c - \delta$, $B < 0$, $A = 0$	P_0 saddle, P_1 saddle, P_2 weak stable focus

3. *The solution* φ_t *corresponds to a solution of system* (5.12) *tending to the singular point* $(0, v)$.
4. *Conversely, any solution of system* (5.12) *tending to the singular point* $(0, v)$ *on the* (x, z)*-plane corresponds to a solution of system* (5.11) *tending to the origin in one of the two angle directions* $\tan\theta = v$.

Proposition 3 *Consider system* (5.11) *and suppose that the origin is a dicritical singular point. Then for every nonsingular direction* θ*, there exists exactly one semipath tending to the origin in the direction* θ *in forward or backward time. If* θ^* *is a singular direction, there may be either no semipaths tending to the origin in the direction* θ^**, or a finite number, or infinitely many.*

5.3.2 Application of the Directional Blow-Ups to Predator-Prey Systems

A Kolmogorov System Obtained from the Rosenzweig-MacArthur System

For system (5.1), we have studied the finite singular points in the positive quadrant, but this study is quite simple as the points turn out to be elementary. For this system, the origin $P_0 = (0, 0)$ and the point $P_1 = (1, 0)$ are singular points for any values of the parameters, and for some values, a third singular point arises: $P_2 = \big(b\delta/(c-\delta), (-bc(\delta + b\delta - c))/(c-\delta)^2\big)$. In summary, we obtain the classification described in Table 5.1 for the finite singular points according the values of the parameters b, c and δ. More details can be found in [11].

However, for the study of infinite singular points, as we mentioned in Sect. 5.2, it is necessary to use some desingularization technique. More precisely, we need to desingularize the origin of the system in chart U_2, i.e., the origin of

$$
\begin{aligned}
\dot{u} &= -u^3 + (\delta + 1 - b - c)u^2 v + b(\delta + 1)uv^2 - uv,\\
\dot{v} &= (\delta - c)uv^2 + b\delta v^3.
\end{aligned}
\tag{5.13}
$$

For doing that, we use the blow-up technique. More precisely, we do a horizontal blow-up introducing the new variable w_1 by means of the variable change $vw_1 = u$, and get the system

$$\begin{aligned} \dot{w_1} &= v^2 w_1^3 + (1-b)v^2 w_1^2 + b w_1 v^2 - w_1 v, \\ \dot{v} &= (\delta - c) w_1 v^3 + b\delta v^3. \end{aligned} \tag{5.14}$$

Now rescaling the time variable, we cancel the common factor v, getting the system

$$\begin{aligned} \dot{w_1} &= v w_1^3 + (1-b)v w_1^2 + b w_1 v - w_1, \\ \dot{v} &= (\delta - c) w_1 v^2 + b\delta v^2. \end{aligned} \tag{5.15}$$

Now the only singular point on $v = 0$ is the origin, which is semi-hyperbolic, so it is not necessary no do more chained blow-ups. We can apply [25, Theorem 2.19], and thus we conclude that it is a saddle-node. Studying the sense of the flow over the axis, we determine that the phase portrait around the origin of system (5.15) is the one on Fig. 5.3a.

Now we start to undo the change. If we multiply by v, all the points of the w_1-axis become singular points, and the sense of all the orbits on the third and fourth quadrants is reversed. With these modifications, we obtain the phase portrait for system (5.14), given in Fig. 5.3b.

After that, we undo the blow-up going back to the (u, v)-plane. We have to swap the fourth and third quadrants and compress the exceptional divisor into the origin. The phase portrait obtained for system (5.8) is not totally determined in the shaded regions of the third and fourth quadrants; see Fig. 5.3c.

As it was explained in the previous subsection, it can be solved by doing a vertical blow-up, but, in the case of this system, it is not necessary because we only need to know the phase portrait of O_2 in the positive quadrant of the Poincaré disc, which corresponds with the positive quadrant in the plane (u, v), in which the phase portrait has been totally determined.

Finally, we have concluded that the local phase portrait at the origin of chart U_2 (which is an infinite singular point) is the same for all the values of the parameters: particularly, the origin of the chart U_2 has only one hyperbolic sector on the positive quadrant of the Poincaré disc being one separatrix at infinity and the other on $x = 0$.

The application of this technique is essential in order to complete the global classification of all the phase portraits of the system, which is included in Fig. 5.4. More results and details about this system can be found in [11].

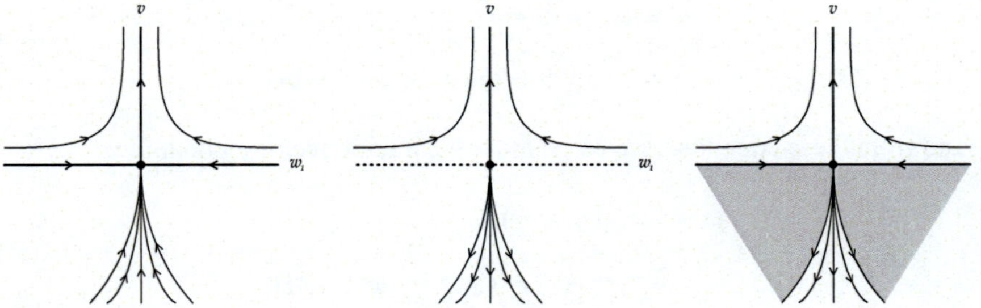

(a) (b) (c)

Fig. 5.3 Desingularization of the origin of system (5.8). **(a)** Origin of system (5.15). **(b)** Origin of system (5.14). **(c)** Origin of system (5.8)

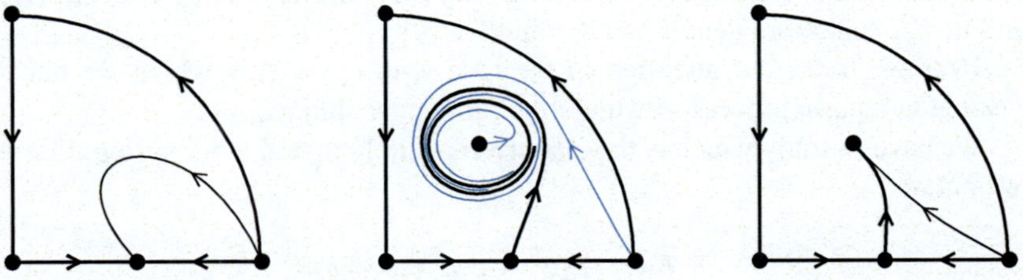

(a) (b) (c)

Fig. 5.4 Phase portraits of system (5.1) in the positive quadrant of the Poincaré disc

A Kolmogorov System Obtained from the Spatial Lotka-Volterra Systems

Let us now deal with system (5.2). For this system, as stated in Sect. 5.2, the singular point at the origin of the chart U_1, which we will name O_1, is linearly zero. For this singular point, we will prove the following result.

Lemma 1 *The origin of the chart U_1 is an infinite singular point of system* (5.2)*, and it has* 12 *distinct local phase portraits described in Fig.* 5.5.

To prove Lemma 1, we use, among other results, the blow-up technique. We illustrate now how is the proof of the result and how we can conclude by using directional blow-ups. It is important to set that we work under the conditions

$$H = \left\{c_2 \neq 0, a_0 \geq 0, c_1 \geq 0, c_3 \geq 0, a_0 \neq c_0, a_0^2 + c_1^2 \neq 0\right\},$$

and the proof that this does not suppose any restriction in order to study all the dynamics can be found in [9].

We recall that the system in chart U_1 has the expression

$$\begin{aligned}\dot{u} &= (c_0 - a_0)uv^2,\\ \dot{v} &= -c_2u^2v - c_3uv^2 - a_0v^3 - c_1v^2,\end{aligned} \tag{5.16}$$

and from these equations, we can remove a common factor v obtaining:

$$\begin{aligned}\dot{u} &= (c_0 - a_0)uv,\\ \dot{v} &= -c_2u^2 - c_3uv - a_0v^2 - c_1v.\end{aligned} \tag{5.17}$$

Then we study the only singular point over the exceptional divisor, i.e., over the line $v = 0$, the origin. We name this singular point $\tilde{O}_1$. Now the eigenvalues of the Jacobian matrix at $\tilde{O}_1$ are zero and $-c_1$, so if $c_1 \neq 0$ the singular point of system (5.17) is semi-hyperbolic and we can study it applying Theorem 2.19 of [25]. The phase portraits corresponding with the semi-hyperbolic case are L_1 to L_4 in Fig. 5.5, and more details can be found in [9].

Here we focus our attention in the case with $c_1 = 0$ in which we must do a desingularization process so we use the blow-up technique.

We have to study which is the characteristic polynomial $\mathcal{F}$ for system (5.17), and we obtain

$$\mathcal{F} = -c_2u^3 - c_3u^2v - c_0uv^2,$$

which cannot be identically zero because $c_2 \neq 0$, so the singular point $\tilde{O}_1$ is nondicritical.

We introduce now a new variable w_1 trying to explode the singular point $\tilde{O}_1$ into a straight line, in particular, the line $u = 0$. We consider the variable change $uw_1 = v$, and making calculations, we obtain the system

$$\begin{aligned}\dot{u} &= (c_0 - a_0)u^2w_1,\\ \dot{w_1} &= -c_0uw_1^2 - c_3uw_1 - c_2u.\end{aligned} \tag{5.18}$$

With this variable change, we have introduced a line filled up with singular points in the line $u = 0$, so, to remove it, we must eliminate a common factor u, obtaining:

$$\begin{aligned}\dot{u} &= (c_0 - a_0)uw_1,\\ \dot{w_1} &= -c_0w_1^2 - c_3w_1 - c_2.\end{aligned} \tag{5.19}$$

Now we focus our attention in the line $u = 0$, which is the exceptional divisor. The singularities over this line are the points with the first coordinate zero and the second one a solution of the equation $-c_0w_1^2 - c_3w_1 - c_2 = 0$. We must study all the singular points, but, as they depend on the parameters, we should distinguish the following cases from (A) to (G).

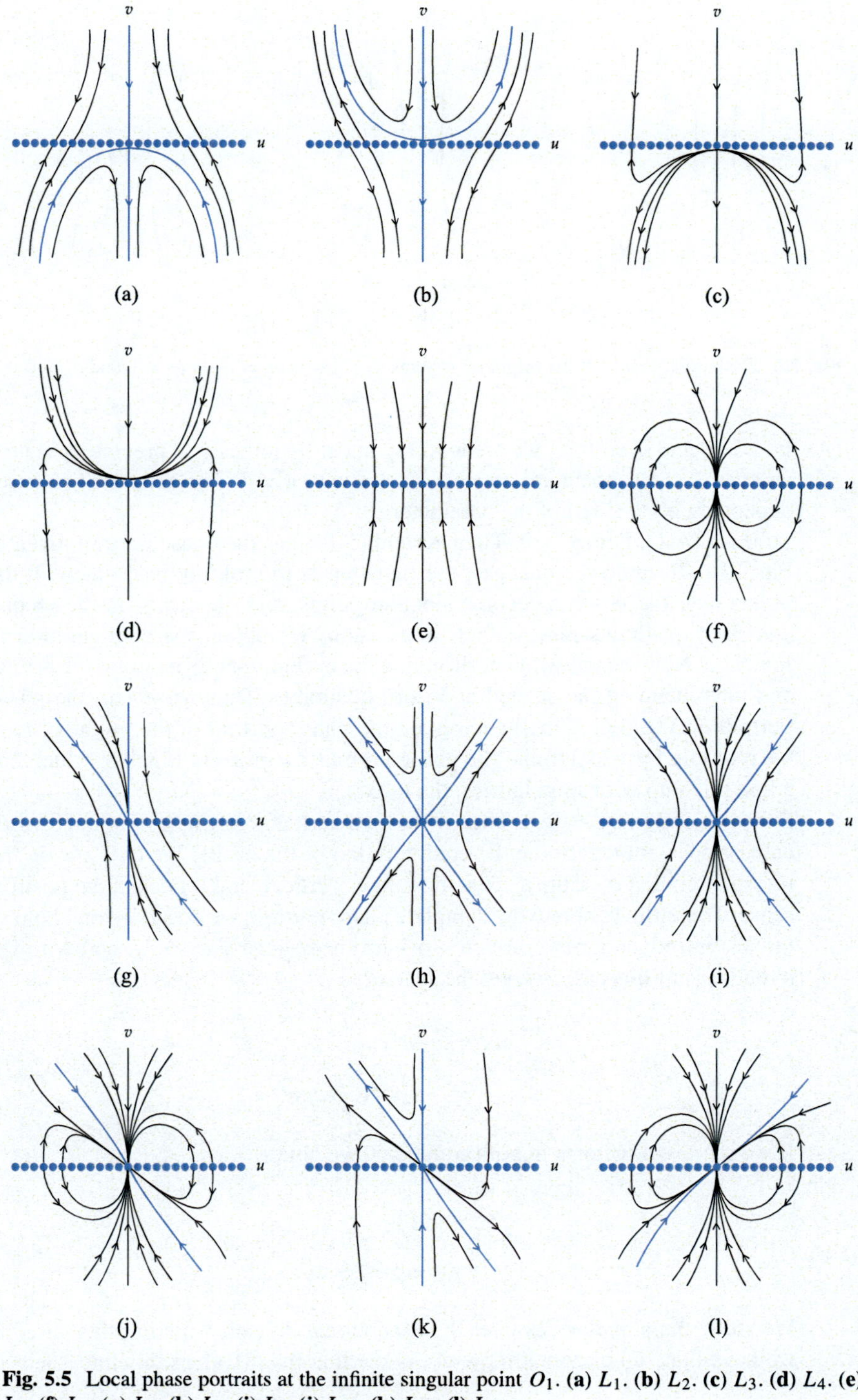

Fig. 5.5 Local phase portraits at the infinite singular point O_1. **(a)** L_1. **(b)** L_2. **(c)** L_3. **(d)** L_4. **(e)** L_5. **(f)** L_6. **(g)** L_7. **(h)** L_8. **(i)** L_9. **(j)** L_{10}. **(k)** L_{11}. **(l)** L_{12}

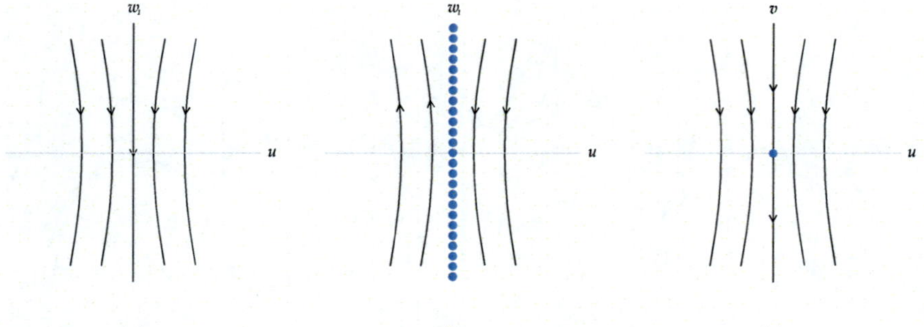

Fig. 5.6 Desingularization of the origin of systems (5.10) with $c_0 = c_1 = c_3 = 0$ and $c_2 > 0$

(A) If $c_0 = 0$ and $c_3 = 0$, as we are working under hypothesis H previously stated, and then $c_2 \neq 0$, there are no singularities with $u = 0$. We consider two cases depending on the sign of the parameter c_2:
Subcase (A.1). Let $c_2 > 0$. Then system (5.19) has the phase portrait given in Fig. 5.6a. To undo the changes, the first step is to multiply by u, then all the points over the w_1-axis become singular points, and the orbits on the second and third quadrants reverse their orientation. We obtain the configuration in Fig. 5.6b. Now we blow down, shrinking the exceptional divisor into the origin, and interchanging the second and third quadrants. Thus we obtain the phase portrait on Fig. 5.6c. Finally, to obtain the phase portrait of the initial system, we multiply by v, obtaining the phase portrait L_5 given in Fig. 5.5, which has a line filled up with singularities, the u-axis.
Subcase (A.2). Let $c_2 < 0$. Then if we do a vertical blow up as in the previous case, it does not determine the configuration of the orbits. We only obtain the information that over the u-axis, the flow is vertical, and it goes in the positive sense. Therefore to obtain the complete phase portrait, we do a horizontal blow-up. We introduce a new variable w_2 with the change $vw_2 = u$, and with the hypothesis of this case, we get the system

$$\begin{aligned} \dot{w_2} &= c_2 w_2^3 v, \\ \dot{v} &= -c_2 w_2^2 v - a_0 v^2. \end{aligned} \tag{5.20}$$

If we remove a factor v in both equations, we obtain

$$\begin{aligned} \dot{w_2} &= c_2 w_2^3, \\ \dot{v} &= -c_2 w_2^2 - a_0 v. \end{aligned} \tag{5.21}$$

We study the singularities over the exceptional divisor, which in this case, as we have done a horizontal blow-up, is the line $v = 0$. Over this line, there is

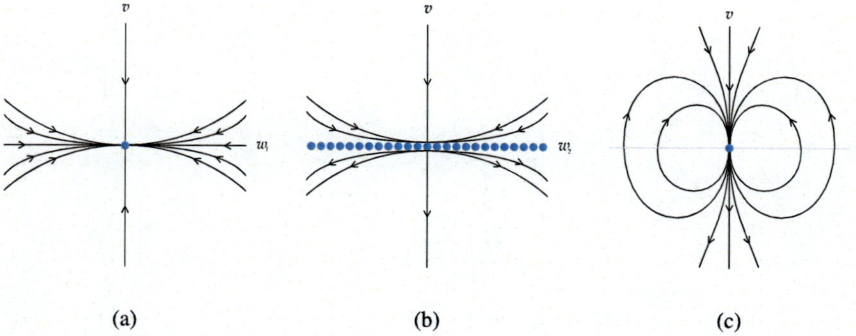

Fig. 5.7 Desingularization of the origin of systems (5.10) with $c_0 = c_1 = c_3 = 0$ and $c_2 < 0$

only one singular point, the origin, and it is a semi-hyperbolic singularity. We can use Theorem 2.19 in [25] to determine the phase portrait, and we obtain that it is a stable topological node. Then the phase portrait near the origin for system (5.21) is the one given in Fig. 5.7a.

If we multiply by the variable v, the phase portrait that we obtain for system (5.20) is the one in Fig. 5.7b.

To blow down, we shrink the exceptional divisor into the origin and swap the third and fourth quadrants, so we get that in the first and second quadrants, the orbits go to the origin tangent to the v-axis, and in the third and fourth quadrants, the orbits leave the origin tangent to the v-axis.

This qualitative behavior together with the information from the vertical blow-up leads to the phase portrait in Fig. 5.7c.

Finally, we obtain the phase portrait for the initial system multiplying by v. The final result is the L_6 of Fig. 5.5.

(B) Let $c_0 = 0$ and $c_3 > 0$. Then over the exceptional divisor we have one singular point: $Q_1 = (0, -c_2/c_3)$. This singularity is hyperbolic and the eigenvalues of the Jacobian matrix are $a_0 c_2/c_3$ and $-c_3$ so we have two subcases depending on their signs:

Subcase (B.1). If $c_2 > 0$ we do a horizontal blow up introducing the variable $vw_2 = u$ in systems (5.17), as with the vertical blow up we obtain some indeterminacies around the exceptional divisor. With the new variable w_2 we get

$$\begin{aligned} \dot{w_2} &= c_2 w_2^3 v + c_3 w_2^2 v, \\ \dot{v} &= -c_2 w_2^2 v + c_3 w_2 v^2 - a_0 v^2. \end{aligned} \tag{5.22}$$

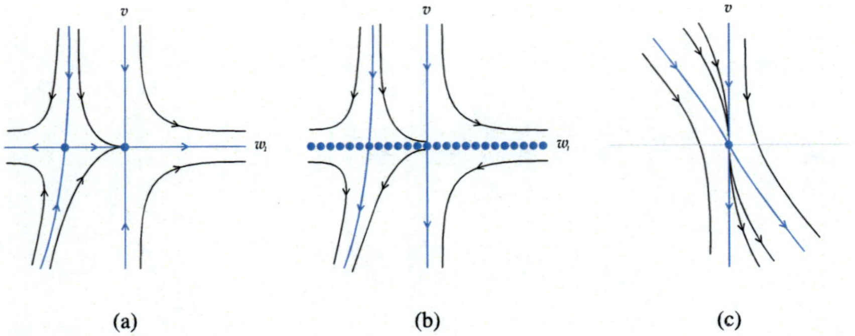

Fig. 5.8 Desingularization of the origin of systems (5.10) with $c_0 = c_1 = 0$, $c_3 > 0$ and $c_2 < 0$

Removing a common factor v, the system becomes:

$$\begin{aligned} \dot{w_2} &= c_2 w_2^3 - c_3 w_2^2, \\ \dot{v} &= -c_2 w_2^2 v - c_3 w_2 v - a_0 v. \end{aligned} \tag{5.23}$$

For this system, we study the existence of singularities on the line $v = 0$, and we found two different points: the origin and the point $(-c_3/c_2, 0)$. The first one is a semi-hyperbolic saddle-node and the second one is a saddle.

Combining the phase portraits of both singularities, the phase portrait for system (5.23) is the one in Fig. 5.8a.

If we multiply by v, the phase portrait for system (5.22) is the one in Fig. 5.8b. If we blow down, we get the phase portrait in Fig. 5.8c, and finally, if we multiply again by v, we obtain for the initial system the phase portrait L_7 of Fig. 5.5.

Subcase (B.2). In the case $c_2 < 0$, the singular point Q_1 is a stable node, and we also need a horizontal blow-up to determine the local phase portrait. The result obtained is the phase portrait L_6 of Fig. 5.5.

(C) Let $c_0 \neq 0$, $c_3 = 0$ and $c_0 c_2 > 0$. In this case, there are no singularities over the exceptional divisor. We distinguish three subcases depending on the behavior of the orbits around that exceptional divisor. If c_0 and c_2 are positive, we obtain the same phase portrait L_5 of Fig. 5.5, and if c_0 and c_2 are negative, we obtain again the phase portrait L_6 of Fig. 5.5, but in this case, to get to the final result, it is also necessary to do a horizontal blow-up.

(D) Let $c_0 \neq 0$, $c_3 = 0$, and $c_0 c_2 < 0$. Then over the exceptional divisor, there are two hyperbolic singularities:

$$Q_2 = \left(0, \sqrt{\frac{-c_2}{c_0}}\right) \quad \text{and} \quad Q_3 = \left(0, -\sqrt{\frac{-c_2}{c_0}}\right).$$

Taking into account the eigenvalues of the Jacobian matrix at both points, we consider three subcases.

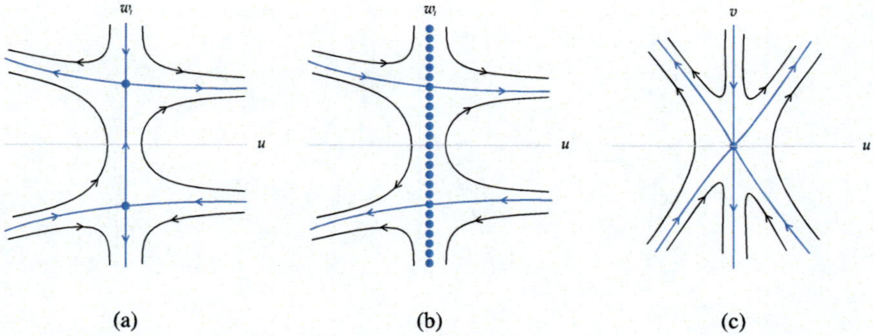

Fig. 5.9 Desingularization of the origin of systems (5.10) with $c_3 = 0$, $c_0 > 0$, $c_2 < 0$ and $a_0 - c_0 < 0$

Subcase (D.1). If $c_0 > 0$ and $a_0 - c_0 > 0$, then the singular point Q_2 is a stable node, and the singular point Q_3 is an unstable node. Undoing the blow-up, we obtain phase portrait L_6 of Fig. 5.5.
Subcase (D.2). If $c_0 > 0$ and $a_0 - c_0 < 0$, then the singular points Q_2 and Q_3 are both saddle points with the orientation of the hyperbolic orbits given in Fig. 5.9a. If we multiply by u and then blow down, we obtain the phase portraits in Fig. 5.9b and c. Multiplying by u again, we obtain that the local phase portrait for O_1 is L_8 of Fig. 5.5.
Subcase (D.3). If $c_0 < 0$ and $a_0 - c_0 > 0$, then the singular points Q_2 and Q_3 are both saddle points but with a different orientation than in case (D.2). Here the vertical blow-up is not enough to determine the behavior of the orbits near the v-axis. We introduce the variable w_2 such that $vw_2 = u$, to explode the origin into a horizontal line. We obtain

$$\begin{aligned} \dot{w}_2 &= c_2 w_2^3 v + c_0 w_2 v, \\ \dot{v} &= -c_2 w_2^2 v - a_0 v^2, \end{aligned} \tag{5.24}$$

and then, removing a common factor v, we obtain

$$\begin{aligned} \dot{w}_2 &= c_2 w_2^3 + c_0 w_2, \\ \dot{v} &= -c_2 w_2^2 - a_0 v. \end{aligned} \tag{5.25}$$

For this system, we study the singularities on the line $v = 0$, which are three: The origin which is a stable node, and two saddle points $(\pm\sqrt{-c_0/c_2}, 0)$. Combining these three singularities, the phase portrait around the w_2-axis for system (5.25) is the one in Fig. 5.10a. If we multiply by v, we get the phase portrait in Fig. 5.10b for systems (5.24). Blowing down, we obtain the phase portrait in Fig. 5.10c. If we multiply again by v, we obtain the local phase portrait for O_1 which is L_9 of Fig. 5.5.

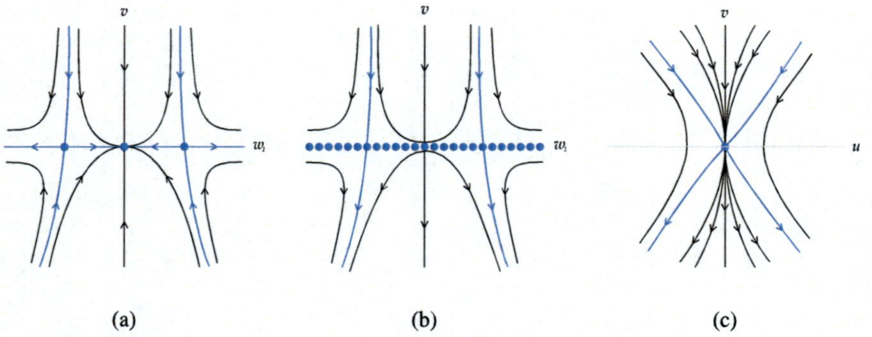

Fig. 5.10 Desingularization of the origin of systems (5.10) with $c_3 = 0$, $c_0 < 0$, $c_2 > 0$ and $a_0 - c_0 > 0$

(E) Let $c_0 \neq 0$, $c_3 \neq 0$ and $c_3^2 - 4c_0c_2 < 0$. In this case there are no singularities over $u = 0$. If $c_2 > 0$ the phase portrait is L_5 of Fig. 5.5, and if $c_2 > 0$, we obtain, by combining with a horizontal blow-up, the phase portrait L_6 of Fig. 5.5.

(F) Let $c_0 \neq 0$, $c_3 \neq 0$ and $c_3^2 - 4c_0c_2 > 0$. Then there are two hyperbolic singularities over $u = 0$:

$$Q_4 = \left(0, -\frac{c_3 + \sqrt{c_3^2 - 4c_0c_2}}{(2c_0)}\right) \quad \text{and} \quad Q_5 = \left(0, -\frac{c_3 - \sqrt{c_3^2 - 4c_0c_2}}{(2c_0)}\right).$$

Subcase (F.1). If $a_0 - c_0 > 0$, $c_0 > 0$ and $c_3^2 - 4c_0c_2 - c_3 > 0$, then Q_4 is an unstable node and Q_5 a stable node. Undoing the blow up we get the phase portrait L_6 of Fig. 5.5.

Subcase (F.2). If $a_0 - c_0 > 0$, $c_0 > 0$ and $c_3^2 - 4c_0c_2 - c_3 < 0$, then the singular point Q_4 is an unstable node and the singular point Q_5 a saddle. Undoing the blow-up, we get the phase portrait L_7 of Fig. 5.5.

Subcase (F.3). If $a_0 - c_0 > 0$, $c_0 < 0$, and $c_3^2 - 4c_0c_2 - c_3 > 0$, then the singular points Q_4 and Q_5 are saddle points, but the vertical blow-up does not determine the behavior of the orbits around the v-axis. Doing a horizontal blow-up, we obtain the phase portrait L_9 of Fig. 5.5.

Subcase (F.4). If $a_0 - c_0 > 0$, $c_0 < 0$ and $R_c - c_3 < 0$, then Q_4 is a saddle, and Q_5 is a stable node. Again the vertical blow-up is not enough to determine the phase portrait. With a horizontal blow-up, we obtain the phase portrait L_{10} of Fig. 5.5.

Subcase (F.5). If $a_0 - c_0 < 0$, $c_0 > 0$, and $R_c - c_3 > 0$, then Q_4 and Q_5 are saddle points. Undoing the blow-up, we get the phase portrait L_8 of Fig. 5.5.

Subcase (F.6). If $a_0 - c_0 < 0$, $c_0 > 0$, and $R_c - c_3 < 0$, then Q_4 is a saddle and Q_5 is a stable node. Undoing the blow-up, we get the phase portrait L_{11} of Fig. 5.5.

(G) Let $c_0 \neq 0$, $c_3 \neq 0$ and $c_3^2 - 4c_0c_2 = 0$. In this case, the singular point on the exceptional divisor is $Q_6 = (0, -c_3/(2c_0))$, and it is a semi-hyperbolic saddle-node. We have three different cases in which the position of the sectors of the saddle-node changes. These cases are determined by the signs of $c_0(a_0 - c_0) < 0$ and $c_0 < 0$. If $c_0(a_0 - c_0) > 0$ and $c_0 > 0$, we obtain the phase portrait L_7, if $c_0(a_0 - c_0) < 0$ and $c_0 > 0$, we obtain the phase portrait L_{11}, and if $c_0(a_0 - c_0) < 0$ and $c_0 < 0$, we get the phase portrait L_{12} of Fig. 5.5.

5.4 Conclusions

In this chapter, we have presented, for two different predator-prey systems, some results about their global dynamics, paying special attention to the dynamics near the infinity and to the configuration of the orbits in a neighborhood of the nonelementary singular points.

First, in Sect. 5.2, we have used the Poincaré compactification, which is a very useful tool to study the behavior of the orbits which go or come from infinity. After a theoretical review of the technique, we have applied it to a Kolmogorov system obtained from a Rosenzweig-MacArthur system and also to a Kolmogorov system obtained from a spatial Lotka-Volterra system. In the first case, we have obtained that there are only isolated singular points at the infinity, while in the second case, the infinity is formed by a continuum of singular points.

Then, in Sect. 5.3, we deal with the desingularization technique consisting on the use of the variable changes called blow-ups. More precisely, we have used directional blow-ups in the horizontal and vertical directions. This allows one to study any type of singularities of analytic systems in dimension two even if they are not elementary. We believe this is important since these types of singular points are not studied in many of the predator-prey systems proposed in the literature. In our case, we apply this technique to the study of singularities in the two planar Kolmogorov systems, giving a description of the process as well as a complete representation of all the results obtained.

Thus, the purpose of this chapter is twofold: on the one hand, to present the results we have obtained for these Kolmogorov systems and, on the other hand, to serve as an illustrative example of how the techniques used can allow a better study of some predator-prey systems proposed in the literature.

References

1. A.J. Lotka, *Elements of Physical Biology* (Waverly Press by the Williams and Wilkins Company, Baltimore, 1925)
2. V. Volterra, Variazione flutuazioni del numero d'individus in specie animali conventi. Atti della Accademia nazionale dei Lincei **2**, 31–113 (1926)
3. É. Diz-Pita, M.V. Otero-Espinar, Predator-prey models: a review on some recent advances. Mathematics **9**, 1783 (2021)
4. M. Rosenzweig, R. MacArthur, Graphical representation and stability conditions of predator-prey interaction. Amer. Natur. **97**, 209–223 (1963)
5. R. Huzak, Predator-prey systems with small predator's death rate. Electron. J. Qualit. Theory Differ. Equ. **86**, 1–16 (2018)
6. A. Kolmogorov, Sulla teoria di Volterra della lotta per l'esistenza. Giornale dell' Istituto Italiano degli Attuari **7**, 74–80 (1936)
7. É. Diz-Pita, J. Llibre, M.V. Otero-Espinar, Phase portraits of a family of Kolmogorv systems depending on six parameters. Electron. J. Differ. Equ. **35**, 1–38 (2021)
8. É. Diz-Pita, J. Llibre, M.V. Otero-Espinar, Planar Kolmogorov systems coming from spatial Lotka-Volterra systems. Int. J. Bifurcat. Chaos **31**(3), 2150201 (2021)
9. É. Diz-Pita, J. Llibre, M.V. Otero-Espinar, Phase portraits of a family of Kolmogorv systems with infinitely many singular points at infinity. Commun. Nonlinear Sci. Numer. Simul. **104**, 106038 (2022)
10. É. Diz-Pita, J. Llibre, M.V. Otero-Espinar, Planar Kolmogorov systems with infinitely many singular points at infinity. Int. J. Bifurcat. Chaos **32**(5), 2250065–4125 (2022)
11. É. Diz-Pita, J. Llibre, M.V. Otero-Espinar, Global phase portraits of a predator-prey system. Electron. J. Qualit. Theory Differ. Equ. **16**, 1–13 (2022)
12. F.H. Busse, Transition to turbulence via the statistical limit cycle route, in *Synergetics*, vol. 39 (Springer, Berlin, 1978)
13. R. Hering, Oscillations in Lotka-Volterra systems of chemical reactions. J. Math. Chem. **5**, 197–202 (1990)
14. G. Gandolfo, *Economic Dynamics*, 4th edn. (Springer, Heidelberg, 2009)
15. G. Gandolfo, Giuseppe Palomba and the Lotka-Volterra equations. Rend. Lincei-Mat. Appl. **19**(4), 347–357 (2008)
16. A.W. Wijeratne, F. Yi, J. Wei, Bifurcation analysis in the diffusive Lotka-Volterra system: an application to market economy. Chaos Solitons Fractals **40**(2), 902–911 (2009)
17. A. Arneodo, P. Coullet, C. Tresser, Occurrence of strange attractors in three-dimensional Volterra equations. Phys. Lett. **79A**, 259–263 (1980)
18. J. Coste, J. Peyraud, P. Coullet, Asymptotic behaviors in the dynamics of competing species. SIAM J. Appl. Math. **36**(3), 516–543 (1979)
19. J. Llibre, D.M. Xiao, Global dynamics of a Lotka-Volterra model with two predators competing for one prey. SIAM K. Appl. Math. **74**(2), 434–453 (2014)
20. C. Lois-Prados, R. Precup, Positive periodic solutions for Lotka-Volterra systems with a general attack rate. Nonlinear Anal.-Real World Appl. **52**, 103024 (2020)
21. S. Smale, On the differential equations of species in competition. J. Math. Biol. **3**, 5–7 (1976)
22. J. Alavez-Ramírez, G. Blé, V. Castellanos, J. Llibre, On the global flow of a 3-dimensional Lotka-Volterra system. Nonlinear Anal.-Theory Methods Appl. **75**, 4114–4125 (2012)
23. J. Llibre, Y.P. Martínez, Dynamics of a family of Lotka-Volterra systems in $\mathbb{R}^3$. Nonlinear Analy. **199**, 111915 (2020)
24. H. Poincaré, Sur l'integration des équations différentielles du premier ordre et du premier degré I. Rendiconti del Circolo Matematico di Palermo **5**, 161–191 (1891)
25. F. Dumortier, J. Llibre, J.C. Artés, *Qualitative Theory of Planar Differential Systems*. UniversiText (Springer, New York, 2006)
26. F. Dumortier, Singularities of vector fields on the plane. J. Differ. Equ. **23**, 53–106 (1977)
27. A.A. Andronov, E.A. Leontovich, I.J. Gordon, A.G. Maier, *Qualitative Theory of 2nd Order Dynamic Systems* (Wiley, Hoboken, 1973)

Chapter 6
A Lotka–Volterra-Type Model Analyzed Through Different Techniques

Jorge Pinto, Sandra Vaz, and Delfim F. M. Torres

6.1 Introduction

The Lotka–Volterra model arose in the middle of 1920s of the twentieth century, by the mathematician Vito Volterra (1860–1940), who intended to explain the oscillatory levels of certain fish catches in the Adriatic sea [18]. At the same time, the biophysicist Alfred Lotka (1880–1949) studied the same interaction predator-prey and published the book Elements of Physical Biology [25]. The Lotka model was similar to the Volterra one, and the predator-prey model is nowadays called the Lotka–Volterra model [7, 11].

The Lotka–Volterra model is a nonlinear population model that has several applications, from epidemiology [2, 22] to biology [18] or economics [9], among others. Sumarti, Nurfitriyana, and Nurwenda, in 2014, applied the continuous Lotka–Volterra model with several interaction laws to the banking system and studied the local stability of the equilibrium points [23]. Here we consider a

J. Pinto
Department of Mathematics, University of Aveiro, Aveiro, Portugal
e-mail: jorge.a.pinto@ua.pt

S. Vaz
Center of Mathematics and Applications (CMA-UBI), Department of Mathematics, University of Beira Interior, Covilhã, Portugal
e-mail: svaz@ubi.pt

D. F. M. Torres (✉)
Center for Research and Development in Mathematics and Applications (CIDMA), Department of Mathematics, University of Aveiro, Aveiro, Portugal

Research Center in Exact Sciences (CICE), Faculty of Sciences and Technology (FCT), University of Cape Verde (Uni-CV), Praia, Cape Verde
e-mail: delfim@ua.pt; delfim@unicv.cv

C. M. A. Pinto, C. M. Ionescu (eds.), *Computational and Mathematical Models in Biology*, Nonlinear Systems and Complexity 38,
https://doi.org/10.1007/978-3-031-42689-6_6

particular case of the Lotka–Volterra model where the growth behavior of one of the species is not exponential but logistic. Precisely, the model that we are going to study is the following one:

$$\begin{cases} {}^{c}D^{\sigma} D(t) = \alpha D(t)\left(1 - \dfrac{D(t)}{C}\right) - pD(t)L(t), \\ {}^{c}D^{\sigma} L(t) = pD(t)L(t) - \beta L(t), \quad 0 < \sigma \leq 1, \end{cases} \tag{6.1}$$

where $D(t)$ represents the prey population in the instant t and $L(t)$ represents the predator population in time t. The value C represents the carrying capacity or the "ideal" capacity of the prey, α represents the growth rate of the prey, and β represents the decay of the predators in the absence of the prey. On the other hand, the parameter p represents the interaction rate between the species. In fractional calculus, several notions of differentiation are available, such as the Grünwald–Letnikov derivative [10], the Riemann–Liouville derivative [10], or the Caputo derivative [4, 5], among others. Here we consider fractional differentiation in the Caputo sense: the model is expressed in terms of the Caputo fractional derivative ${}^{c}D^{\sigma}$. If $\sigma = 1$, then we recover the standard derivative of order one: ${}^{c}D^{1}D(t) = \dot{D}(t)$ [3].

For most nonlinear models, it is not possible to obtain the exact solution, so numerical schemes are important to obtain numerical approximations and to obtain a graphical representation of the solutions. There are several numerical methods, some of them standard, while others are nonstandard. For instance, Euler and Heun methods are standard methods of discretization, while Mickens' method is an example of a nonstandard scheme [8, 13–16].

In Sect. 6.2, we analyze the equilibrium points of (6.1) and their local stability in the case $\sigma = 1$. Afterward, in Sect. 6.3, we discretize the model using the Euler numerical scheme [8], and in Sect. 6.4, the Mickens' numerical scheme [13–17], analyzing the dynamical consistency with the continuous model (6.1). We proceed with Sect. 6.5, using fractional calculus to study (6.1) for $0 < \sigma < 1$: we prove the existence and uniqueness of a positive solution (Proposition 2) and the conservation law (Theorem 4); we determine the local stability of the equilibrium points of the system (Sect. 6.5.5); and we represent the solutions graphically (Sect. 6.5.6). For the fractional derivative, it is necessary to use suitable numerical schemes. Here we use the modified trapezoidal method that involves the modified Euler scheme [1, 19]. We end our work representing graphically some of the solutions for the different methods, comparing them, and with Sect. 6.6 of conclusion. All our numerical simulations were done using Mathematica version 12.1.

6.2 The Modified Lotka–Volterra Model

In this section, we determine the feasible region for model (6.1) with $\sigma = 1$, the equilibrium points, and their local stability.

6.2.1 Model Description

In the model, we have two populations. The variable $D(t)$ represents the prey population that, in the absence of predation, grows in the logistic way. The predator population, $L(t)$, in case of the absence of any prey for sustenance, dies with exponential decay. The model is reasonable since in the physical world the populations are not infinite, so using the logistic function makes sense instead of exponential growth:

$$\begin{cases} \dot{D} = \alpha D\left(1 - \frac{D}{C}\right) - pDL, \\ \dot{L} = pDL - \beta L. \end{cases} \tag{6.2}$$

Throughout the text, we consider two hypothesis:

(P_1) $0 < \alpha < \beta < pC < 1$;
(P_2) all the initial conditions are positive.

The first condition is necessary for the existence of nonnegative equilibrium points. We use the parameter C, the carrying capacity, in our computations, but it will be replaced by $C = 1$ when necessary. In that case D and L represent a fraction of C.

In our model, if $\alpha = 0$, then

$$\begin{cases} \dot{D} = -pDL, \\ \dot{L} = pDL - \beta L, \end{cases}$$

and the prey population tends to extinction, which implies that the predator population will also be extinct because they do not have what to eat.

Assuming the prey is the only food resource for the predators, and the predators do not migrate looking for other place to live, if $p = 0$, then

$$\begin{cases} \dot{D} = \alpha D\left(1 - \frac{D}{C}\right), \\ \dot{L} = -\beta L, \end{cases}$$

and there is no interaction between the two species. This means that the predator population will be extinct and the prey will approach the carrying capacity C.

6.2.2 Nonnegativity of Solutions and Conservation Law

Model (6.2) is a populational model so it has to satisfy some properties, namely, the nonnegativity of the solutions and the conservation law.

Roughly speaking, Proposition 1 asserts that the prey and predator populations are always nonnegative.

Proposition 1 *Under hypotheses* (P_1) *and* (P_2), *the solutions of system* (6.2) *are nonnegative for all* $t > 0$, *that is, the solutions are in*

$$\Omega_+ = \left\{(D, L) \in (\mathbb{R}_0^+)^2 : D \geqslant 0, L \geqslant 0\right\}.$$

Proof Considering the first equation of (6.2),

$$\frac{dD}{dt} = \alpha D\left(1 - \frac{D}{C}\right) - pDL \Leftrightarrow \frac{dD}{D} = \left[\alpha\left(1 - \frac{D}{C}\right) - pL\right]dt,$$

and integrating both sides we get

$$D(T) = D_1 \exp^{\displaystyle\int_0^T \alpha\left(1 - \frac{D}{C}\right) - pL\,dt} \geqslant 0, \quad D_1 = e^{D_0}.$$

Analogously, considering the second equation of (6.2),

$$\frac{dL}{dt} = L(pD - \beta),$$

and we have

$$L(T) = L_1 \exp^{\displaystyle\int_0^T pD - \beta\,dt} \geqslant 0, \quad L_1 = e^{L_0}.$$

Therefore, $(D, T) \in \Omega_+$. □

Theorem 1 *Let* $M = \max\{D(0), C\}$. *If hypotheses* (P_1) *and* (P_2) *hold, then the feasible region is given by*

$$\Omega = \left\{(D, L) \in (\mathbb{R}_0^+)^2 : 0 \leq D \leq M \text{ and } 0 \leq D + L \leq \left(\frac{\alpha + 4\beta}{4\beta}\right) M\right\},$$

that is, any solution that starts in Ω *remains in* Ω.

Proof Considering the first equation of (6.2),

$$\frac{dD}{dt} = \alpha D\left(1 - \frac{D}{C}\right) - pDL \Rightarrow \frac{dD}{dt} \leq \alpha D\left(1 - \frac{D}{C}\right) \Leftrightarrow \frac{dD}{D\left(1 - \frac{D}{C}\right)} \leq \alpha dt.$$

Integrating both sides,

$$\int \frac{1}{D} dD - \int \frac{-\frac{1}{C}}{1 - \frac{D}{C}} dD \leq \alpha t + k \Leftrightarrow D(t) \leq \frac{C_1 \exp^{\alpha t}}{1 + \frac{C_1}{C} \exp^{\alpha t}}.$$

If $D(0) = D_0$, then $C_1 = \dfrac{D_0}{1 - \frac{D_0}{C}}$. Thus,

$$D(t) \leqslant \frac{D_0 \exp^{\alpha t}}{1 - \frac{D_0}{C}(1 - \exp^{\alpha t})},$$

that is, $\lim_{t\to\infty} D(t) = C$ and $\lim_{t\to 0} D(t) = D_0$. Let $W(t) = D(t) + L(t)$. Then,

$$\begin{aligned}
\frac{dW}{dt} &= \frac{dD}{dt} + \frac{dL}{dt} \\
&= \alpha D\left(1 - \frac{D}{C}\right) + \beta D - \beta D - \beta L \\
&= \alpha D\left(1 - \frac{D}{C}\right) + \beta D - \beta W.
\end{aligned}$$

If $f(D) = \alpha D\left(1 - \frac{D}{C}\right)$, then $\overline{D} = \frac{C}{2}$, which is a critical point that is the maximum. Therefore, $f(\overline{D}) = \frac{\alpha C}{4}$ and

$$\begin{aligned}
\frac{dW}{dt} &\leq \frac{\alpha C}{4} + \beta D - \beta W, \\
\frac{dW}{dt} &\leq \left(\frac{\alpha + 4\beta}{4}\right) M - \beta W, \\
\frac{dW}{dt} + \beta W &\leq \left(\frac{\alpha + 4\beta}{4}\right) M.
\end{aligned}$$

Solving the differential equation,

$$W(t) \leq W_0 \exp^{-\beta t} + \left(\frac{\alpha + 4\beta}{4\beta}\right) M(1 - \exp^{-\beta t})$$

and we have $\lim_{t\to+\infty} W(t) = \left(\dfrac{\alpha + 4\beta}{4\beta}\right) M.$ □

6.2.3 Stability Analysis

The equilibrium points of (6.2) are

$$\begin{cases} \dot{D} &= 0, \\ \dot{L} &= 0, \end{cases} \Leftrightarrow \begin{cases} \overline{D} &= 0, \\ \overline{L} &= 0, \end{cases} \vee \begin{cases} \overline{D} &= C, \\ \overline{L} &= 0, \end{cases} \vee \begin{cases} \overline{D} &= \frac{\beta}{p}, \\ \overline{L} &= \frac{\alpha}{p}\left(1 - \frac{\beta}{pC}\right). \end{cases}$$

We obtain three equilibrium points:

$$e_1 = (0, 0), \quad e_2 = (C, 0), \quad e_3 = \left(\frac{\beta}{p}, \frac{\alpha}{p}\left(1 - \frac{\beta}{pC}\right)\right).$$

The equilibrium e_1 means both populations tend to extinction, and e_2 means the prey population tend to C, and the predators population tend to extinction. It is important to note that the equilibrium point e_3 only exists if $1 - \frac{\beta}{pC} > 0$, which is one of the relations in (P_1), meaning that the prey population tends to $\frac{\alpha}{\beta}$ and the predator population tends to $\frac{\alpha}{p}\left(1 - \frac{\beta}{pC}\right)$.

To study the local stability, we linearize the system. The Jacobian matrix is

$$J = \begin{bmatrix} \alpha\left(1 - 2\frac{D}{C}\right) - pL & -pD \\ pL & pD - \beta \end{bmatrix}.$$

In $e_1 = (0, 0)$,

$$J(e_1) = \begin{bmatrix} \alpha & 0 \\ 0 & -\beta \end{bmatrix}$$

and so

$$P(\lambda) = \det(J(e_1) - \lambda Id) = (\alpha - \lambda)(-\beta - \lambda).$$

The eigenvalues are $\lambda_1 = \alpha$ and $\lambda_2 = -\beta$. By hypothesis (P_1), $\lambda_1 > 0$ and $\lambda_2 < 0$. This means e_1 is a *saddle point*.

In $e_2 = (C, 0)$,

$$J(e_2) = \begin{bmatrix} -\alpha & -pC \\ 0 & pC - \beta \end{bmatrix}$$

and

$$P(\lambda) = \det(J(e_2) - \lambda Id) = (-\alpha - \lambda)(pC - \beta - \lambda).$$

The eigenvalues are $\lambda_1 = -\alpha$ and $\lambda_2 = -\beta + pC$. From (P_1) it can be seen that $\lambda_1 < 0$ and $\lambda_2 > 0$. Once again, e_2 is a *saddle point*.

For $e_3 = \left(\frac{\beta}{p}, \frac{\alpha}{p}\left(1 - \frac{\beta}{pC}\right)\right)$,

$$J(e_3) = \begin{bmatrix} -\frac{\alpha\beta}{pC} & -\beta \\ \alpha\left(1 - \frac{\beta}{pC}\right) & 0 \end{bmatrix}$$

and

$$P(\lambda) = \det(J(e_3) - \lambda Id) = \lambda^2 + \frac{\alpha\beta}{pC}\lambda + \beta\alpha\left(1 - \frac{\beta}{pC}\right).$$

By the Routh–Hurwitz criterion and hypothesis (P_1), all coefficients are greater than zero and we conclude that all the roots are negative. Therefore, e_3 is a *sink* or *asymptotically stable*.

6.2.4 *Graphical Analysis*

Throughout the work, we represent graphically the solutions, under different initial conditions, and compare the different methods, visualizing the similarities and differences. In our simulations, for comparison reasons, the parameters will always be $C = 1$, $a = 0.05$, $b = 0.3$, and $p = 0.4$.

In Fig. 6.1, the continuous model (6.1) is represented using different initial conditions. We see that if $(D_0, L_0) = (0.2, 0.3)$, then the predators tend to extinction and the prey population increase. When the prey tend to the carrying capacity, then the predators population increase, converging to e_3. On the other hand, if $(D_0, L_0) = (0, 0.5)$, and since the prey are extinct, the predators tend to extinction because of the absence of food. Finally, if $(D_0, L_0) = (0.85, 0.1)$, then the population of prey and predators tend to e_3.

Figure 6.2 corresponds to initial conditions $(D_0, L_0) = (0.2, 0.3)$ and will be used for comparison with the other models in the sequel.

6.3 Euler's Numerical Scheme

Leonhard Euler (1707–1783) presented his method, now known as the Euler method, in *Institutionum calculi integralis* (1768–1770). Euler's method is an

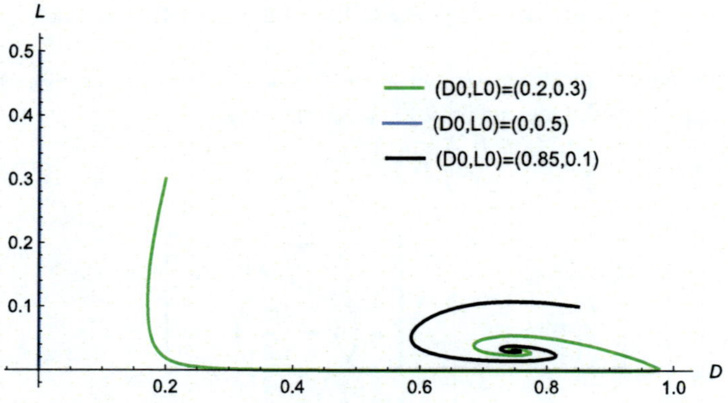

Fig. 6.1 Continuous model (6.1) with different initial conditions

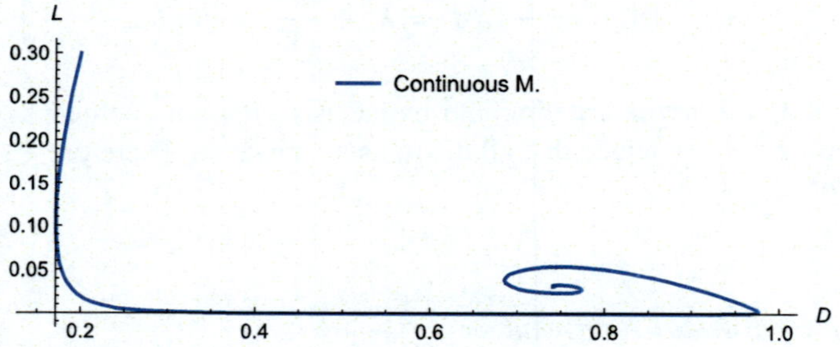

Fig. 6.2 Continuous model (6.1) with $(D_0, L_0) = (0.2, 0.3)$

iterative process for approximating the solution of a problem. In other words, given an ordinary first-order differential equation or system, with known initial condition, Euler's method gives us a way to determine an approximated numerical solution. Let us consider $y' = f(t, y)$, $y(t_0) = y_0$, and $t \in [a, b]$, and let us compute the Taylor series for $t = t_1 = t_0 + h$:

$$y(t_1) \approx y_1 = y_0 + hf(t_0, y_0) + \frac{h^2}{2} f'(t_0, y_0) + \ldots + \frac{h^k}{k!} f^{(k-1)}(t_0, y_0),$$

where $f^{(j)}(t_0, y_0) = \dfrac{d^j f}{dt^j}(t_0, y_0)$. The following explicit method with uniform step-size h allows us to obtain approximated solutions $y_i \approx y(t_i)$:

$$\begin{aligned} y_0 &= y(t_0), \\ y(t_{i+1}) \approx y_{i+1} &= y_i + hf(t_i, y_i), \quad i = 0, \ldots, n. \end{aligned} \tag{6.3}$$

The scheme (6.3) is precisely the Euler method. The step-size $h = t_{i+1} - t_i$, $i = 0, \ldots, n$, is uniform and $t_i = t_0 + ih$, $i = 0, \ldots, n$, where $h = \frac{b-a}{n}$. For physical models, this method is not frequently used because it may be difficult to prove the nonnegativity of the solutions. Moreover, it is also known to cause numerical instabilities and present dynamical inconsistency with the continuous model [12, 14, 21].

6.3.1 Model Discretization

Using the concepts just introduced,

$$\begin{cases} \dot{D}(t) \approx \dfrac{D(t+h) - D(t)}{h}, \\ \dot{L}(t) \approx \dfrac{L(t+h) - L(t)}{h}, \end{cases}$$

and we get

$$\begin{cases} \dfrac{D_{i+1} - D_i}{h} = \alpha D_i \left(1 - \dfrac{D_i}{C}\right) - p D_i L_i, \\ \dfrac{L_{i+1} - L_i}{h} = p D_i L_i - \beta L_i, \end{cases}$$

$$\Leftrightarrow \begin{cases} D_{i+1} = D_i \left(\alpha h \left(1 - \dfrac{D_i}{C}\right) - phL_i + 1\right), \\ L_{i+1} = L_i \left(phD_i - \beta h + 1\right), \end{cases} \tag{6.4}$$

$i \geq 0$. System (6.4) is the explicit discrete model by the standard Euler's scheme.

6.3.2 Nonnegativity of Solutions and Conservation Law

Now we prove that the discrete system (6.4) satisfies the nonnegativity condition and the conservation law. First we prove an auxiliary lemma under the following hypothesis:

(H_1) $1 - \beta h > 0$.

Lemma 1 *Let h be the step-size and suppose that (P_1) and (H_1) hold. Then,*

$$\frac{1 + \alpha h}{ph} \geq C.$$

Proof From hypotheses (P_1) and (H_1), $0 > 1 + \alpha h - \beta h > 1 + \alpha h - Cph$ because $\beta h < Cph$. □

Theorem 2 *If hypotheses* (P_1), (P_2) *and* (H_1) *hold, then all solutions of system* (6.4) *are nonnegative for all* $n \geq 0$ *with the feasible region being given by*

$$\Omega = \left\{(D, L) \in (\mathbb{R}_0^+)^2 : W_n = D_n + L_n \leq C\right\}.$$

Proof By hypotheses (P_1) and (H_1), we easily see that the second equation of (6.4),

$$L_{n+1} = L_n\left(1 - \beta h + phD_n\right),$$

satisfy $L_{n+1} \geqslant 0$ for all n. Regarding the first equation, by (P_1) it follows that

$$\begin{aligned} D_{n+1} &= D_n\left(1 + \alpha h - h\left(\frac{\alpha}{C}D_n + pL_n\right)\right) \\ &> D_n(1 + \alpha h - ph(D_n + L_n)) \\ &> D_n(1 + \alpha h - phW_n). \end{aligned}$$

Therefore, $D_{n+1} \geq 0$ if, and only if, $1 + \alpha h - phW_n \geq 0$, that is, $W_n \leq \frac{1+\alpha h}{ph}$. Let $W_n = D_n + L_n$. By (6.4), we have

$$\begin{aligned} &\frac{W_{n+1} - W_n}{h} = \alpha D_n\left(1 - \frac{D_n}{C}\right) - \beta L_n \\ \Leftrightarrow &\frac{W_{n+1} - W_n}{h} = \alpha D_n\left(1 - \frac{D_n}{C}\right) - \beta L_n - \beta D_n + \beta D_n \\ \Leftrightarrow &\frac{W_{n+1} - W_n}{h} = \alpha D_n\left(1 - \frac{D_n}{C}\right) - \beta W_n + \beta D_n \end{aligned}$$

and from (P_1) it follows that

$$\begin{aligned} W_{n+1} &\leq (1 - \beta h)W_n + \beta hD_n\left(1 - \frac{D_n}{C}\right) + \beta hD_n \\ &= (1 - \beta h)W_n + \beta hD_n\left(2 - \frac{D_n}{C}\right). \end{aligned}$$

Let $f(D) = \beta hD\left(2 - \frac{D}{C}\right)$. Then the critical point is $\overline{D} = C$, which is a maximum, and $f(\overline{D}) = \beta hC$. Moreover, $W_{n+1} \leq (1 - \beta h)W_n + \beta hC$, that is,

$$W_{n+1} \leq (1 - \beta h)^{n+1}W_0 + C\left(1 - (1 - \beta h)^n\right).$$

Thus, $\lim_{n\to+\infty} W_{n+1} = C$, which means that $W_n \leq C \leq \dfrac{1 + \alpha h}{ph}$ for all $n \geqslant 0$. We conclude that the feasible region is

$$\Omega = \left\{(D, L) \in \mathbb{R}_0^{+2} : 0 \leq W_n = D_n + L_n \leq C\right\}$$

and that (6.4) satisfies the nonnegativity condition and the conservation law if (H_1) is verified. □

6.3.3 *Stability Analysis*

The stationary points of the Euler discrete system (6.4) are obtained from

$$\begin{cases} F(D^*) = D^*, \\ F(L^*) = L^*, \end{cases} \Leftrightarrow \begin{cases} D^* = 0, \\ L^* = 0, \end{cases} \vee \begin{cases} D^* = C, \\ L^* = 0, \end{cases} \vee \begin{cases} D^* = \dfrac{\beta}{p}, \\ L^* = \dfrac{\alpha}{p}\left(1 - \dfrac{\beta}{pC}\right), \end{cases}$$

that is,

$$e_1 = (0, 0), \quad e_2 = (C, 0), \quad e_3 = \left(\frac{\beta}{p}, \frac{\alpha}{p}\left(1 - \frac{\beta}{pC}\right)\right),$$

which are equal to the ones of the continuous model (6.2).

To study the local stability, we linearize the system. The Jacobian matrix is given by

$$J = \begin{bmatrix} \alpha h\left(1 - \dfrac{2D}{C}\right) - phL + 1 & -phD \\ phL & phD - \beta h + 1 \end{bmatrix}.$$

For $e_1 = (0, 0)$, the Jacobian matrix is

$$J(e_1) = \begin{bmatrix} 1 + \alpha h & 0 \\ 0 & 1 - \beta h \end{bmatrix}$$

and

$$P(\lambda) = \det(J(e_1) - \lambda Id) = (1 + \alpha h - \lambda)(1 - \beta h - \lambda).$$

The eigenvalues are $\lambda_1 = 1 + \alpha h$ and $\lambda_2 = 1 - \beta h$. By (P_1), $\lambda_1 > 1$, and by (H_1), $\lambda_2 < 1$. We conclude that e_1 is a *saddle point*.

For $e_2 = (C, 0)$, the Jacobian matrix is

$$J(e_2) = \begin{bmatrix} 1 - \alpha h & -phC \\ 0 & phC - \beta h + 1 \end{bmatrix}$$

and

$$P(\lambda) = \det(J(e_2) - \lambda Id) = (1 - \alpha h - \lambda)(phC - \beta h + 1 - \lambda).$$

The eigenvalues are $\lambda_1 = 1 - \alpha h$ and $\lambda_2 = -\beta h + phC + 1$. By (P_1), we can see that $\lambda_2 > 1$ and $\lambda_1 < 1$, and thus e_2 is a *saddle point*.

For $e_3 = \left(\frac{\beta}{p}, \frac{\alpha}{p}\left(1 - \frac{\beta}{pC}\right)\right)$, the Jacobian matrix is

$$J(e_3) = \begin{bmatrix} 1 - \dfrac{\alpha h \beta}{pC} & -\beta h \\ \alpha h \left(1 - \dfrac{\beta}{pC}\right) & 1 \end{bmatrix}$$

and

$$P(\lambda) = \det(J(e_3) - \lambda Id) = \left(1 - \frac{\alpha\beta h}{pC} - \lambda\right)(1 - \lambda) + \alpha\beta h^2\left(1 - \frac{\beta}{pC}\right).$$

From Schur–Cohn criterion for quadratic polynomials $P(x)$, if $P(1) > 0$, $P(-1) > 0$, and $| P(0) | < 1$, then all the roots are inside the unit circle (see [6, 12]). Here we have $P(1) = \alpha\beta h^2\left(1 - \frac{\beta}{pC}\right)$, so by (P_1) we can conclude that $P(1) > 0$; while

$$P(-1) = 2\left(2 - \frac{\alpha\beta h}{pC}\right) + \alpha\beta h^2\left(1 - \frac{\beta}{pC}\right)$$

and by (P_1), since $\dfrac{\beta}{pC} < 1 \Leftrightarrow 2 - \dfrac{\alpha\beta h}{pC} > 2 - \alpha h$, one has $P(-1) > 0$. Moreover,

$$P(0) = 1 - \frac{\alpha\beta h}{pC} + \alpha\beta h^2\left(1 - \frac{\beta}{pC}\right).$$

We need to show that $P(0) > -1$. By (P_1), and since

$$1 - \frac{\alpha\beta h}{pC} > 1 - \alpha h > 1 - \beta h > 0,$$

we have

$$P(0) = 1 - \frac{\alpha\beta h}{pC} + \alpha\beta h^2\left(1 - \frac{\beta}{pC}\right) > 1 - \beta h + \alpha\beta h^2\left(1 - \frac{\beta}{pC}\right) > 0 > -1$$

and to have $P(0) < 1$ we need to show that

$$-\frac{\alpha\beta}{pC}h + \alpha\beta\left(1 - \frac{\beta}{pC}\right)h^2 < 0. \tag{6.5}$$

Because (6.5) is a polynomial $g(h)$ of degree 2 with roots

$$h_1 = 0 \quad \vee \quad h_2 = \frac{1}{pC - \beta},$$

we have $P(0) < 1$ as long as $0 < h < h_2$. Therefore, we have shown that $\mid P(0) \mid < 1$ as long as $0 < h < h_2$ and, in this case, e_3 is a *sink* or *asymptotically stable*.

Under Euler's numerical scheme, the conservation law is not straightforward and we need to impose some conditions to have dynamical consistency with the continuous model. In order for e_3 to be asymptotically stable, the step-size must be smaller than $h_2 = \dfrac{1}{pC - \beta}$. By (P_1), all parameters are less than 1, so it is not a difficult condition to be attained.

6.3.4 *Graphical Analysis*

For the numerical scheme of Euler, we take the time interval to be [0, 300] and the step-size as $h = 0.25$.

In Fig. 6.3, some solutions of (6.4), for different initial conditions, are plotted. Figure 6.3 is similar to Fig. 6.1.

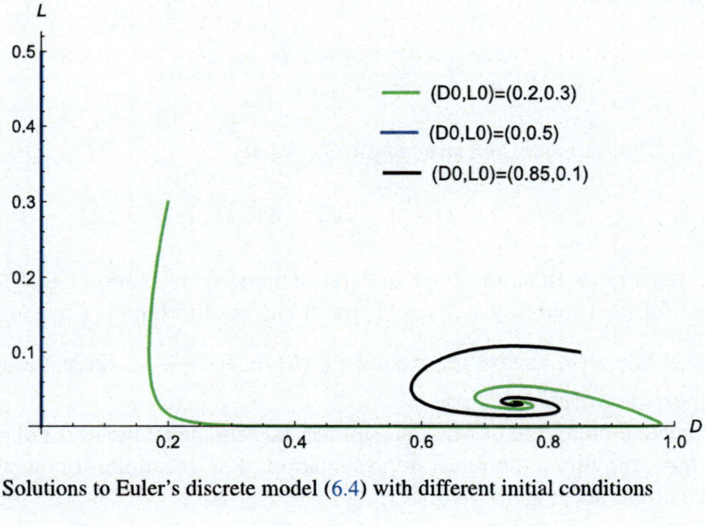

Fig. 6.3 Solutions to Euler's discrete model (6.4) with different initial conditions

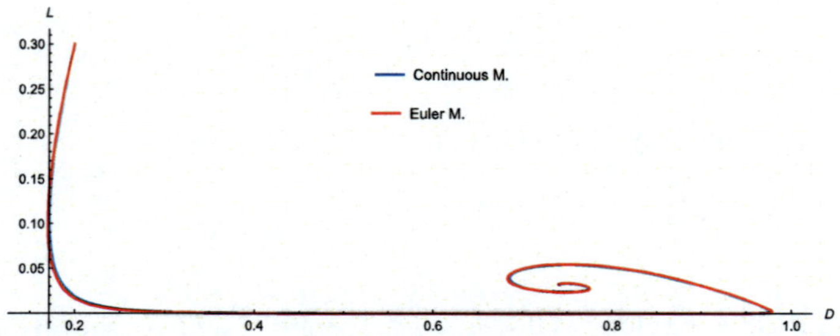

Fig. 6.4 Euler's method compared with the continuous model

In Fig. 6.4, we compare the solution of the continuous model (6.2) with the one obtained by the Euler method with initial conditions $(D_0, L_0) = (0.2, 0.3)$. There are only a few mild differences between the plots, which are qualitatively the same.

6.4 Mickens' Numerical Scheme

The nonstandard finite difference numerical scheme (NSFD), started by Mickens [13], is based on Mickens' work [14–17]. It was created with the goal of solving some problems produced by Euler's method, namely, numerical instabilities and the difficulty of showing the nonnegativity of the solutions for physical models or dynamical inconsistency [14].

Mickens' numerical scheme has two main rules [14]. The first is the derivative is approximated by

$$\frac{dx}{dt} \approx \frac{x_{k+1} - \varphi(h)x_k}{\psi(h)}, \quad h = \Delta t,$$

where h is the step-size, and $\varphi(h)$ and $\psi(h)$ satisfy

$$\varphi(h) = 1 + O(h^2) \quad \text{and} \quad \psi(h) = h + O(h^2).$$

The numerator function $\varphi(h)$ and the denominator function $\psi(h)$ may take different forms. Generally, $\varphi(h) = 1$, but it can be different, for instance, $\varphi(h) = \cos(\lambda h)$, while $\psi(h)$ can be, for example, $\psi(h) = \dfrac{1 - e^{-\lambda h}}{\lambda}$, where λ is a parameter that appears in the model.

The second main rule of Mickens' numerical scheme is: the linear and nonlinear terms may require a nonlocal representation. For examples of such nonlocal representations see, e.g., [14–16].

6.4.1 Model Discretization

Applying the Mickens' rules,

$$\begin{cases} \dot{D}(t) & \approx \dfrac{D((i+1)h) - D(ih)}{\phi(h)} = \dfrac{D_{i+1} - D_i}{\phi(h)}, \\ \dot{L}(t) & \approx \dfrac{L((i+1)h) - L(ih)}{\phi(h)} = \dfrac{L_{i+1} - L_i}{\phi(h)}. \end{cases}$$

By Mickens and Washington [17], if the populations have to satisfy a conservation law, then the denominator function is obtained by the conservation law of the continuous model (6.1). We have seen that

$$\dot{W} = \left(\frac{\alpha + 4\beta}{4\beta}\right) M - \beta W,$$

so the denominator function is chosen to be $\phi(h) = \dfrac{1 - e^{-\beta h}}{\beta}$. For simplification in writing, in this work we denote $\phi := \phi(h)$.

The discrete model obtained from Mickens' method is then given by

$$\begin{cases} \dfrac{D_{i+1} - D_i}{\phi} & = \alpha D_i \left(1 - \dfrac{D_{i+1}}{C}\right) - p D_{i+1} L_i, \\ \dfrac{L_{i+1} - L_i}{\phi} & = p D_{i+1} L_i - \beta L_{i+1}, \end{cases}$$

$$\Leftrightarrow \begin{cases} D_{i+1} & = \dfrac{(\alpha\phi + 1) D_i}{1 + p\phi L_i + \dfrac{\alpha\phi D_i}{C}}, \\ L_{i+1} & = \dfrac{(p\phi D_{i+1} + 1) L_i}{1 + \beta\phi}, \end{cases} \tag{6.6}$$

for $i \geq 0$. The explicit discrete model is

$$\begin{cases} D_{i+1} & = \dfrac{(\alpha\phi + 1) D_i}{1 + p\phi L_i + \dfrac{\alpha\phi D_i}{C}}, \\ \\ L_{i+1} & = \left(\dfrac{p\phi(\alpha\phi + 1) D_i}{1 + p\phi L_i + \dfrac{\alpha\phi D_i}{C}} + 1\right) \dfrac{L_i}{1 + \beta\phi}, \end{cases} \tag{6.7}$$

for $i \geq 0$.

6.4.2 Nonnegativity of Solutions and Conservation Law

If the initial conditions are nonnegative, by (P_1), and the direct observation of (6.6), we have the nonnegativity of the solutions. In the next result, for practical reasons, we consider $C = 1$.

Theorem 3 *Under hypotheses* (P_1) *and* (P_2), *if* $C = 1$ *and* $\xi = 1 + \alpha\phi$, *then*

$$\Omega = \left\{ (D, L) \in (\mathbb{R}_0^+)_+^2 : 0 \leq D_n \leq 1 \quad and \quad W_n \leq 1 \leq \frac{4\alpha^2 + \xi\beta^2}{4\alpha\beta} \right\},$$

where $W_n = D_n + L_n$.

Proof By the first equation of (6.6), we have

$$D_{n+1} = \frac{(1+\alpha\phi)D_n}{1 + p\phi L_n + \alpha\phi D_n} < \frac{(1+\alpha\phi)D_n}{1+\alpha\phi D_n}$$

for some $n > 0$ and

$$\lim_{n\to+\infty} \frac{(1+\alpha\phi)D_n}{1+\alpha\phi D_n} = 1.$$

Thus, $0 \leq D_n \leq 1$. Adding the equations of (6.6), we have

$$\begin{aligned}
\frac{W_{n+1} - W_n}{\phi} &= \alpha D_n - \alpha D_n D_{n+1} - \beta L_{n+1} - \beta D_{n+1} + \beta D_{n+1} \\
&= \alpha D_n + D_{n+1}(-\alpha D_n + \beta) - \beta W_{n+1} \\
&\leq \alpha D_n + \beta D_{n+1}\left(1 - \frac{\alpha}{\xi\beta} D_{n+1}\right) - \beta W_{n+1}.
\end{aligned}$$

Let $f(D) = \beta D\left(1 - \frac{D}{K}\right)$, where $K = \frac{\xi\beta}{\alpha}$. Then, $\overline{D} = \frac{K}{2}$ is the critical point and maximizer, so that $f(D)$ attains its maximum at $f(\overline{D}) = \frac{K\beta}{4}$. Therefore,

$$\begin{aligned}
\frac{W_{n+1} - W_n}{\phi} &\leq \alpha + \frac{K\beta^2}{4} - \beta W_{n+1}, \\
W_{n+1} &\leq \frac{A}{1+\beta\phi} + \frac{W_n}{1+\beta\phi},
\end{aligned}$$

where $A = \alpha + \frac{\beta K}{4} = \frac{4\alpha^2 + \beta^2\xi}{4\alpha}$. We conclude that

$$W_{n+1} \le \left(\frac{1}{1+\beta\phi}\right)^{n+1} W_0 + \frac{A\phi}{1+\beta\phi}\left(\frac{1-\left(\frac{1}{1+\beta\phi}\right)^n}{1-\frac{1}{1+\beta\phi}}\right)$$

$$\le \left(\frac{1}{1+\beta\phi}\right)^{n+1} W_0 + \frac{A}{\beta}\left(1-\left[\frac{1}{1+\beta\phi}\right]^n\right)$$

and $\lim_{n\to+\infty} W_{n+1} = \dfrac{4\alpha^2+\xi\beta^2}{4\alpha\beta}$. □

Considering hypothesis (P_1), $\phi(h) = h + O(h^2)$ and $\xi = 1+\alpha\phi$, it is reasonable to consider $1 < \xi < 2$.

6.4.3 Stability Analysis

The stationary points of the discrete system (6.6) are:

$$\begin{cases} F(D^*) = D^* \\ F(L^*) = L^* \end{cases} \Leftrightarrow \begin{cases} D^* = 0, \\ L^* = 0, \end{cases} \vee \begin{cases} D^* = C, \\ L^* = 0, \end{cases} \vee \begin{cases} D^* = \dfrac{\beta}{p}, \\ L^* = \dfrac{\alpha}{p}\left(1-\dfrac{\beta}{pC}\right). \end{cases}$$

The stationary points are equal to the ones of the continuous model (6.1):

$$e_1 = (0,0), \quad e_2 = (C,0), \quad e_3 = \left(\frac{\beta}{p}, \frac{\alpha}{p}\left(1-\frac{\beta}{pC}\right)\right).$$

Once more, to study the local stability, we linearize the system. The Jacobian matrix of (6.7) is given by

$$J(D,L) = \begin{bmatrix} \dfrac{(1+\alpha\phi)(1+p\phi L)}{\left(1+p\phi L+\dfrac{\alpha\phi D}{C}\right)^2} & -\dfrac{(1+\alpha\phi)p\phi D}{\left(1+p\phi L+\dfrac{\alpha\phi D}{C}\right)^2} \\ \dfrac{p\phi L(1+\alpha\phi)(1+p\phi L)}{(1+\beta\phi)\left(1+p\phi L+\dfrac{\alpha\phi D}{C}\right)^2} & \dfrac{1}{1+\beta\phi}\left(1+\dfrac{p\phi(1+\alpha\phi)D\left(1+\dfrac{\alpha\phi D}{C}\right)}{\left(1+p\phi L+\dfrac{\alpha\phi D}{C}\right)^2}\right) \end{bmatrix}.$$

For $e_1 = (0,0)$, the Jacobian matrix is

$$J(e_1) = \begin{bmatrix} 1+\alpha\phi & 0 \\ 0 & \dfrac{1}{1+\beta\phi} \end{bmatrix}$$

and

$$P(\lambda) = \det(J(e_1) - \lambda Id) = (1+\alpha\phi - \lambda)\left(\frac{1}{1+\beta\phi} - \lambda\right).$$

In this way, the eigenvalues are $\lambda_1 = 1+\alpha\phi$ and $\lambda_2 = \dfrac{1}{1+\beta\phi}$. By (P_1), we have $\lambda_1 > 1$ and $\lambda_2 < 1$, that is, e_1 is a *saddle point*.

For $e_2 = (C, 0)$, the Jacobian matrix is

$$J(e_2) = \begin{bmatrix} \dfrac{1}{1+\alpha\phi} & -\dfrac{Cp\phi}{1+\alpha\phi} \\ 0 & \dfrac{1+Cp\phi}{1+\beta\phi} \end{bmatrix}$$

and the characteristic polynomial is

$$P(\lambda) = \det(J(e_2) - \lambda Id) = \left(\frac{1}{1+\alpha\phi} - \lambda\right)\left(\frac{1+Cp\phi}{1+\beta\phi} - \lambda\right).$$

The eigenvalues are $\lambda_1 = \dfrac{1}{1+\alpha\phi}$ and $\lambda_2 = \dfrac{1+Cp\phi}{1+\beta\phi}$. Once more, by (P_1), $\lambda_1 < 1$ and $\lambda_2 > 1$. We conclude that e_2 is a *saddle point*.

For $e_3 = \left(\dfrac{\beta}{p}, \dfrac{\alpha}{p}\left(1 - \dfrac{\beta}{pC}\right)\right)$, we have

$$J(e_3) = \begin{bmatrix} \dfrac{1+\alpha\phi\left(1-\frac{\beta}{pC}\right)}{1+\alpha\phi} & -\dfrac{\beta\phi}{1+\alpha\phi} \\ \dfrac{\alpha\phi\left(1-\frac{\beta}{pC}\right)\left(1+\alpha\phi\left(1-\dfrac{\beta}{pC}\right)\right)}{(1+\alpha\phi)(1+\beta\phi)} & \dfrac{1}{1+\beta\phi}\left(1+\dfrac{\beta\phi(1+\frac{\alpha\phi\beta}{pC})}{1+\alpha\phi}\right) \end{bmatrix}$$

and, after some computations, it can be seen that

$$P(\lambda) = \det(J(e_3) - \lambda Id)$$

$$= \lambda^2 - \left(1 + \frac{(1+\beta\phi) + \alpha\phi\left(1 - \frac{\beta}{pC}\right)}{(1+\beta\phi)(1+\alpha\phi)}\right)\lambda + \frac{1 + \alpha\phi\left(1 - \frac{\beta}{pC}\right)}{1+\alpha\phi}.$$

By the Schur–Cohn criterion for quadratic polynomials, if $P(1) > 0$, $P(-1) > 0$ and $| \, P(0) \, | < 1$, then all of its roots are in the unit circle [6, 12]. By (P_1),

- $P(1) = \dfrac{\alpha\phi}{1+\alpha\phi}\dfrac{\beta\phi}{1+\beta\phi}\left(1 - \dfrac{\beta}{pC}\right) > 0;$
- $P(-1) = 2 + \left(\dfrac{1 + \alpha\phi\left(1 - \dfrac{\beta}{pC}\right)}{1+\alpha\phi}\right) + \dfrac{1 + \beta\phi + \alpha\phi\left(1 - \dfrac{\beta}{pC}\right)}{(1+\alpha\phi)(1+\beta\phi)} > 0.$

Regarding

$$P(0) = \frac{1 + \alpha\phi\left(1 - \dfrac{\beta}{pC}\right)}{1+\alpha\phi},$$

by (P_1), $0 < P(0) < 1$.

So, we conclude that e_3 is a *sink* or an *asymptotically stable point*.

We proved the boundedness of the solutions to Mickens' numerical scheme, considering $C = 1$, and the dynamical consistency with the continuous model without further restrictions.

6.4.4 Graphical Analysis

Figure 6.5 presents solutions to the Mickens' discrete model (6.6) with different initial conditions.

Figure 6.5 is similar to Figs. 6.1 and 6.3.

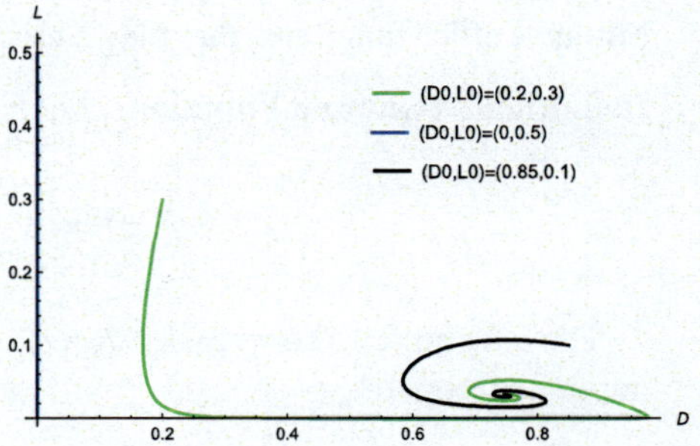

Fig. 6.5 Solutions to Mickens' discrete model (6.6) with different initial conditions

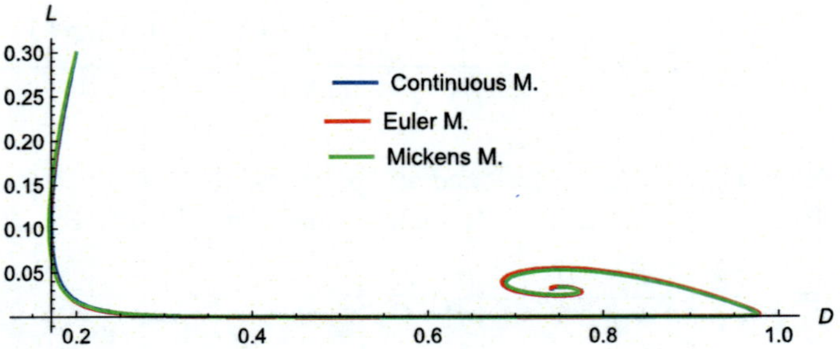

Fig. 6.6 Solutions of Mickens' model (6.6) versus solutions of previous models (6.1) and (6.4).

Figure 6.6 compares the three previous models (6.1), (6.4) and (6.6). As we can see, the differences between them are only mild.

6.5 Fractional Calculus

The calculus of non-integer order, known as Fractional Calculus (FC), is the branch of mathematics that studies the extension of the derivative and integral concepts to an arbitrary order, not necessarily of a fraction order [20].

Considering $f(x) = \frac{1}{2}x^2$, the first and second derivative is $f'(x) = x$ and $f''(x) = 1$. But what about the derivative of order $n = \frac{1}{2}$? This was the question that the father of fractional calculus, L'Hôpital, considered, asking it, by letter, to Leibniz, in 1695. Since then, several mathematicians worked in such kind of calculus, namely, Grünwald, Letnikov, Riemann, Liouville, Caputo, among many others [10, 24].

6.5.1 Preliminaries on FC

In fractional calculus, there are some functions, namely, the Gamma, Beta, and Mittag–Leffler functions, that play a crucial role.

Definition 1 (Gamma Function) Let $z \in \mathbb{C}$. The gamma function is defined by

$$\Gamma(z) = \int_0^{\infty} e^{-t} t^{z-1} \, dt. \tag{6.8}$$

The integral (6.8) converges if $Re(z) > 0$. The gamma function has the following important property:

$$\Gamma(z+1) = z\Gamma(z).$$

Definition 2 (Beta Function) Let $z, w \in \mathbb{C}$. The beta function is defined by

$$B(z, w) = \int_0^1 \tau^{z-1}(1-\tau)^{w-1}\, d\tau \tag{6.9}$$

for $Re(w) > 0$.

Using the Laplace transform, we can rewrite (6.9) as

$$B(z, w) = \frac{\Gamma(z)\Gamma(w)}{\Gamma(z+w)}. \tag{6.10}$$

By (6.10), we conclude that $B(z, w) = B(w, z)$.

It is known that e^z,

$$e^z = \sum_{k=0}^{\infty} \frac{z^k}{\Gamma(k+1)}, \tag{6.11}$$

has an important role in the integration of ordinary differential equations. Similar role has the Mittag–Leffler function for fractional differential equations. The Mittag–Leffler function is a generalization of the exponential function (6.11).

Definition 3 (Mittag–Leffler Function) The Mittag–Leffler function of two parameters is defined by

$$E_{\alpha,\beta} = \sum_{k=0}^{\infty} \frac{z^k}{\Gamma(\alpha k + \beta)},$$

where $\alpha, \beta \in \mathbb{C}$ and $Re(\alpha) > 0$.

The Mittag–Leffler function is uniformly convergent in every compact subset of $\mathbb{C}$. If $\alpha = \beta = 1$, then $E_{1,1}(z) = e^z$.

The fractional calculus has several formulations. The two most well-known are the Riemann–Liouville and Caputo approaches. Here we use fractional derivatives in Caputo's sense.

To define the Riemann–Liouville fractional order integral, let us begin by considering the integral of first order of a function f:

$$J^1 f(x) = \int_0^x f(t)\, dt. \tag{6.12}$$

Analogously, let us consider the integral

$$J^2 f(x) = \int_0^x \int_0^{t_1} f(t)\, dt\, dt_1$$

$$= \int_0^x \int_t^x f(t)\, dt_1\, dt$$
$$= \int_0^x f(t) \int_t^x 1\, dt_1\, dt$$
$$= \int_0^x (x - t) f(t)\, dt.$$

By induction, it follows that

$$J^{n+1} f(x) = \frac{1}{n!} \int_0^x (x - t)^n f(t)\, dt. \tag{6.13}$$

The fractional order integral is obtained taking n in (6.13) to be a real number.

Definition 4 (Riemann–Liouville Fractional Order Integral) Let $\alpha > 0$. The Riemann–Liouville fractional order integral of order α is given by

$$J^\alpha f(x) = \frac{1}{\Gamma(\alpha)} \int_0^x (x - t)^{\alpha-1} f(t)\, dt.$$

The fractional derivative of Riemann–Liouville is obtained with the help of the arbitrary order integral of Riemann–Liouville and the integer order derivative.

Definition 5 (Riemann–Liouville Fractional Order Derivative) Let $\alpha > 0$, $m = \lceil \alpha \rceil$, and $v = m - \alpha$. The Riemann–Liouville fractional order derivative of order α of function f is defined by

$$\begin{aligned} D^\alpha f(x) &= D^m [J^v f(x)] \\ &= D^m \left(\frac{1}{\Gamma(\alpha)} \int_0^x (x - t)^{\alpha-1} f(t)\, dt \right) \\ &= \frac{d^m}{dt^m} \left(\frac{1}{\Gamma(\alpha)} \int_0^x (x - t)^{\alpha-1} f(t)\, dt \right) \\ &= \frac{1}{\Gamma(\alpha)} \frac{d^m}{dt^m} \int_0^x (x - t)^{\alpha-1} f(t)\, dt. \end{aligned}$$

When one uses the Laplace transform of the Riemann–Liouville fractional-order derivative, initial values with integer-order derivatives are not obtained. This fact has no physical meaning [5]. For this reason, in many applications, the Caputo fractional derivative is preferred.

The definition of Caputo's fractional derivative is similar to the Riemann–Liouville definition, the difference being the order one takes the operations of Riemann–Liouville integration and integer-order differentiation. Indeed, in

Caputo's definition, first we compute the derivative of integer order and only then the fractional-order integral.

Definition 6 (Caputo Fractional Order Derivative) Let $\alpha > 0$, $m = \lceil \alpha \rceil$, and $v = m - \alpha$. The Caputo fractional order derivative of order α of function f is defined by

$$
\begin{aligned}
{}^{c}D^{\alpha} f(x) &= J^{v}[D^{m} f(x)] = J^{v}\left[\frac{d^{m}}{dx^{m}} f(x)\right] = J^{v} f^{(m)}(x) \\
&= \frac{1}{\Gamma(v)} \int_{0}^{x} (x-t)^{v-1} f^{(m)}(t)\, dt.
\end{aligned}
$$

Caputo's definition has two advantages with respect to the Riemann–Liouville definition: (i) in applications that involve fractional differential equations, the presence of initial values are physical meaningful; (ii) the derivative of a constant is zero, in contrast with the Riemann–Liouville.

6.5.2 *Model Description*

Let $0 < \sigma < 1$, $(D(0), L(0)) \in (\mathbb{R}_0^+)^2$, and

$$
\begin{cases}
{}^{c}D^{\sigma} D(t) = \alpha D\left(1 - \dfrac{D}{C}\right) - pDL, \\
{}^{c}D^{\sigma} L(t) = pDL - \beta L.
\end{cases}
\tag{6.14}
$$

To write (6.14) in a compact way, let

$$
(\mathbb{R}_0^+)^2 = \{X \in \mathbb{R}^2 : X \geqslant 0\},
$$

$X(t) = (D(t)\ L(t))^T$, and

$$
{}^{c}D^{\sigma} X(t) = F(X(t)), \quad X(0) = (D(0), L(0)) \in (\mathbb{R}_0^+)^2,
\tag{6.15}
$$

where

$$
F(X) = \begin{pmatrix} \alpha D\left(1 - \dfrac{D}{C}\right) - pDL \\ pDL - \beta L \end{pmatrix}.
\tag{6.16}
$$

6.5.3 Existence and Uniqueness of Nonnegative Solutions

Before the stability analysis, it is necessary to show the existence and uniqueness of nonnegative solutions. The following lemma is important.

Lemma 2 (See [22]) *Let us assume that F in (6.16) satisfies the following conditions:*

- *$F(X)$ and $\frac{\partial F(X)}{\partial X}$ are continuous in $X \in \mathbb{R}^n$;*
- *$\|F(X)\| \leqslant w + \lambda \|X\|$, $\forall X \in \mathbb{R}^n$, where w and λ are positive constants.*

Then a solution to (6.15)–(6.16) exists and is unique.

Proposition 2 (Existence and Uniqueness of Nonnegative Solutions) *There exists only one solution to the IVP (6.15) in*

$$(\mathbb{R}_0^+)^2 = \left\{(D, L) \in \mathbb{R}^2 : (D(t), L(t)) \geqslant 0, \forall t > 0\right\}.$$

Proof The existence and uniqueness of solution follows from Lemma 2. The vector function (6.16) is a polynomial so it is continuously differentiable. Let

$$Z = \begin{pmatrix} -\frac{\alpha}{C} & -p \\ 0 & p \end{pmatrix}, \quad B = \begin{pmatrix} \alpha & 0 \\ 0 & -\beta \end{pmatrix}.$$

Then,

$$F(X) = DZX + BX. \tag{6.17}$$

Using the sup norm,

$$\begin{aligned} \|F(X)\| &\leq \|ZX\| + \|BX\| \\ &= (\|Z\| + \|B\|)\, \|X\| \\ &< \epsilon + (\|Z\| + \|B\|)\, \|X\| \end{aligned}$$

for some $\epsilon > 0$. So, by Lemma 2, system (6.15) has a unique solution. The proof of the nonnegative of the solutions follows the same idea of [2]. To prove that $(D_k^*(t), L_k^*(t)) \in (\mathbb{R}_0^+)^2$ for all $t \geqslant 0$, let us consider the following auxiliary fractional differential system:

$$\begin{cases} {}^cD^\sigma D(t) = \alpha D\left(1 - \frac{D}{C}\right) - pDL + \frac{1}{k}, \\ {}^cD^\sigma L(t) = pDL - \beta L + \frac{1}{k}, \end{cases}$$

with $k \in \mathbb{N}$. By contradiction, let us assume that exists a time instant where the condition fails. Let

$$t_0 = \inf\left\{t > 0 : (D_k^*(t), L_k^*(t)) \notin (\mathbb{R}_0^+)^2\right\}.$$

Then $(D_k^*(t_0), L_k^*(t_0)) \in (\mathbb{R}_0^+)^2$ and one of the quantities $D_k^*(t_0)$ or $L_k^*(t_0)$ is zero. Let us suppose that $D_k^*(t_0) = 0$. Then

$${}^cD^\sigma D_k^*(t_0) = 0 + \frac{1}{k} > 0.$$

By continuity of ${}^cD^\sigma D_k^*$, we conclude that ${}^cD^\sigma D_k^*([t_0, t_0 + \zeta]) \subseteq \mathbb{R}^+$ so D_k^* is nonnegative. Analogously, we can do the same for ${}^cD^\sigma L_k^*$, obtaining the intended contradiction. It follows by Lemma 1 of [2], when $k \to \infty$, that $(D^*(t), L^*(t)) \in (\mathbb{R}_0^+)^2$ for all $t \geqslant 0$. □

6.5.4 The Conservation Law

Considering Proposition 2, there exists only one solution to the IVP (6.15). The proof of our next result uses some auxiliary results found in [2, 22].

Theorem 4 (Conservation Law) *The solution to the IVP* (6.15) *is in*

$$\Omega = \left\{(D, L) \in (\mathbb{R}_0^+)^2 : W(t) = D(t) + L(t) \leqslant W(0) + \frac{A}{\beta}\right\},$$

where $A = \dfrac{\alpha + 4\beta}{4} M$.

Proof Let

$$\begin{aligned}{}^cD^\sigma W(t) &\leqslant \alpha D(t)(1 - D(t)) - \beta L(t) - \beta D(t) + \beta D(t)\\ &\leqslant \frac{\alpha}{4} M + \beta M - \beta W(t) \leqslant \left(\frac{\alpha + 4\beta}{4}\right) M - \beta W(t).\end{aligned}$$

It is known that

$${}^cD^\sigma W(t) = J^{1-\alpha}\dot{W}(t) \quad \text{and} \quad \phi_\alpha(t) = \frac{t^{\alpha-1}}{\Gamma(\alpha)}, \qquad \text{for } t > 0.$$

Then

$$\mathcal{L}\{{}^cD^\sigma W(t)\} = \mathcal{L}\{\phi_{1-\alpha}(t) * \dot{W}(t)\} = \mathcal{L}\{\phi_{1-\alpha}(t)\} \cdot \mathcal{L}\{\dot{W}(t)\}$$

$$= s^{\alpha-1}(sW(s) - W(0)) = s^{\alpha}W(s) - s^{\alpha-1}W(0),$$

so that

$$\begin{aligned} {}^{c}D^{\sigma}W(t) + \beta W(t) \leq A &\Leftrightarrow \mathcal{L}\{{}^{c}D^{\sigma}W(t) + \beta W(t)\} \leq \mathcal{L}\{A\} \\ &\Leftrightarrow s^{\alpha}W(s) - s^{\alpha-1}W(0) + \beta W(s) \leq \frac{A}{s} \\ &\Leftrightarrow W(s) \leq \frac{A}{s(s^{\alpha}+\beta)} + \frac{s^{\alpha-1}}{s^{\alpha}+\beta}W(0). \end{aligned}$$

Therefore,

$$\begin{aligned} W(t) &= \mathcal{L}^{-1}\{W(s)\} \\ &\leq \mathcal{L}^{-1}\left\{\frac{A}{s(s^{\alpha}+\beta)}\right\} + \mathcal{L}^{-1}\left\{\frac{s^{\alpha-1}}{s^{\alpha}+\beta}W(0)\right\} \\ &= A\mathcal{L}^{-1}\left\{\frac{1}{s(s^{\alpha}+\beta)}\right\} + W(0)\mathcal{L}^{-1}\left\{\frac{s^{\alpha-1}}{s^{\alpha}+\beta}\right\} \\ &= \frac{A}{\beta}(1 - E_{\alpha}(-\beta t^{\alpha})) + W(0)E_{\alpha}(-\beta t^{\alpha}). \end{aligned}$$

Since $0 \leqslant E_{\alpha}(-\beta t^{\alpha}) \leqslant 1$, we have $W(t) \leqslant W(0) + \frac{A}{\beta}$. □

6.5.5 *Stability Analysis*

The equilibrium points of the fractional system (6.14) are

$$\begin{cases} {}^{c}D^{\alpha}D(t) = 0 \\ {}^{c}D^{\alpha}L(t) = 0 \end{cases} \Leftrightarrow \begin{cases} D = 0, \\ L = 0, \end{cases} \vee \begin{cases} D = C, \\ L = 0, \end{cases} \vee \begin{cases} D = \frac{\beta}{p}, \\ L = \frac{\alpha}{p}\left(1 - \frac{\beta}{pC}\right). \end{cases}$$

They are similar to the ones of the continuous model (6.2):

$$e_1 = (0, 0), \quad e_2 = (C, 0), \quad e_3 = \left(\frac{\beta}{p}, \frac{\alpha}{p}\left(1 - \frac{\beta}{pC}\right)\right),$$

and the study of their local stability is equal to the continuous model. Precisely:

- e_1 and e_2 are *saddle points*;
- e_3 is a *sink* or an *asymptotically stable point*.

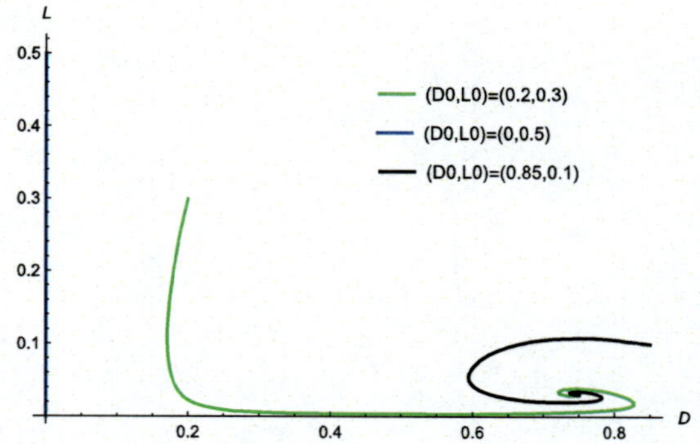

Fig. 6.7 Solutions to the fractional model (6.14) with different initial conditions

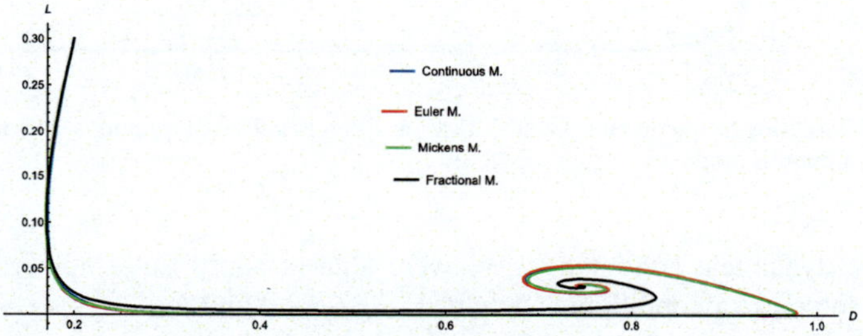

Fig. 6.8 Solution to fractional model compared with solutions to continuous and discrete Euler's and Micken's models

6.5.6 Graphical Analysis

We now present some graphics obtained by the modified trapezoidal method, for $\sigma = 0.95$, step-size $h = 0.25$, and the time interval $[0, 300]$.

In Fig. 6.7, we observe that the orbit with initial conditions $(D_0, L_0) = (0.2, 0.3)$ tend to the equilibrium point faster than the other methods. The same happens with $(D_0, L_0) = (0.85, 0.1)$.

Figure 6.8 shows that the equilibrium point is attained faster and Fig. 6.9 illustrates, graphically, that when σ tends to 1 the solution of (6.14) tends to the solution of the continuous model (6.2), showing that the modified trapezoidal method is a good discretization method to our problem.

6.6 Conclusion

In this work, we considered a predator-prey Lotka–Volterra-type model and analyzed it through different methods. The use of the logistic function, to explain the growth of the prey population in the absence of predation, and the study of the model from different perspectives, allow us to know when extra conditions are needed to

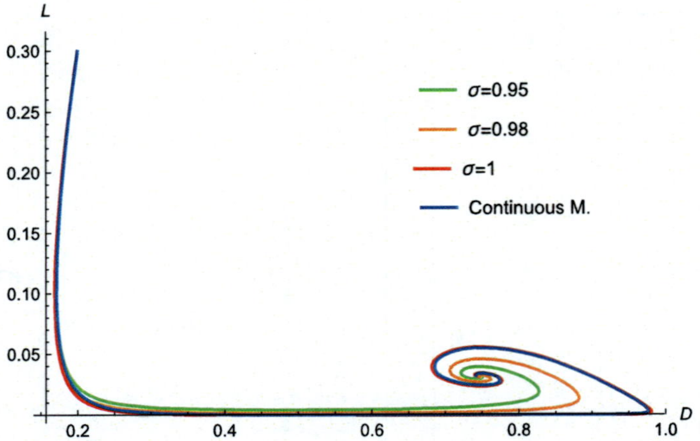

Fig. 6.9 Solution to fractional system (6.14), for different values of σ, compared with the solution to the continuous model (6.2)

achieve dynamical consistency. We were able to apply Euler numerical scheme, under some extra conditions, namely, $1 - \beta h > 0$ and $h \in]0, \frac{1}{pC-\beta}[$. Regarding Mickens' nonstandard finite difference numerical scheme, the proofs were carried out using the carrying capacity $C = 1$. Writing the model using fractional calculus, and using the same initial conditions, our graphical analysis allowed us to conclude that with $\sigma = 0.95$, the equilibrium point is attained faster.

Acknowledgments The authors were partially supported by the Portuguese Foundation for Science and Technology (FCT): Vaz through the Center of Mathematics and Applications of *Universidade da Beira Interior* (CMA-UBI), project UIDB/00212/2020; Torres through the Center for Research and Development in Mathematics and Applications (CIDMA), grants UIDB/04106/2020 and UIDP/04106/2020, and within the project "Mathematical Modelling of Multi-scale Control Systems: Applications to Human Diseases" (CoSysM3), reference 2022.03091.PTDC, financially supported by national funds (OE) through FCT/MCTES.

References

1. H.F. Ahmed, Fractional Euler method; an effective tool for solving fractional differential equations. J. Egypt. Math. Soc. **26**(1), 38–43 (2018)
2. R. Almeida, A.M.C. Brito da Cruz, N. Martins, M.T.T. Monteiro, An epidemiological MSEIR model described by the Caputo fractional derivative. Int. J. Dyn. Control **7**(2), 776–784 (2019)
3. R. Almeida, S. Pooseh, D.F.M. Torres, *Computational Methods in the Fractional Calculus of Variations* (Imperial College Press, London, 2015)
4. M. Caputo, Linear models of dissipation whose Q is almost frequency independent—II. Geophys. J. Int. **13**(5), 529–539 (1967)
5. M. Dias de Carvalho, J.E. Ottoni, Introdução ao cálculo fracionário com aplicações. Rev. Mat. Ouro Preto **1**, 50–77 (2018)

6. S. Elaydi, *An Introduction to Difference Equations*, 3rd edn. Undergraduate Texts in Mathematics (Springer, New York, 2005)
7. G. Garcia Lorenzana, A. Altieri, Well-mixed Lotka-Volterra model with random strongly competitive interactions. Phys. Rev. E **105**(2), 024307, 15 pp. (2022)
8. W. Gautschi, *Numerical Analysis*, 2nd edn. (Birkhäuser Boston, Boston, 2012)
9. A. Kamimura, G.F. Burani, H.M. França, The economic system seen as a living system: a Lotka–Volterra framework. Emergence: Complexity Organ. **13**(3), 80–93 (2011)
10. D.E. Koning, *Fractional Calculus*. Bachelor Project Mathematics, Faculty of Mathematics and Natural Sciences (University of Groningen, Netherlands, 2015)
11. M. Lemos-Silva, D.F.M. Torres, A note on a prey-predator model with constant-effort harvesting, in *Dynamic Control and Optimization* (Springer Nature Switzerland AG, 2022), pp. 201–209
12. P. Liu, S.N. Elaydi, Discrete competitive and cooperative models of Lotka–Volterra type, J. Comput. Anal. Appl. **3**(1), 53–73 (2001)
13. R.E. Mickens, *Nonstandard Finite Difference Models of Differential Equations* (World Scientific Publishing, River Edge, 1994)
14. R.E. Mickens, Nonstandard finite difference schemes for differential equations, J. Differ. Equ. Appl. **8**(9), 823–847 (2002)
15. R.E. Mickens, Dynamic consistency: a fundamental principle for constructing nonstandard finite difference schemes for differential equations. J. Differ. Equ. Appl. **11**(7), 645–653 (2005)
16. R.E. Mickens, Calculation of denominator functions for nonstandard finite difference schemes for differential equations satisfying a positivity condition. Numer. Methods Partial Differ. Equ. **23**(3), 672–691 (2007)
17. R.E. Mickens, T.M. Washington, NSFD discretizations of interacting population models satisfying conservation laws. Comput. Math. Appl. **66**(11), 2307–2316 (2013)
18. J.D. Murray, *Mathematical Biology. I*, 3rd edn. Interdisciplinary Applied Mathematics, 17, (Springer, New York, 2002)
19. Z.M. Odidat, S. Momami, An algorithm for the numerical solution of differential equations of fractional order. J. Appl. Math Inf. **26**(1–2), 15–27 (2008)
20. B. Ross, The development of fractional calculus 1695–1900, Historia Math. **4**, 75–89 (1977)
21. P. Shi, L. Dong, Dynamical behaviors of a discrete HIV-1 virus model with bilinear infective rate. Math. Methods Appl. Sci. **37**(15), 2271–2280 (2014)
22. M.R. Sidi Ammi, M. Tahiri, D.F.M. Torres, Global stability of a Caputo fractional SIRS model with general incidence rate. Math. Comput. Sci. **15**(1), 91–105 (2021)
23. N. Sumarti, R. Nurfitriyana, W. Nurwenda, A dynamical system of deposit and loan volumes based on the Lotka–Volterra model. AIP Conf. Proc. **1587**, 92–94 (2014)
24. D. Valério, J. Tenreiro Machado, V. Kiryakova, Some pioneers of the applications of fractional calculus. Fract. Calc. Appl. Anal. **17**(2), 552–578 (2014)
25. J. Véron, Alfred J. Lotka and the mathematics of population. J. Électron. Hist. Probab. Stat. **4**(1), 10 pp. (2008)

Chapter 7
From Duffing Equation to Bio-oscillations

Felix Sadyrbaev and Inna Samuilik

7.1 Introduction

Systems of ordinary differential equations (ODE), exhibiting irregular behavior of solutions, appear in various fields, including engineering, mechanics, hydrodynamics, and biology [1–7]. The Duffing differential equation

$$x'' + \delta x' + \alpha x + \beta x^3 = h \cos \omega t \tag{7.1}$$

had attracted much attention due to its natural appearance in mechanics. It is known that it is the "minimal" scalar ODE, which exhibits chaotic behavior of solutions. Equation (7.1) can be written also as a three-dimensional system:

$$\begin{cases} x' = y, \\ y' = -\delta y - \alpha x - \beta x^3 - h \cos z, \\ z' = \omega. \end{cases} \tag{7.2}$$

The authors of [1] have considered the similar system:

F. Sadyrbaev (✉)
Institute of Mathematics and Computer Science, Riga, Latvia
e-mail: felix@latnet.lv

I. Samuilik
Riga Technical University, Riga, Latvia
e-mail: Inna.Samuilika@rtu.lv

C. M. A. Pinto, C. M. Ionescu (eds.), *Computational and Mathematical Models in Biology*, Nonlinear Systems and Complexity 38,
https://doi.org/10.1007/978-3-031-42689-6_7

$$\begin{cases} x' = y, \\ y' = by + x - x^3 - kz, \\ z' = w(y - z) \end{cases} \tag{7.3}$$

In system, (7.3) z is treated as "the third independent dynamical variable, w is its characteristic rate, and k is the feedback coefficient." The damping term in (7.3) has an opposite sign, compared to (7.2). The authors suggest that the negative damping yields additional spiral instability. A specific electrical circuit, imitating solutions of (7.3), is proposed in [1]. The simulation and experimental results were obtained for the oscillator (7.3) and the corresponding electronic device. The authors have observed the irregular behavior of time series and were led to the conclusion that system (7.3) can be chaotic.

We provide our own analysis of system (7.3) around the specific values of parameters. Our observations are used later for the study of some systems from the theory of genetic networks.

Let $b = 1.9$, $k = 2.5$, $w \in (2.5, 3.85)$. The system (7.3) has exactly three critical points at $(-1, 0, 0)$, $(0, 0, 0)$, $(1, 0, 0)$. We have examined the characteristics of these points under the change of the variable w from 3.85 to 2.55. Starting from the value $w = 3.55$, all three critical points are non-attractive. There are periodic trajectories around the two side critical points $(-1, 0, 0)$ and $(+1, 0, 0)$. The attractivity properties of these points are weakened, as the parameter w diminishes, and finally, trajectories start to travel between the side critical points, exhibiting irregular behavior and sensitivity to initial conditions.

The evolution of two orbits in system (2.3) to the chaotic behavior is illustrated by Figs. 7.1 and 7.2.

Parameter w	Critical points $\pm(1, 0, 0)$	Critical point $(0, 0, 0)$
3.85	$-1.86, -0.045 \pm 2.03i$	$0.948, -1.45 \pm 1.4i$
3.75	$-1.8, -0.024 \pm 2.04i$	$0.954, -1.4 \pm 1.4i$
3.55	$-1.69, 0.02 \pm 2.05i$	$0.968, -1.3 \pm 1.4i$
3.25	$-1.53, 0.09 \pm 2.05i$	$0.992, -1.17 \pm 1.38i$
3.00	$-1.41, 0.057 \pm 2.054i$	$1.016, -1.06 \pm 1.35i$
2.90	$-1.36, 0.183 \pm 2.05$	$1.027, -1.01 \pm 1.34i$
2.80	$-1.32, 0.21 \pm 2.047i$	$1.039 \pm -0.97 \pm 1.32$
2.65	$-1.25, 0.252 \pm 2.039i$	$1.058 \pm -0.90 \pm 1.29$
2.55	$-1.21, 0.28 \pm 2.032i$	$1.072 \pm -0.86 \pm 1.28$

(7.4)

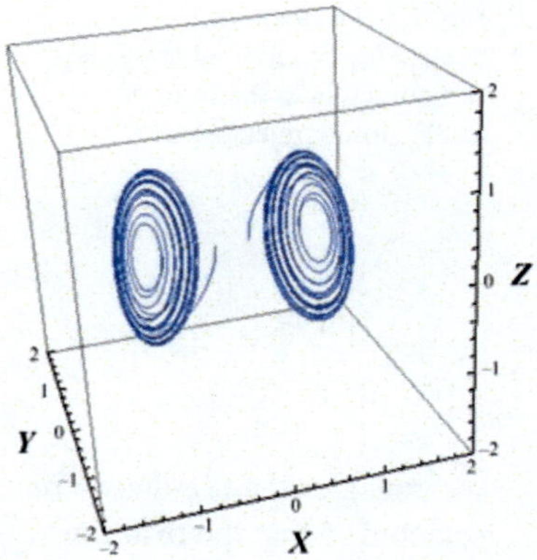

Fig. 7.1 Two orbits in the system (7.3), $b = 1.9$, $k = 2.5$, $w = 3.25$

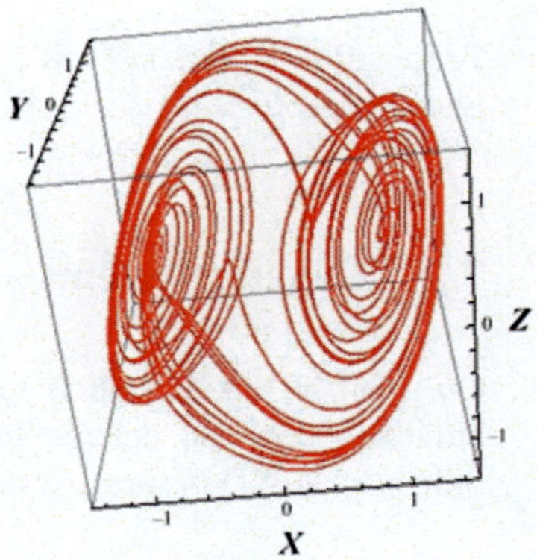

Fig. 7.2 Chaotic behavior of trajectories in the system (7.3), $b = 1.9$, $k = 2.5$, $w = 2.55$

7.2 Genomic System

Consider the system:

$$\begin{cases} x' = \frac{1}{1+e^{-\mu_1(w_{11}x+w_{12}y+w_{13}z-\theta_1)}} - x, \\ y' = \frac{1}{1+e^{-\mu_2(w_{21}x+w_{22}y+w_{23}z-\theta_2)}} - y, \\ z' = \frac{1}{1+e^{-\mu_3(w_{31}x+w_{32}y+w_{33}z-\theta_3)}} - z \end{cases} \tag{7.5}$$

which arises in the theory of gene regulatory networks (GRN). It describes the three-element network with the elements x, y, and z. They can influence each other, developing the collective response to some external threats. The system (7.5) has an invariant unit cube in the positive octant of the phase space. Therefore, the system has an attractor. The knowledge of attractors, their locations, and properties is of great theoretical and practical interest. We would like to claim that there are two stable periodic trajectories that attract other trajectories of the system. To show this,

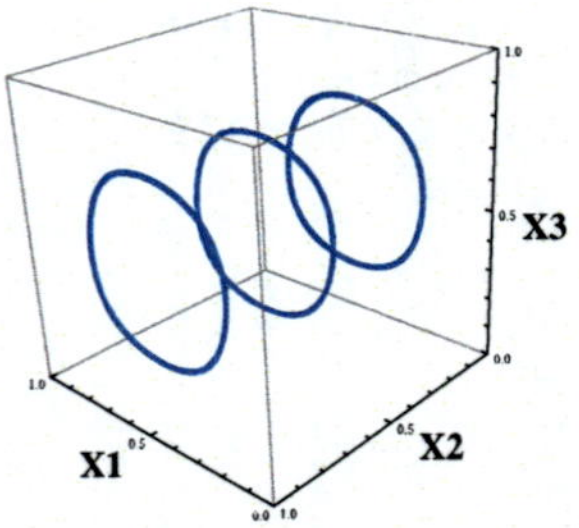

Fig. 7.3 Proposition 1 System (5) can have at least two attractors in the form of stable closed trajectories

we consider the regulatory matrix of the special form. All other parameters can be selected so that the resulting system (7.5) has exactly three periodic trajectories like in Fig. 7.3.

Proposition 1 *System (7.5) can have at least two attractors in the form of stable closed trajectories*

7.3 Genetic Systems of Low Dimensionality

Our goal in this section is to construct several examples, which show how the attracting sets can emerge in GRN systems. We start with the simplest two-dimensional (2D in short) system.

7.3.1 Two-Dimensional Systems

We consider the system:

$$\begin{cases} \dfrac{dx_1}{dt} = \dfrac{1}{1+e^{-\mu_1(w_{11}x_1+w_{12}x_2-\theta_1)}} - v_1x_1, \\ \\ \dfrac{dx_2}{dt} = \dfrac{1}{1+e^{-\mu_2(w_{21}x_1+w_{22}x_2-\theta_2)}} - v_2x_2, \end{cases} \tag{7.6}$$

where μ_i and v_i are positive. System (7.6) contains ten parameters w, μ, θ, v. Changing any of these parameters can essentially affect properties of the system and solutions.

The argument z of a sigmoidal function is transformed by the regulatory (coefficient) matrix:

$$W = \begin{pmatrix} w_{11} & w_{12} \\ w_{21} & w_{22} \end{pmatrix}. \tag{7.7}$$

This matrix describes interrelation of elements x_i of a network. If GRN are studied, then the structure of W affects properties of the system and their solutions. The nullclines are given by the equations:

$$\begin{cases} x_1 = \dfrac{1}{v_1} \dfrac{1}{1 + e^{-\mu_1 (w_{11}x_1 + w_{12}x_2 - \theta_1)}}, \\ \\ x_2 = \dfrac{1}{v_2} \dfrac{1}{1 + e^{-\mu_2 (w_{21}x_1 + w_{22}x_2 - \theta_2)}}. \end{cases} \tag{7.8}$$

The function $f(z) = \frac{1}{1+e^{-\mu z}}$ is sigmoidal and takes values in the interval (0, 1). Therefore, the first nullcline is in the strip $\{(x_1, x_2) : 0 < x_1 < \frac{1}{v_1},\ x_2 \in R\}$, and the second one is in the strip $\{(x_1, x_2) : x_1 \in R,\ 0 < x_2 < \frac{1}{v_2}\}$. Therefore, all critical points are located in the bounded box. There exists at least one critical point.

For analysis of critical points, we need the linearized system. It is

$$\begin{cases} u_1' = -v_1 u_1 + \mu_1 w_{11} g_1 u_1 + \mu_1 w_{12} g_1 u_2, \\ u_2' = -v_2 u_2 + \mu_2 w_{21} g_2 u_1 + \mu_2 w_{22} g_2 u_2, \end{cases} \tag{7.9}$$

where

$$g_1 = \frac{e^{-\mu_1 (w_{11}x_1^* + w_{12}x_2^* - \theta_1)}}{[1 + e^{-\mu_1 (w_{11}x_1^* + w_{12}x_2^* - \theta_1)}]^2}, \tag{7.10}$$

$$g_2 = \frac{e^{-\mu_2 (w_{21}x_1^* + w_{22}x_2^* - \theta_2)}}{[1 + e^{-\mu_2 (w_{21}x_1^* + w_{22}x_2^* - \theta_2)}]^2}, \tag{7.11}$$

where (x_1^*, x_2^*) is a critical point under consideration. Notice that $0 < g_i < 0.25$ for $i = 1, 2$:

$$A = \begin{vmatrix} \mu_1 w_{11} g_1 - v_1 & \mu_1 w_{12} g_1 \\ \mu_2 w_{21} g_2 & \mu_2 w_{22} g_2 - v_2 \end{vmatrix} \tag{7.12}$$

$$A - \lambda I = \begin{vmatrix} \mu_1 w_{11} g_1 - v_1 - \lambda & \mu_1 w_{12} g_1 \\ \mu_2 w_{21} g_2 & \mu_2 w_{22} g_2 - v_2 - \lambda \end{vmatrix} \tag{7.13}$$

and the characteristic equation is

$$\begin{aligned} \det|A - \lambda I| &= (\mu_1 w_{11} g_1 - v_1 - \lambda)(\mu_2 w_{22} g_2 - v_2 - \lambda) - (\mu_2 w_{21} g_2)(\mu_1 w_{12} g_1) \\ &= \mu_1 \mu_2 w_{11} w_{22} g_1 g_2 - \mu_1 w_{11} g_1 v_2 - \mu_1 w_{11} g_1 \lambda - \mu_2 w_{22} g_2 v_1 + v_1 v_2 \\ &\quad + v_1 \lambda - \mu_2 w_{22} g_2 \lambda + v_2 \lambda + \lambda^2 - \mu_1 \mu_2 w_{12} w_{21} g_1 g_2 = \lambda^2 + (v_1 + v_2 \\ &\quad - \mu_1 w_{11} g_1 - \mu_2 w_{22} g_2)\lambda + \mu_1 \mu_2 w_{11} w_{22} g_1 g_2 - \mu_1 w_{11} g_1 v_2 \\ &\quad - \mu_2 w_{22} g_2 v_1 - \mu_1 \mu_2 w_{12} w_{21} g_1 g_2 + v_1 v_2 = 0. \end{aligned} \tag{7.14}$$

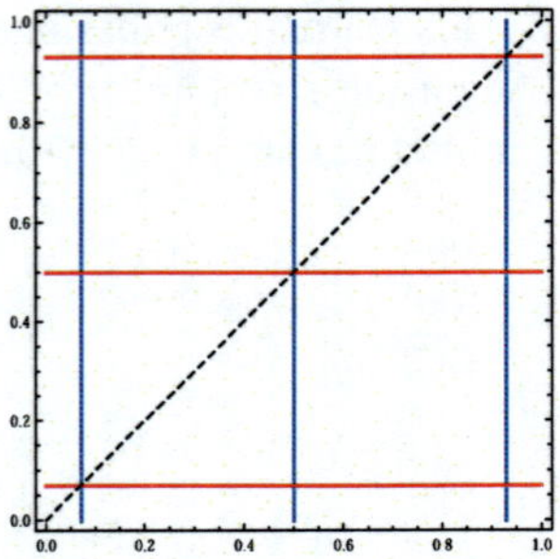

Fig. 7.4 Self-activation case, matrix W in (7.18)

To simplify, we can write the characteristic equation as

$$\lambda^2 + B\lambda + C = 0, \tag{7.15}$$

$$B = v_1 + v_2 - \mu_1 w_{11} g_1 - \mu_2 w_{22} g_2, \tag{7.16}$$

$$C = \mu_1\mu_2 w_{11} w_{22} g_1 g_2 - \mu_1 w_{11} g_1 v_2 - \mu_2 w_{22} g_2 v_1 - \mu_1\mu_2 w_{12} w_{21} g_1 g_2 + v_1 v_2. \tag{7.17}$$

Simple analysis of the quadratic equation shows that a critical point is the stable node if $B > 0$ and $B^2/4 - C > 0$. Similarly, a critical point is a stable focus if $B > 0$ and $B^2/4 - C < 0$.

Proposition 2 *The system (7.6) can have attractors in the form of stable critical points and stable periodic solutions.*

Proof by constructing examples.

Self-Activation Consider the system (7.6) with the regulatory matrix:

$$W = \begin{pmatrix} 1.5 & 0 \\ 0 & 1.5 \end{pmatrix}, \tag{7.18}$$

where the elements on the main diagonal are positive, $\mu_1 = \mu_2 = 4$, $\theta_1 = \theta_2 = 2$, $v_1 = v_2 = 1$. The attractor consists of four stable nodes at the corners (Fig. 7.4).

Cross-Activation Consider the system (7.6) with the regulatory matrix:

$$W = \begin{pmatrix} 1 & 1 \\ 1 & 1 \end{pmatrix}, \tag{7.19}$$

where $\mu_1 = \mu_2 = 4$, $\theta_1 = \theta_2 = 1$, $v_1 = v_2 = 1$. The attractor is a pair of stable nodes (Fig. 7.5).

Inhibition The inhibition appears as the negative influence on genes. The matrix W in this case contains negative entries. The appropriate references are [8, 9].

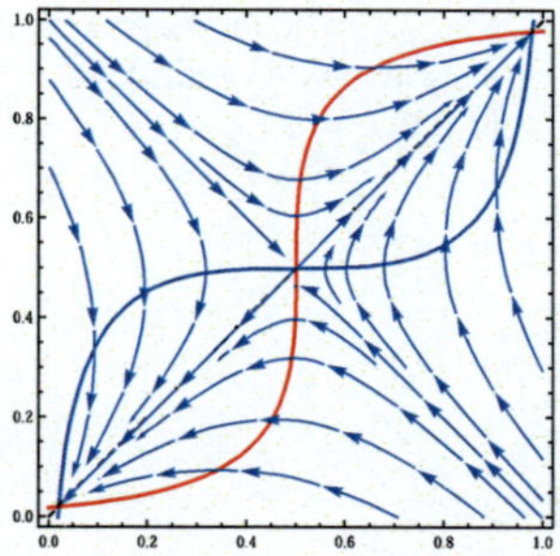

Fig. 7.5 Cross-activation case, matrix W in (7.19)

Oscillation Consider the system (7.6) with the regulatory matrix:

$$W = \begin{pmatrix} k & 1 \\ -1 & k \end{pmatrix}. \tag{7.20}$$

For $k = 0$, the coefficients in the characteristic Eq. (7.15) are $B = v_1 + v_2$, $C = 4\mu_1\mu_2 g_1 g_2 + v_1 v_2$. Therefore, the roots

$$\lambda_{1,2} = -\frac{B}{2} \pm \sqrt{\frac{B^2}{4} - C} = -\frac{B}{2} \pm \sqrt{(v_1 - v_2)^2 - 4\mu_1\mu_2 g_1 g_2}.$$

If $v_1 = v_2$, the characteristic numbers $\lambda_{1,2}$ are complex conjugate with the negative real part. The unique critical point is a stable focus.

For $k \neq 0$, $B = v_1 + v_2 - k(\mu_1 g_1 + \mu_2 g_2)$, $C = k^2\mu_1\mu_2 g_1 g_2 - k(\mu_1 g_1 v_2 + \mu_2 g_2 v_1) + 4\mu_1\mu_2 g_1 g_2 + v_1 v_2$.

It follows from (7.8) that

$$\begin{aligned} 1 + e^{-\mu_1 (w_{11}x_1 + w_{12}x_2 - \theta_1)} &= \frac{1}{v_1 x_1}, \\ 1 + e^{-\mu_2 (w_{21}x_1 + w_{22}x_2 - \theta_2)} &= \frac{1}{v_2 x_2} \end{aligned} \tag{7.21}$$

and, therefore, using (7.10) and (7.11)

$$\begin{aligned} g_1 &= \frac{\frac{1}{v_1 x_1} - 1}{\frac{1}{v_1^2 x_1^2}} = (1 - v_1 x_1) v_1 x_1 \\ g_2 &= \frac{\frac{1}{v_2 x_2} - 1}{\frac{1}{v_2^2 x_2^2}} = (1 - v_2 x_2) v_2 x_2. \end{aligned} \tag{7.22}$$

Suppose $v_1 = v_2 = v$. Denote $Y_1 = \mu_1(1 - v x_1) v x_1$, $Y_2 = \mu_2(1 - v x_2) v x_2$. Then

$$B = 2v - k(Y_1 + Y_2), \quad C = k^2 Y_1 Y_2 - kv(Y_1 + Y_2) + 4Y_1 Y_2 + v^2. \tag{7.23}$$

Fig. 7.6 The limit cycle (red) in system (7.6) with W in (7.20), $k = 1$

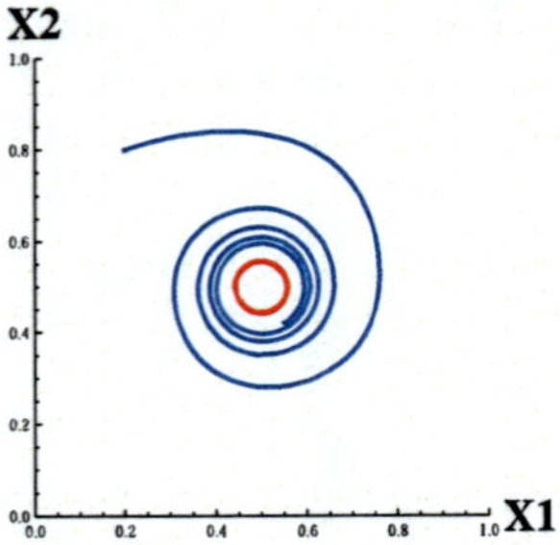

Fig. 7.7 The nullclines in system (7.6) with W in (7.20) meet for $k = 2.63$

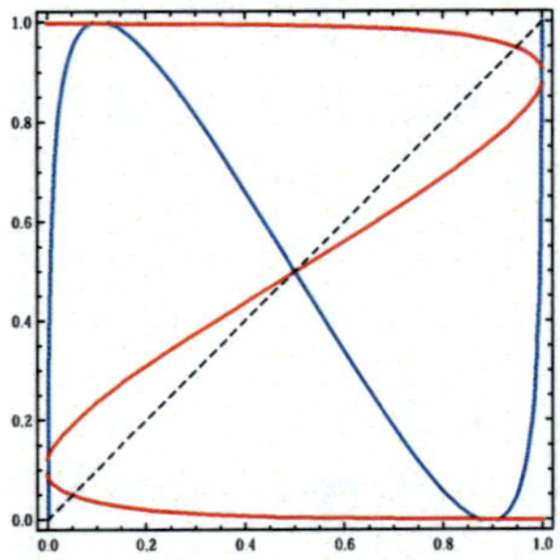

One has performing calculations that

$$B^2 - 4C = k^2(Y_1 - Y_2)^2 - 16Y_1Y_2. \tag{7.24}$$

Suppose now that:

1. Nullclines

$$\begin{cases} x_1 = \dfrac{1}{v}\dfrac{1}{1 + e^{-\mu_1(kx_1 + 2x_2 - \theta_1)}}, \\ x_2 = \dfrac{1}{v}\dfrac{1}{1 + e^{-\mu_2(-2x_1 + kx_2 - \theta_2)}}. \end{cases} \tag{7.25}$$

   intersect only once at the point $(\bar{x}_1, (\bar{x}_2)$, which depends on k.
2. $B = 2v - k(Y_1 + Y_2) < 0$.
3. $B^2 - 4C = k^2(Y_1 - Y_2)^2 - 16Y_1Y_2 < 0$,

where $Y_i = \mu_i(1 - v\bar{x}_i)v\bar{x}_i$, $i = 1, 2$.

If these conditions are satisfied, the single critical point $(\bar{x}_1, \bar{x}_2)$ is unstable focus, and it is surrounded by a stable periodic solution (limit cycle). The limit cycle appears as the result of Andronov–Hopf bifurcation (Fig. 7.6).

As an illustration, consider the system with the matrix: W as in (7.20), $k = 1$, $\mu_1 = \mu_2 = 4$, $\theta_1 = 1$, $\theta_2 = 0$, $v_1 = v_2 = 1$. The limit cycle exists, while $k < 2.63$. After that, for greater k, the limit cycle does not exist. Several critical points appear instead, nodes and saddle points (Fig. 7.7).

7.4 Three-Dimensional Systems

Consider the three-element GRN together with the modeling system:

$$\begin{cases} x_1' = \frac{1}{1+e^{-\mu_1(w_{11}x_1+w_{12}x_2+w_{13}x_3-\theta_1)}} - v_1x_1, \\ x_2' = \frac{1}{1+e^{-\mu_2(w_{21}x_1+w_{22}x_2+w_{23}x_3-\theta_2)}} - v_2x_2, \\ x_3' = \frac{1}{1+e^{-\mu_3(w_{31}x_1+w_{32}x_2+w_{33}x_3-\theta_3)}} - v_3x_3, \end{cases} \tag{7.26}$$

where the regulatory matrix is

$$W = \begin{pmatrix} w_{11} & w_{12} & w_{13} \\ w_{21} & w_{22} & w_{23} \\ w_{31} & w_{32} & w_{33} \end{pmatrix}. \tag{7.27}$$

Proposition 3 *The system (7.26) can have attractors in the form of stable critical points and stable periodic solutions.*

Proof by Constructing Examples The first one easily can be proved if the matrix (7.27) is filled by units, $w_{ij} = 1$ for all i, j, $v_i = 1$, $\mu_i = 1$ for $i = 1, 2, 3$:

$$W = \begin{pmatrix} w_{11} & w_{12} & w_{13} \\ w_{21} & w_{22} & w_{23} \\ w_{31} & w_{32} & w_{33} \end{pmatrix}. \tag{7.28}$$

This is true also for the diagonal matrix:

$$W = \begin{pmatrix} w_{11} & 0 & 0 \\ 0 & w_{22} & 0 \\ 0 & 0 & w_{33} \end{pmatrix}, \tag{7.29}$$

where the elements w_{ii} are such that the equations

$$v_i x_i = \frac{1}{1 + e^{-\mu_i(w_{11}x_1+w_{12}x_2+w_{13}x_3-\theta_1)}}$$

have three zeros, $i = 1, 2, 3$. The system of nullclines is then a group of nine planes, which is the 3D version of Fig. 7.4.

The second part of the assertion of proposition 3 follows from the next one.

Proposition 4 *The system (7.26) can have two limit cycles.*

Proof by Example Consider system (7.26) with the matrix:

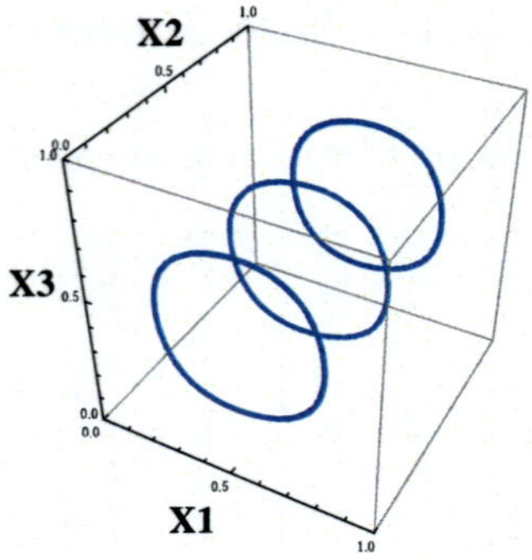

Fig. 7.8 Three periodic solutions in (7.26) with the matrix (7.30), $k = 1.2$

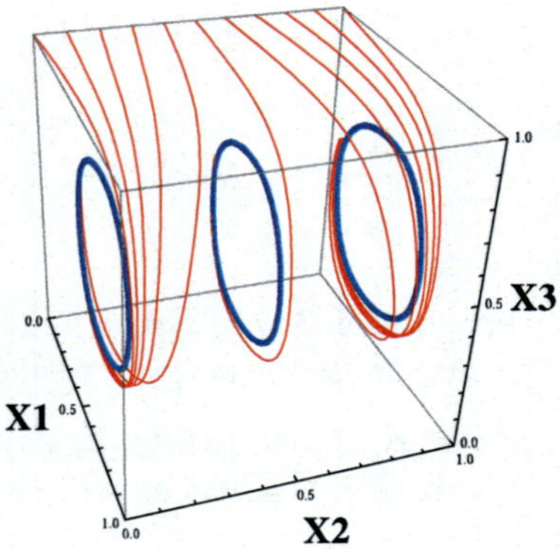

Fig. 7.9 The same periodic solutions with tending to them solutions

$$W = \begin{pmatrix} k & 0 & 1 \\ 0 & 1.5 & 0 \\ -1 & 0 & k \end{pmatrix}, \tag{7.30}$$

where k is the bifurcation parameter. For $k = 1.2$, the system has three periodic solutions, which locate in the three components (parallel planes) of the second nullcline, given by (Figs. 7.8 and 7.9)

$$x_2 = \frac{1}{1 + e^{-4(1.5x_2 - 0.75)}}. \tag{7.31}$$

Stable Periodic Solutions Consider the system (7.26) with the regulatory matrix:

$$W = \begin{pmatrix} k & 0 & 2 \\ 0 & w_{22} & 0 \\ -2 & 0 & k \end{pmatrix}, \tag{7.32}$$

where w_{22} is the same (so the Eq. (7.31) has three roots $r_1 < r_2 < r_3$) and $k = 1.2$.

For the matrix

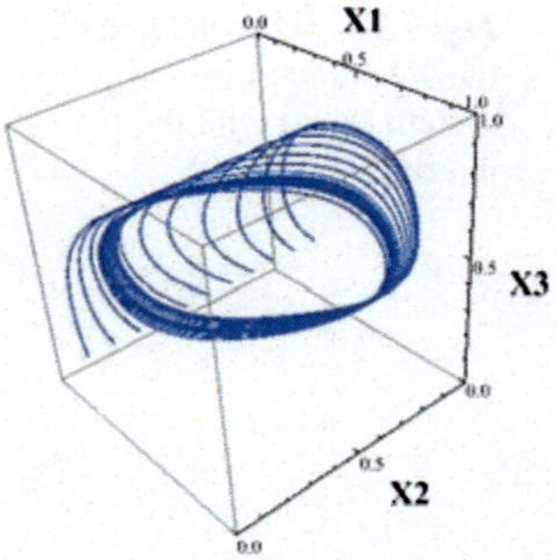

Fig. 7.10 3D attractor in (7.26) with the matrix (7.39), $k = 1.2$

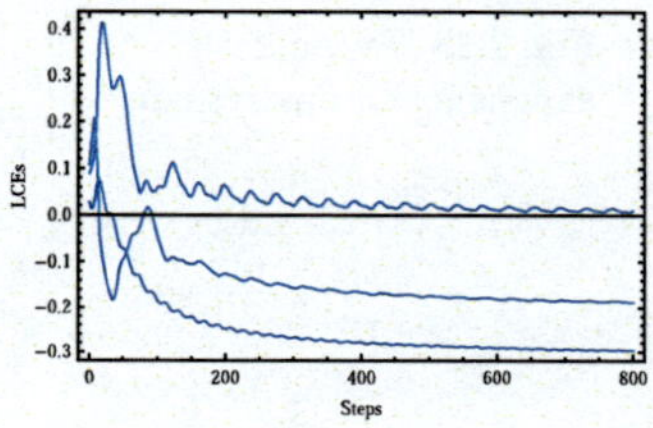

Fig. 7.11 Lyapunov exponents for this system

$$W = \begin{pmatrix} 1.2 & 0 & 1 \\ 0 & 1.5 & 1.5 \\ -1 & 0 & 1.2 \end{pmatrix} \tag{7.33}$$

with the same values of other parameters, one obtained quasi-periodic solutions, exhibiting chaotic behavior as indicated by the respective Lyapunov exponents (Figs. 7.10 and 7.11).

7.5 Four-Dimensional System

We consider the four-dimensional system:

$$\begin{cases} x_1' = \frac{1}{1+e^{-\mu_1(w_{11}x_1+w_{12}x_2+w_{13}x_3+w_{14}x_4-\theta_1)}} - v_1x_1, \\ x_2' = \frac{1}{1+e^{-\mu_2(w_{21}x_1+w_{22}x_2+w_{23}x_3+w_{24}x_4-\theta_2)}} - v_2x_2, \\ x_3' = \frac{1}{1+e^{-\mu_3(w_{31}x_1+w_{32}x_2+w_{33}x_3+w_{34}x_4-\theta_3)}} - v_3x_3, \\ x_4' = \frac{1}{1+e^{-\mu_4(w_{41}x_1+w_{42}x_2+w_{43}x_3+w_{44}x_4-\theta_4)}} - v_4x_4, \end{cases} \tag{7.34}$$

The coefficient matrix

$$W = \begin{pmatrix} w_{11} & w_{12} & w_{13} & w_{14} \\ w_{21} & w_{22} & w_{23} & w_{24} \\ w_{31} & w_{32} & w_{33} & w_{34} \\ w_{41} & w_{42} & w_{43} & w_{44} \end{pmatrix}. \tag{7.35}$$

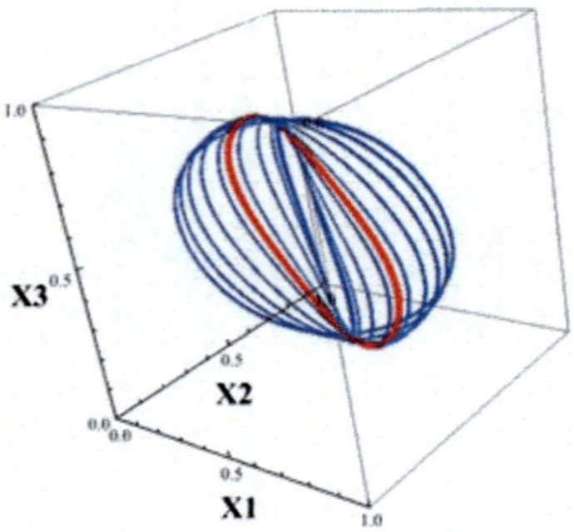

Fig. 7.12 3D projection of the 4D attractor in system (7.34) with the matrix (7.36)

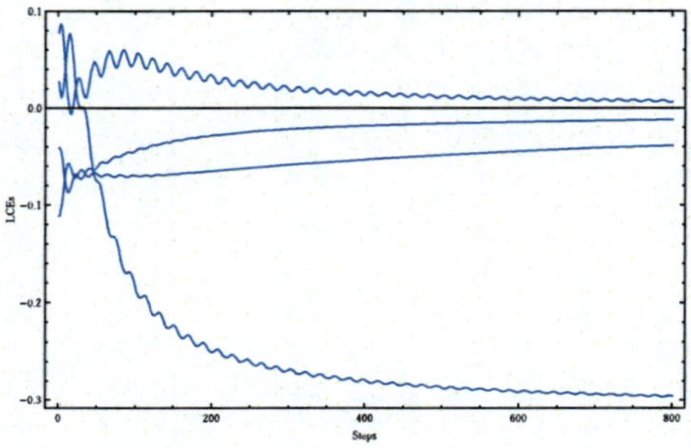

Fig. 7.13 Lyapunov exponents for this system

We construct examples of 4D attractors, continuing arguing as before. First, we take a two-dimensional system with the regulatory matrix, say, $W1$, and then combine it with another 2D system with another matrix $W2$, thus obtaining 4D system. Some of them exhibit chaotic behavior.

Let $\mu_i = 4$, $v_i = 1$ for $i = 1, 2, 3, 4$. The first regulatory matrix is

$$W = \begin{pmatrix} 1.2 & 0 & 0 & 1 \\ 0 & 1 & 1 & 0 \\ 0 & -1 & 1 & 0 \\ -1 & 0 & 0 & 1.2 \end{pmatrix}. \tag{7.36}$$

Thus, the oscillatory 2D system is combined with the cross-activating one. The resulting behavior of trajectories and the corresponding Lyapunov curves are visualized (Figs. 7.12 and 7.13).

Proposition 5 *Solutions of the system (7.34) with matrices of the type (7.36) can be sensitive dependent to initial conditions.*

Let the regulatory matrix be:

$$W = \begin{pmatrix} 1.2 & 0 & 0 & 1 \\ 0 & 1.2 & 1 & 0 \\ 0 & -1 & 1.2 & 0 \\ -1 & 0 & 0 & 1.2 \end{pmatrix}. \tag{7.37}$$

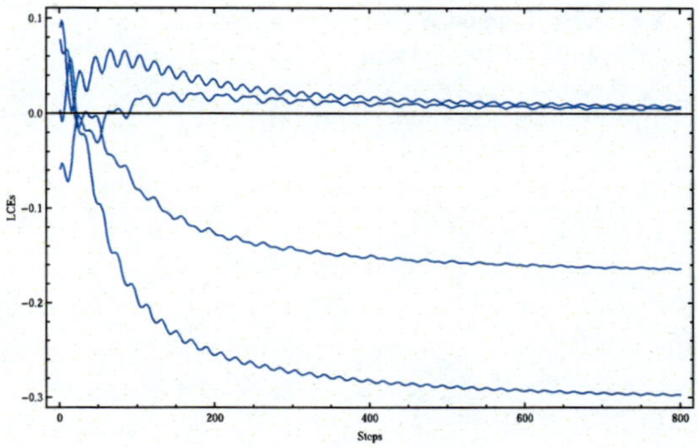

Fig. 7.14 Lyapunov exponents for 4D system with the regulatory matrix (7.37)

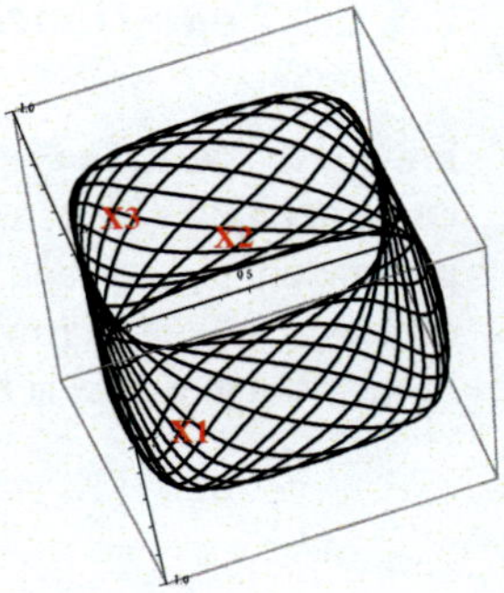

Fig. 7.15 3D projection of the 4D attractor in system (7.34) with the matrix (7.38)

Here, two identical oscillatory 2D systems are combined to produce the 4D system. The respective Lyapunov curves are depicted in (Fig. 7.14).

Proposition 6 *Solutions of the system (7.34) with matrices of the type (7.37) can have two positive Lyapunov exponents.*

Remark Such behavior is called sometimes *hyperchaotic* [4].

Let the regulatory matrix be:

$$W = \begin{pmatrix} 1 & 0 & 0 & 1 \\ 0 & 1.76 & 1 & 0 \\ 0 & -1 & 1.76 & 0 \\ -1 & 0 & 0 & 1 \end{pmatrix}. \tag{7.38}$$

Here, two oscillatory 2D systems are combined to produce the 4D system. The periods $T_1 = 6.25$ and $T_2 = 12.5$ of the limit cycles in both 2D systems relate as $T_2/T_1 = 2$. Equalities are approximate. The behavior of trajectories and the corresponding Lyapunov curves are illustrated by (Figs. 7.15 and 7.16).

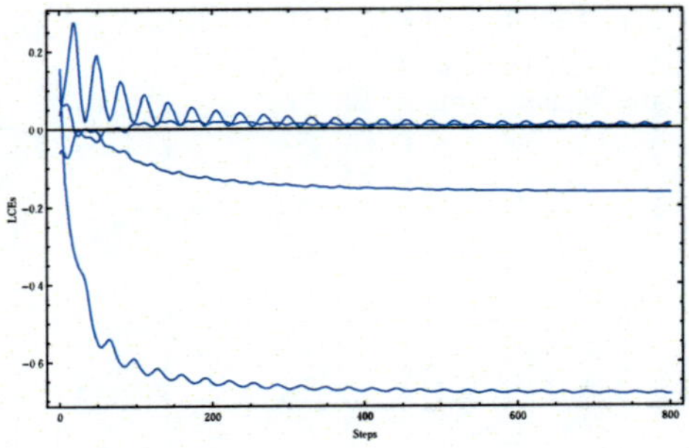

Fig. 7.16 Lyapunov exponents for this system

7.5.1 Four-Dimensional System: 3 + 1

Previously, we have constructed 4D systems by combining two 2D systems. We will construct now 4D system from one 3D system and a single equation. Suppose that parameters $\mu_i = 4$ and $v_i = 1$ are as above. Thetas θ_i are taken as the half-sum of elements in the i-th row of the corresponding regulatory matrix. This will keep one of the critical points at the central location (0.5, 0.5, 0.5). Let the 3D matrix W be:

$$W = \begin{pmatrix} 1.5 & 0 & 2 \\ 0 & 1.5 & 0 \\ -2 & 0 & 1.5 \end{pmatrix}, \tag{7.39}$$

Let the 4D system be:

$$W = \begin{pmatrix} 1.5 & 0 & 2 & 0 \\ 0 & 1.5 & 0 & 0 \\ -2 & 0 & 1.5 & 0 \\ 0 & 0 & 0 & 1.2 \end{pmatrix}. \tag{7.40}$$

If other parameters are the same, as before, this system has the 3D limit cycle, associated with the matrix (7.39). On the other hand, the x_4-nullcline is again a union of three 4D hyperplanes.

7.6 Artificial Neural Networks

An artificial neural network (ANN) is a computational architecture for processing complex data using multiple interconnected processors and computational paths. Artificial neural networks, created by analogy with the human brain, can train and analyze large and complex data sets that are extremely difficult to process using more linear algorithms.

In 1943, American neurophysiologist and cybernetician Warren Sturgis McCulloch and American logician Walter Harry Pitts modeled a neuron as a switch that receives input from other neurons and, depending on the total weighted input, is either activated or remains inactive [10, 11].

In 1958, ANN was created by psychologist Frank Rosenblatt. It was called perceptron and was designed to simulate the activity of the human brain in processing visual data and in learning to recognize objects. Subsequently, similar artificial neural networks were developed to study the process of cognition. Over time, it became clear that in addition to analyzing the activity of the human brain, it can perform other very useful functions. Because of their ability to pattern match and learn, these networks have been used to analyze many problems that are extremely difficult or impossible to solve using traditional computational or statistical methods [12].

In the late 1980s, artificial neural networks began to be actively used for a variety of purposes [11]. For example, they are used in the economy to increase the productivity of discount calculation, targeted marketing, and credit evaluation.

The principle of operation of an artificial neural network is to form connections between many different processing elements, each of which serves as an analog of one neuron in the brain of a biological being. Neurons can be physically reproduced or simulated using a digital computer. Each neuron receives a set of input signals and then, taking into account the internal system of weight coefficients, generates one output signal, which, as a rule, serves as input for another neuron. Neurons are closely interconnected with each other and are organized into several different levels. The input layer receives the input data, and the output layer generates the final result. Typically, there are one or more hidden levels between these two levels. In such a structure, it is impossible to predict or know exactly how data is transmitted.

Definition 1 An artificial neural network is a mathematical model that tries to simulate the structure and functionalities of biological neural networks. A basic building block of every artificial neural network is an artificial neuron, that is, a simple mathematical model (function) [13].

Neural networks consist of neurons interconnected, so the neuron is the main part of the neural network. The neurons only do two things: multiply the inputs by the weights and sum them up and add the bias, and the second action is the activation.

The input data is the data that the neuron receives from previous neurons or the user. Weights are assigned to each input of the neuron; initially, they are assigned random numbers. When training a neural network, the value of neurons and biases changes. The weights are multiplied by the input data that is fed to the neuron. The biases are assigned to each neuron; just like the initial bias weights, these are random numbers. Bias makes it easier and faster to train a neural network [12]. Typically, this transformation involves the use of a sigmoid, hyperbolic tangent, or other nonlinear functions [14].

One example of which is

$$x_i' = \tanh \sum_{j=1}^{N} a_j x_j - b_i x_i, \tag{7.41}$$

where N is the number of neurons, each of which represents a dimension of the system [15]. The hyperbolic tangent is a sigmoid function.

Consider the system:

$$\begin{cases} x_1' = \tanh(x_1 + x_2 + x_3 + x_4) - bx_1, \\ x_2' = \tanh(x_1 + x_2 + x_3 + x_4) - bx_2, \\ x_3' = \tanh(x_1 + x_2 + x_3 + x_4) - bx_3, \\ x_4' = \tanh(x_1 + x_2 + x_3 + x_4) - bx_4 \end{cases} \tag{7.42}$$

with the regulatory matrix

$$W = \begin{pmatrix} 0 & -1 & 0 & 1 \\ 1 & 0 & 0 & 1 \\ 1 & 1 & 0 & -1 \\ 0 & -1 & 1 & 0 \end{pmatrix} \tag{7.43}$$

and $b = 0.03$.

The initial conditions are

$$x_1(0) = 1.2;\ x_2(0) = 0.4;\ x_3(0) = 1.2;\ x_4(0) = -1.$$

The graph of the regulatory matrix (7.43) is presented in Fig. 7.17.

The attractor is shown in Figs. 7.18 and 7.19 and the solutions in Figs. 7.20 and 7.21.

For specific parameters, the solution of the system has a chaotic trajectory as shown in Figs. 7.22 and 7.23. The minimal dissipative artificial neural network that exhibits chaos has $N = 4$ and is given by system (7.42) with the regulatory matrix (7.43) and $b = 0.043$ and an attractor as shown in Figs. 7.24 and 7.25 [15].

The dynamics of Lyapunov exponents is shown in Fig. 7.26.

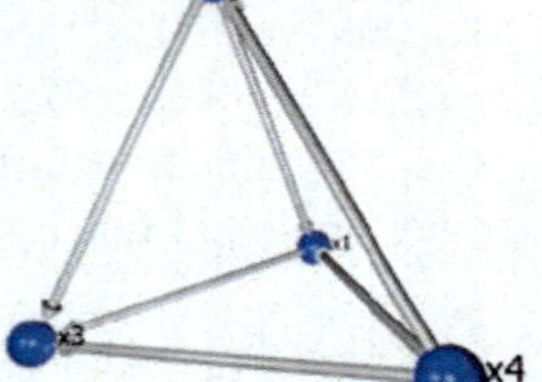

Fig. 7.17 The graph, corresponding to the system (7.42), with the regulatory matrix (7.43)

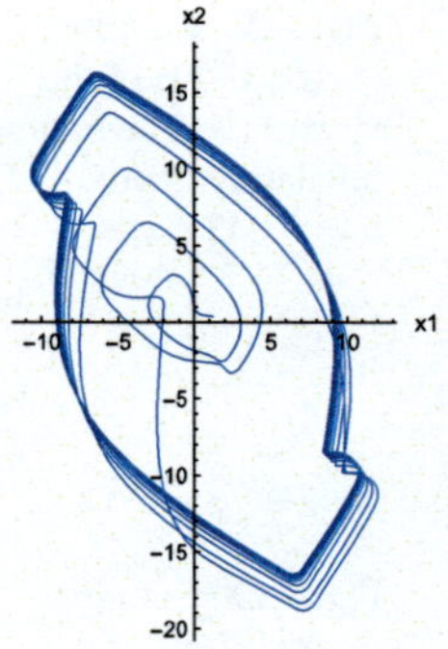

Fig. 7.18 The projection of the attractor on 2D subspace $(x_1(t), x_2(t))$

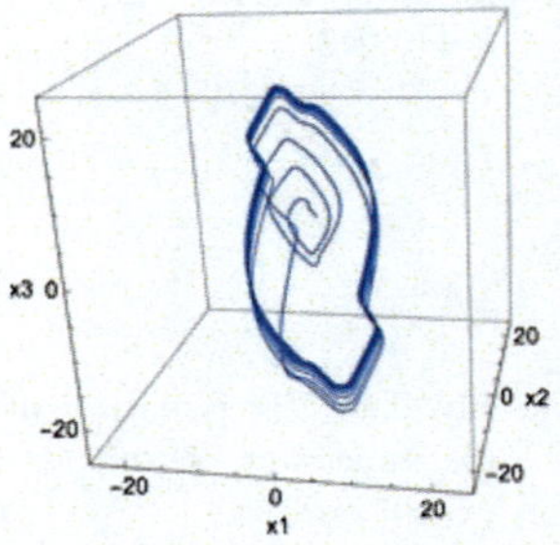

Fig. 7.19 The projection of the attractor on 3D subspace $(x_1(t), x_2(t), x_3(t))$

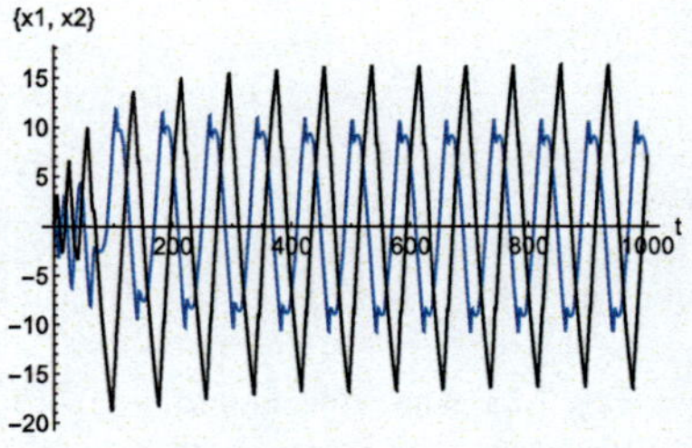

Fig. 7.20 Solutions $(x_1(t), x_2(t))$ of the system (7.42) with the regulatory matrix (7.43)

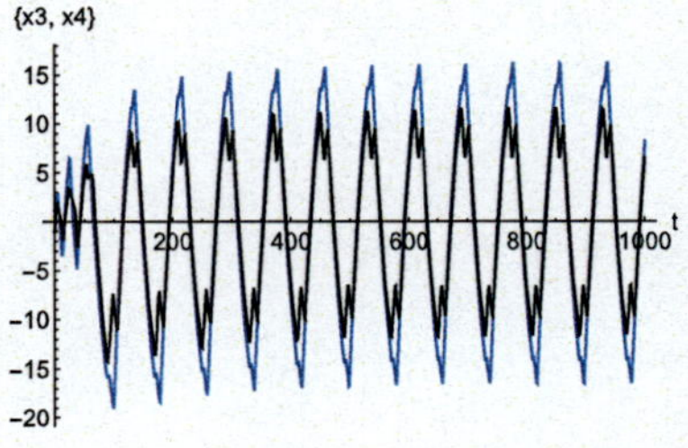

Fig. 7.21 Solutions $(x_3(t), x_4(t))$ of the system (7.42) with the regulatory matrix (7.43)

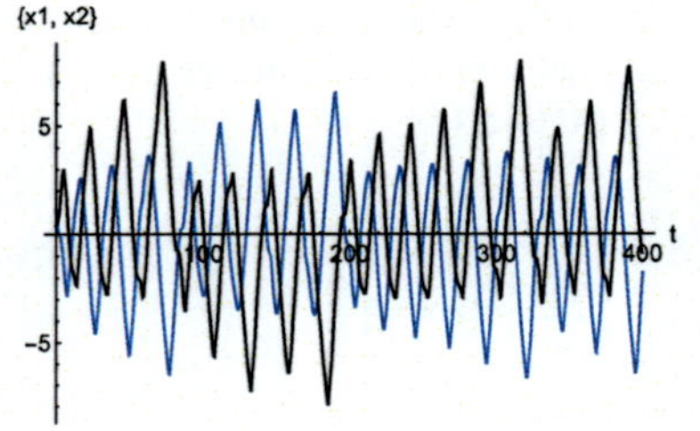

Fig. 7.22 Solutions $(x_1(t), x_2(t))$ of the system (7.42) with the regulatory matrix (7.43), $b = 0.043$

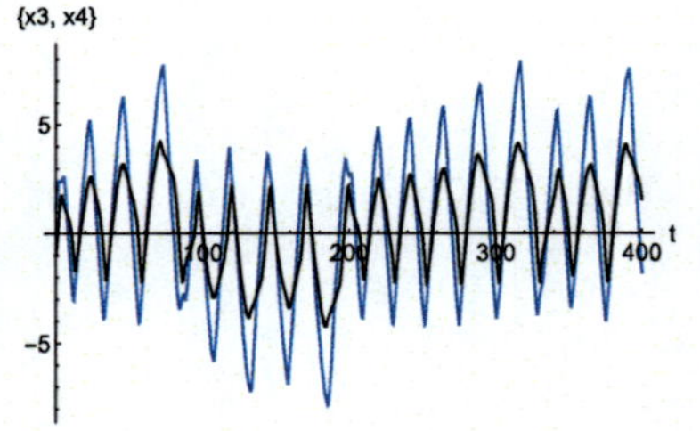

Fig. 7.23 Solutions $(x_3(t), x_4(t))$ of the system (7.42) with the regulatory matrix (7.43), $b = 0.043$

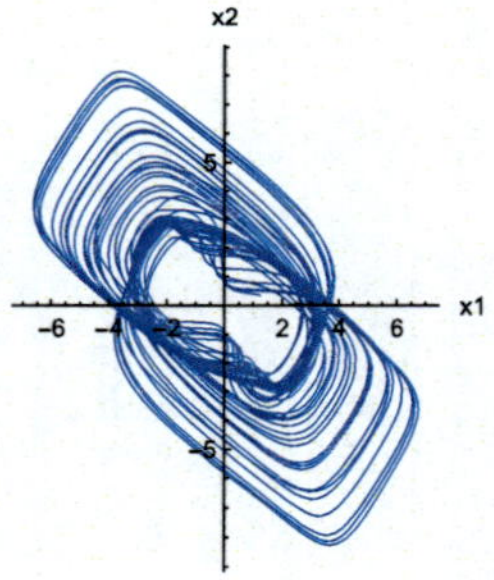

Fig. 7.24 The projection of the attractor on 2D subspace $(x_1(t), x_2(t))$

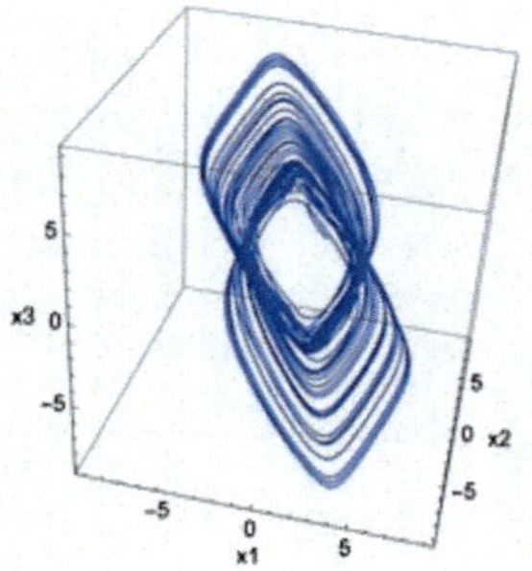

Fig. 7.25 The projection of the attractor on 3D subspace $(x_1(t), x_2(t), x_3(t))$

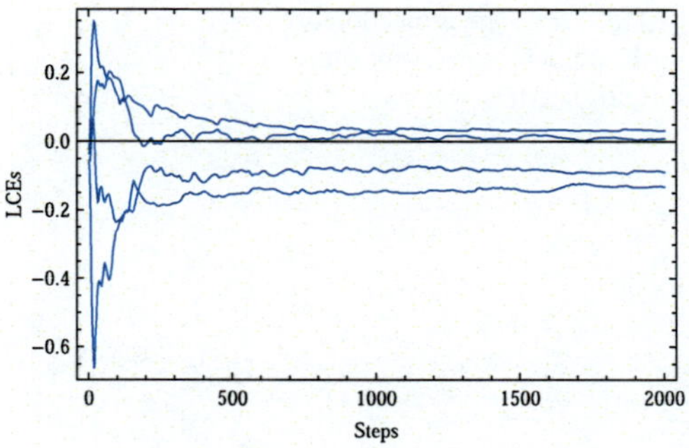

Fig. 7.26 $LE_1 = 0.03, LE_2 = 0.01, LE_3 = -0.09, LE_4 = -0.13$

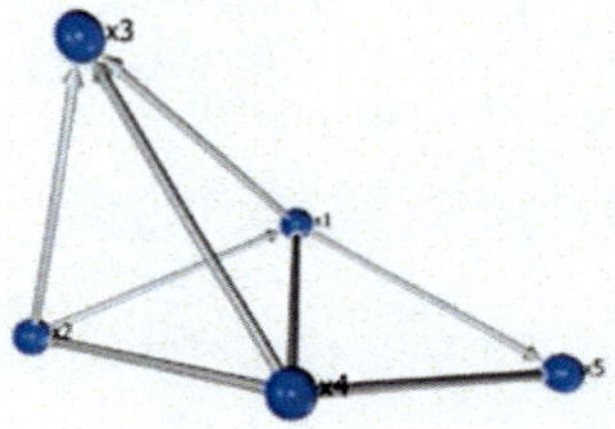

Fig. 7.27 The graph, corresponding to the system (7.44)

7.7 Five-Dimensional (5D) Systems

Consider the system:

$$\begin{cases} x_1' = \tanh(x_4 - x_2) - bx_1, \\ x_2' = \tanh(x_1 + x_4) - bx_2, \\ x_3' = \tanh(x_1 + x_2 - x_4) - bx_3, \\ x_4' = \tanh(x_3 - x_2) - bx_4, \\ x_5' = \tanh(x_1 + x_4 - x_5) - bx_5 \end{cases} \tag{7.44}$$

and $b = 0.043$.

The initial conditions are

$$x_1(0) = 1.2;\ x_2(0) = 0.4;\ x_3(0) = 1.2;\ x_4(0) = -1;\ x_5(0) = -1.$$

The graph of the system (7.44) is depicted in Fig. 7.27. This system has an attractor as shown in Figs. 7.28 and 7.29. The irregular behavior of the three solutions can be seen in Figs. 7.30 and 7.31.

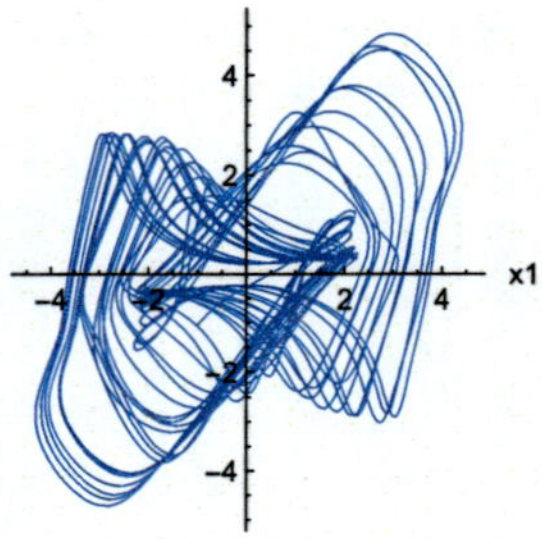

Fig. 7.28 The projection of the attractor on 2D subspace $(x_4(t), x_5(t))$

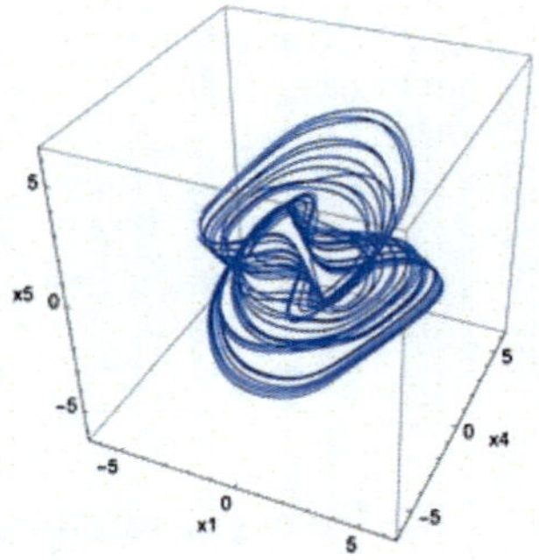

Fig. 7.29 The projection of the attractor on 3D subspace $(x_1(t), x_4(t), x_5(t))$

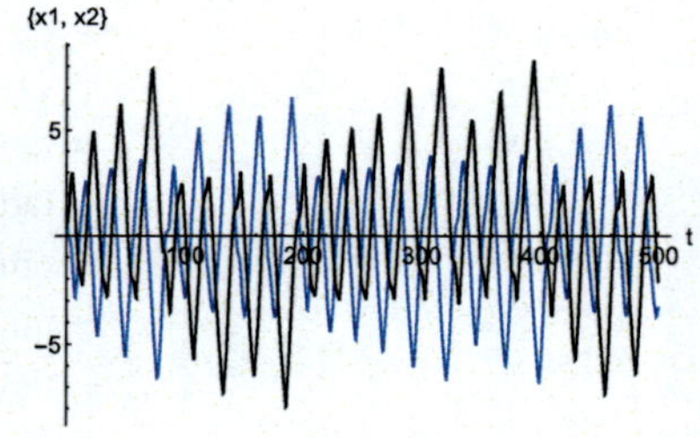

Fig. 7.30 The graphs of solutions $(x_1(t), x_2(t))$ of the system (7.44)

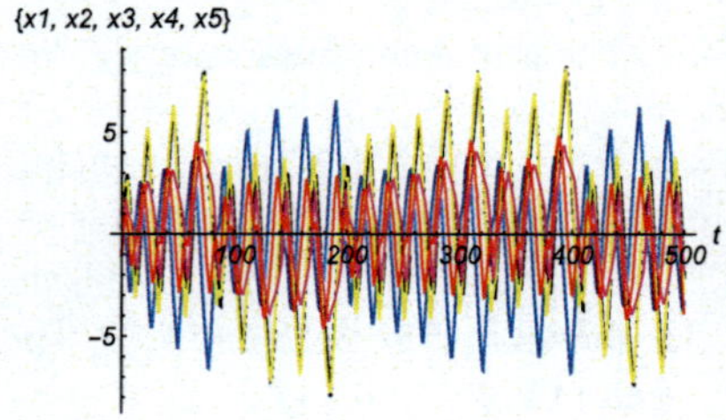

Fig. 7.31 The graphs of solutions $(x_1(t), x_2(t), x_3(t), x_4(t), x_5(t))$ of the system (7.44)

7.8 Six-Dimensional (6D) Systems

Consider the system:

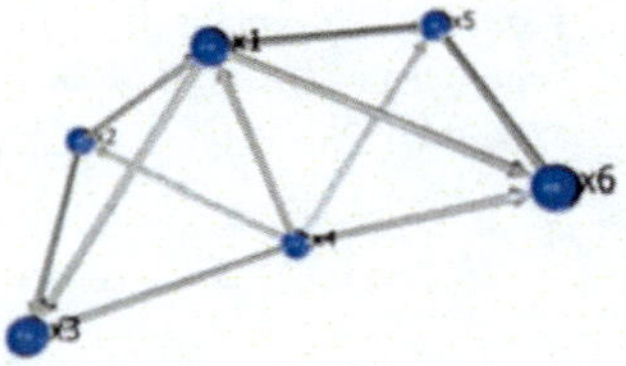

Fig. 7.32 The graph, corresponding to the case of the regulatory matrix (7.45)

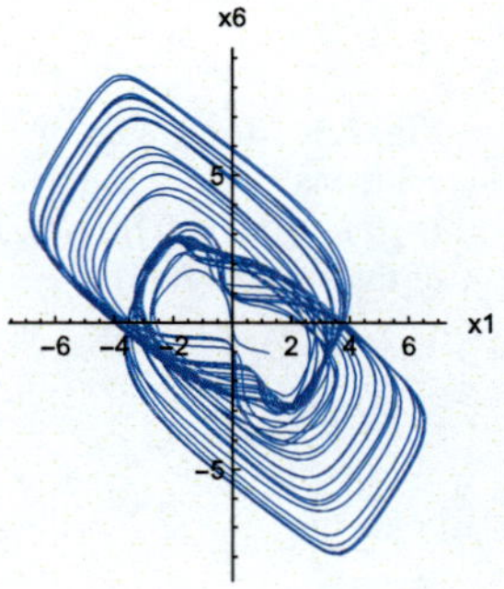

Fig. 7.33 The projection of the attractor on 2D subspace $(x_1(t), x_6(t))$

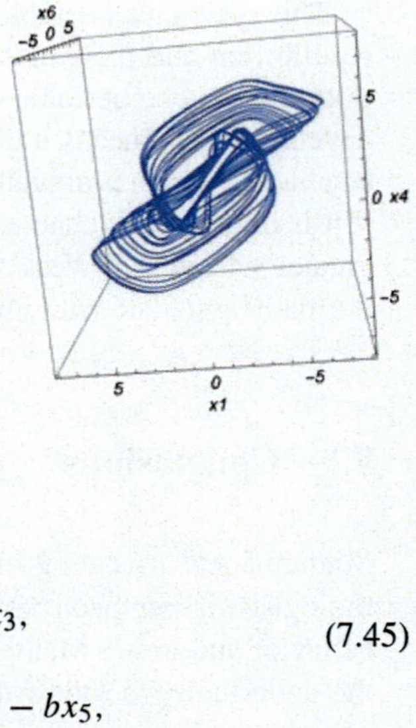

Fig. 7.34 The projection of the attractor on 3D subspace $(x_1(t), x_4(t), x_6(t))$

$$\begin{cases} x_1' = \tanh(x_4 - x_2) - bx_1, \\ x_2' = \tanh(x_1 + x_4) - bx_2, \\ x_3' = \tanh(x_1 + x_2 - x_4) - bx_3, \\ x_4' = \tanh(x_3 - x_2) - bx_4, \\ x_5' = \tanh(x_1 + x_4 - x_5 + x_6) - bx_5, \\ x_6' = \tanh(x_1 + x_4) - bx_6 \end{cases} \tag{7.45}$$

and $b = 0.043$.

The initial conditions are

$$x_1(0) = 1.2;\ x_2(0) = 0.4;\ x_3(0) = 1.2;\ x_4(0) = -1;\ x_5(0) = -1;\ x_6(0) = -1.$$

The graph of the system (7.45) is depicted in Fig. 7.32.

This system has an attractor as shown in Figs. 7.33 and 7.34. The irregular behavior of solutions can be seen in Figs. 7.35 and 7.36.

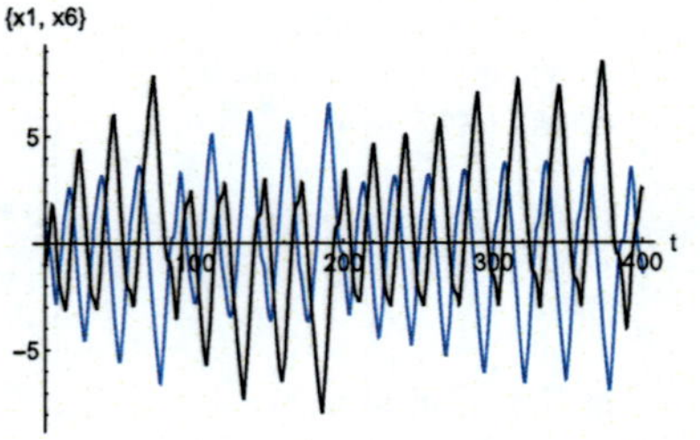

Fig. 7.35 The graphs of solutions $(x_1(t), x_6(t))$ of the system (7.45)

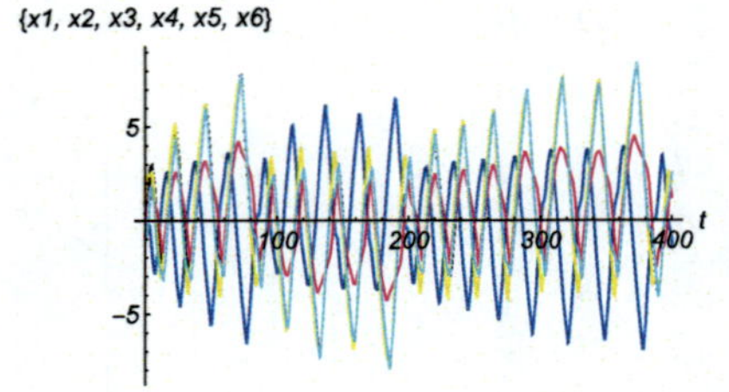

Fig. 7.36 The graphs of solutions $(x_1(t), x_2(t), x_3(t), x_4(t), x_5(t), x_6(t))$ of the system (7.45)

This system can exhibit three types of dynamics. Solutions can approach a static equilibrium and thereafter remain forever. It means that the human brain is dead. Solutions can be periodic or quasi-periodic. In this case, the human brain does not develop, which means it does not have the ability to think creatively. Solutions can be chaotic, which is arguably the most healthy state for a natural network, especially if it is only weakly chaotic so that it retains some memory but can explore a vastly greater state space. Weakly chaotic networks exhibit the complex behavior that we normally associate with intelligent living systems [16].

7.9 Conclusions

Mathematical modeling of experimental data (time series), obtained by research biologists, is perspective. It allows strictly formulating observed cases of regular behavior and order. Mathematical modeling, which uses dynamical systems, gives the opportunity to follow the evolution of underlying biological networks. Systems of ordinary differential equations of quasi-linear form provide convenient apparatus for description of gene regulatory networks. These systems contain a great number of parameters and they are not very particular ones. The linear part describes the natural decay of networks without interaction. The nonlinear part is associated with the responses of GRN to fluctuations in the environment. GRN have many other important for the survival and development of living organisms functions. Any knowledge about the principles of the functioning of GRN is useful. It allows to maintain and change the behavior of cells. In particular, some diseases can be treated, studying first their mathematical models.

Self-organization of genetic networks gives rise to attempts to copy them in various fields. Designing the architecture of telecommunication networks can use knowledge the mechanisms of functioning of GRN, which can quickly adapt to harmful events in the environment.

Neuronal networks are another object of investigation for mathematicians. In this chapter, we have studied systems of ODE, which have much in common with systems arising in the theory of GRN. These systems are quasi-linear again. The nonlinearity is also the sigmoidal function hyperbolic tangent, which takes values of both signs, in contrast with the logistic function, which is usually used in models of GRN. Therefore, much of the theory of GRN is valid for models of neuronal networks and vice versa.

The purely mathematical problems, which are met, when studying models of GRN and neuronal networks (artificial neuronal networks, ANN), are challenging and important for further progress in the theory of dynamical systems. Among the important problems mentioned are the description of attractors, studying the structures of the phase spaces for higher-dimensional systems, investigating methods of control, and maintaining these systems. The inverse problems require to build systems with prescribed properties, and this is the way to GRN engineering.

References

1. A. Tamaševičius, S. Bumelienė, R. Kirvaitis, G. Mykolaitis, E. Tamaševičiūtė, E. Lindberg, Autonomous Duffing-Holmes type chaotic oscillator. Elektronika ir Elektrotechnika. **5**(93), 43–46 (2009). https://www.eejournal.ktu.lt/index.php/elt/article/download/10178/5036
2. Y. Koizumi, T. Miyamura, S. Arakawa, E. Oki, K. Shiomoto, and M. Murata, Application of attractor selection to adaptive virtual network topology control, in *Proceedings of BIONETICS* (2008), pp. 1–8
3. L.-Z. Wang, R.-Q. Su, Z.-G. Huang, X. Wang, W.-X. Wang, C. Grebogi, Y.-C. Lai, A geometrical approach to control and controllability of nonlinear dynamical networks. Nat. Commun. **7**, 11323 (2016). https://doi.org/10.1038/ncomms11323
4. O.E. Rössler, Chaotic oscillations: an example of hyperchaos, in *Nonlinear Oscillations in Biology*. Lectures in Applied Mathematics, vol. 17 (AMS, Providence, 1979)
5. P. Holmes, A nonlinear oscillator with a strange attractor. Philos. Trans. R. Soc. Lond. A: Math. Phys. Sci. **292**, 419–448 (1979). https://royalsocietypublishing.org/doi/10.1098/rsta.1979.0068
6. D.K. Arrowsmith, C.M. Place, *Dynamical Systems*. Differential Equations, Maps and Haotic Behavior (Chapman and Hall/CRC, London, 1992)
7. V.W. Noonburg, *Differential Equations: From Calculus to Dynamical Systems*, 2nd edn. (MAA Press, Providence, 2019)
8. F. Sadyrbaev F., V. Sengileyev, Remarks on inhibition. Int. J. Math. Comput. Methods **7**, 11–17 (2022)
9. D. Ogorelova, F. Sadyrbaev F., V. Sengileyev, Control in inhibitory genetic regulatory network models. Contemp. Math. **1**(5), 421–428
10. A. Krogh, What are artificial neural networks? Nat. Biotechnol. **26**(2), 195–197 (2008). https://doi.org/10.1038/nbt1386
11. S. Haykin, Neural networks expand SP's horizons. IEEE Signal Process. Mag. **13**(2), 24–49 (1996). https://doi.org/10.1109/79.487040
12. F. Gafarov, A. Galimyanov, *Artificial Neural Networks and Their Applications* (Kazan, 2018).

13. K. Suzuki, *Artificial Neural Networks*. Methodological Advances and Biomedical Applications (InTech, Rijeka, 2011)
14. S. Walczak, N. Cerpa, *Artificial Neural Networks*, 3rd edn. Encyclopedia of Physical Science and Technology (2003)
15. J.C. Sprott, *Elegant Chaos Algebraically Simple Chaotic Flows* (World Scientific Publishing Company, Singapore, 2010), 302 p. https://doi.org/10.1142/7183
16. J.C. Sprott, Chaotic dynamics on large networks. Chaos **18**(2), 023135 (2008). https://doi.org/10.1063/1.2945229

Chapter 8
Impact of Travel on Spread of Infection

Kjetil Holm, Josef Rebenda, and Yuriy Rogovchenko

8.1 Introduction

The study of infectious diseases was initiated in 1662 by Graunt in the book titled Natural and Political Observations Made Upon the Bills of Mortality [5]. In 1906, Hamer suggested that the spread of infection should depend on the number of susceptible and infected individuals [12]. This was an important step toward the use of compartmental models where the individuals within a closed population N are split into mutually exclusive compartments according to their disease status. Each individual can be only in one compartment at a given time but can move to other compartments. The flow between different compartments is usually described by differential equations.

K. Holm (✉)
Faculty of Engineering and Science, Department of Mathematical Sciences, University of Agder, Kristiansand, Norway
Oslo University Hospital, Oslo, Norway
e-mail: hokjet@ous-hf.no

J. Rebenda
Faculty of Electrical Engineering and Communication, Department of Mathematics, Brno University of Technology, Brno, Czech Republic
e-mail: rebenda@vutbr.cz

Y. Rogovchenko
Faculty of Engineering and Science, Department of Mathematical Sciences, University of Agder, Kristiansand, Norway

Faculty of Electrical Engineering and Communication, Department of Mathematics, Brno University of Technology, Brno, Czech Republic
e-mail: yuriy.rogovchenko@uia.no

C. M. A. Pinto, C. M. Ionescu (eds.), *Computational and Mathematical Models in Biology*, Nonlinear Systems and Complexity 38,
https://doi.org/10.1007/978-3-031-42689-6_8

In 1910, Ross suggested a system of differential equations to model the spread of malaria concluding that it can be controlled by killing a fraction of mosquitoes and keeping their number under a certain threshold [16]. This situation is typical for many models in epidemiology which have a threshold parameter, known as the basic reproduction number, the average number of infected contacts per infected individual. It is related to stability of a disease-free equilibrium, spread of disease, and its eradication. In 1927, almost a decade after the 1918 influenza pandemic, Kermack and McKendrick [14] proposed a compartmental model to describe the dynamics of an infectious disease in population with the three cohorts, susceptible $S(t)$, infected $I(t)$, and recovered $R(t)$. The original system of integro-differential equations was reduced to a form known in the literature as an SIR model:

$$\frac{dS}{dt} = -\frac{\beta IS}{N}, \qquad \frac{dI}{dt} = \frac{\beta IS}{N} - \gamma I, \qquad \frac{dR}{dt} = \gamma I. \tag{8.1}$$

We refer to excellent surveys by Brauer [5, 6] where further details can be found.

The SIR model is simple and easy to work with, but it oversimplifies complex disease processes leaving aside many important details. For instance, according to the model's design, all individuals become immune to disease after recovery, which is the case for malaria but not for viral infections like influenza or coronavirus. Furthermore, (8.1) does not account for the effect of travel which was essential for the analysis of the recent Covid-19 epidemics. Both features are important for a model developed in this paper.

In this chapter, we suggest a modification of the classical SIR model characterized by two distinctive features. Firstly, the susceptible cohort is split into two sub-cohorts with respect to their travel patterns where individuals which travel more frequently are assumed to have a higher infection rate. Secondly, we allow repeated reinfection for those recovered individuals that return to the susceptible cohort. To the best of our knowledge, such models were not considered yet in the literature, and existing models exhibit different, often controversial, scenarios. The research question we address is how do active travel and a possibility for repeated infection contribute to the spread of disease?

8.2 SIR Model with Two Different Travel Patterns

Expanding further the classical SIR model (8.1), we keep the notation $S(t)$, $I(t)$, and $R(t)$ for the number of susceptible, infected, and recovered individuals. Our fundamental assumption is that all susceptible individuals S are divided into two subgroups, S_1 and S_2, in accordance with their travel patterns. The former, larger group S_1, comprises less frequent travelers, and the latter, smaller group S_2, accounts for those who travel more often. We do not distinguish between business and leisure trips and include in S_2 both types. Given the total number S of susceptible individuals in the basic model (8.1), we split the cohort into two

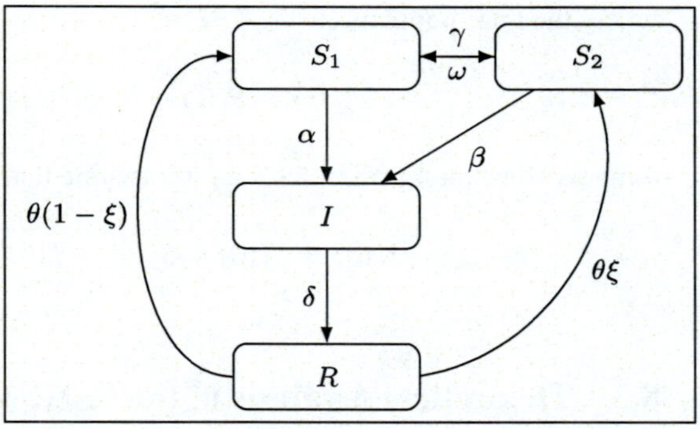

Fig. 8.1 The flowchart for an SSIR model

subgroups defining S_1 and S_2 by

$$S_1 = (1-q)S, \qquad S_2 = qS,$$

where the coefficient $q \in (0, 1)$ determines the initial proportion of active travelers within the susceptible cohort. Due to the presence of two different sub-compartments of susceptible, we call the modified model an SSIR model.

To the best of our knowledge, no similar models were discussed in the literature yet. Our SSIR model is a natural extension of the classical SIR model [5, 6] where one differential equation for susceptible cohort S is replaced with two equations for sub-cohorts S_1 and S_2. Susceptible individuals move between two subgroups when they start or stop traveling intensively. This process is described with the help of the coefficients γ and ω representing transition rates from a nonmobile cohort to a mobile one and from the mobile back to a nonmobile cohort, respectively. Both S_1 and S_2 interact with the infected cohort I, but the chances of getting infected differ. The infection rates for individuals in nonmobile and mobile groups are denoted by α and β, respectively. We assume that infected individuals in both mobile and nonmobile groups recover at the same rate δ and leave the infected cohort with part of them gaining immunity from infection afterward. Some individuals may become infected again; these are removed from the recovered cohort at the rate θ and become susceptible again. Individuals from the latter group are distributed between S_1 and S_2; this distribution is regulated by the coefficient ξ. Schematically, our SSIR model is presented in Fig. 8.1.

The system of four differential equations governing the spread of infectious disease in the SSIR model assumes the form

$$\dot{S_1} = -\alpha I(t)S_1(t) - \gamma S_1(t) + \omega S_2(t) + \theta(1-\xi)R(t), \tag{8.2}$$

$$\dot{S_2} = -\beta I(t)S_2(t) + \gamma S_1(t) - \omega S_2(t) + \xi\theta R(t), \tag{8.3}$$

$$\dot{I} = \alpha I(t)S_1(t) + \beta I(t)S_2(t) - \delta I(t), \tag{8.4}$$

$$\dot{R} = \delta I(t) - \theta R(t), \tag{8.5}$$

where the total population

$$N(t) = S_1(t) + S_2(t) + I(t) + R(t) \tag{8.6}$$

remains constant for all times, and we assume that $N(t)$ is normalized so that

$$N(0) = S_1(0) + S_2(0) + I(0) + R(0) = 1. \tag{8.7}$$

8.3 Theoretical Analysis of the SSIR Model

8.3.1 Positivity of Solutions

First, we prove that solutions with nonnegative initial values remain nonnegative for all times. In what follows, inequalities between vectors should be understood component-wise.

Theorem 1 *Assume that*

$$\mathbf{x}(0) = (S_1(0), S_2(0), I(0), R(0))^{\mathrm{T}} \geq (0, 0, 0, 0)^{\mathrm{T}}. \tag{8.8}$$

Then, the solution $\mathbf{x}(t) = (S_1(t), S_2(t), I(t), R(t))^{\mathrm{T}}$ *to the SSIR system* (8.2)–(8.5) *satisfying the initial condition* (8.8) *is nonnegative for all* $t > 0$.

Proof Let $\mathbf{x}(t)$ be a solution to the system (8.2)–(8.5) satisfying the initial condition (8.8). If $S_1(t_1) = 0$ for some $t_1 > 0$ and both $S_2(t_1)$ and $R(t_1)$ are nonnegative, then the derivative of S_1 is nonnegative:

$$\begin{aligned}\dot{S}_1(t_1) &= -\alpha I(t_1)S_1(t_1) - \gamma S_1(t_1) + \omega S_2(t_1) + \theta(1-\xi)R(t_1)\\ &= \omega S_2(t_1) + \theta(1-\xi)R(t_1) \geq 0.\end{aligned}$$

Thus, if $S_1(t_1) = 0$, the function $S_1(t)$ is nondecreasing afterward and cannot become negative as long as other components are nonnegative. Similarly, assuming that $S_2(t_2) = 0$ for some $t_2 > 0$, and both values $S_1(t_2)$ and $R(t_2)$ are nonnegative, we get

$$\dot{S}_2(t_2) = -\beta I(t_2)S_2(t_2) + \gamma S_1(t_2) - \omega S_2(t_2) + \xi\theta R(t_2) = \gamma S_1(t_2) + \xi\theta R(t_2) \geq 0.$$

Direct integration of Eq. (8.4) over $[0, t)$ yields

$$I(t) = I(0)\exp\left(\int_0^t (\alpha S_1(u) + \beta S_2(u) - \delta)\, du\right).$$

Since $I(0) \geq 0$, we conclude that $I(t) \geq 0$ for all $t > 0$. Finally, assuming that $R(t_3) = 0$ for some $t_3 \geq 0$, we get

$$\dot{R} = \delta I(t_3) - \theta R(t_3) = \delta I(t_3) \geq 0$$

because $I(t_3) \geq 0$. Therefore, if (8.8) holds, solutions to the system (8.2)–(8.5) cannot eventually become negative. □

Corollary 1 *Assume that*

$$(0, 0, 0, 0)^{\mathrm{T}} \leq (S_1(0), S_2(0), I(0), R(0))^{\mathrm{T}} \leq (1, 1, 1, 1)^{\mathrm{T}}.$$

Then, for all $t \geq 0$, *one has*

$$(0, 0, 0, 0)^{\mathrm{T}} \leq (S_1(t), S_2(t), I(t), R(t))^{\mathrm{T}} \leq (1, 1, 1, 1)^{\mathrm{T}}.$$

Proof The estimate from below follows from Theorem 1. On the other hand, condition (8.7) and the relation

$$\dot{N}(t) = \dot{S}_1(t) + \dot{S}_2(t) + \dot{I}(t) + \dot{R}(t) = 0, \qquad \text{for all} \qquad t \geq 0,$$

yield that

$$1 = N(t) = S_1(t) + S_2(t) + I(t) + R(t) \tag{8.9}$$

for all $t \geq 0$, and thus, none of the components of the solution vector $\mathbf{x}(t)$ can exceed the unity. □

8.3.2 Equilibrium Points

Equilibria of the system (8.2)–(8.5) satisfy the system of nonlinear algebraic equations

$$-\alpha I(t)S_1(t) - \gamma S_1(t) + \omega S_2(t) + \theta(1 - \xi)R(t) = 0, \tag{8.10}$$

$$-\beta I(t)S_2(t) + \gamma S_1(t) - \omega S_2(t) + \xi\theta R(t) = 0, \tag{8.11}$$

$$\alpha I(t)S_1(t) + \beta I(t)S_2(t) - \delta I(t) = 0, \tag{8.12}$$

$$\delta I(t) - \theta R(t) = 0. \tag{8.13}$$

First, assuming that $I = 0$, we find the disease-free equilibrium E_0. Equation (8.13) immediately yields $R = 0$. Substituting $I = 0$ and $R = 0$ into Eqs. (8.9) and (8.10), we first relate S_1 and S_2 by

$$S_1 = \frac{\omega}{\gamma} S_2,$$

and then deduce that

$$S_1 = \frac{\omega}{\gamma + \omega} \qquad \text{and} \qquad S_2 = \frac{\gamma}{\gamma + \omega}.$$

The disease-free equilibrium E_0 of the system (8.2)–(8.5) is therefore given by

$$E_0 = \left(\frac{\omega}{\gamma + \omega}, \frac{\gamma}{\gamma + \omega}, 0, 0 \right), \tag{8.14}$$

and its location depends on the transfer rates γ and ω between the sub-cohorts of mobile and nonmobile susceptible.

To find the endemic equilibrium E_*, assume now that $I(0) \neq 0$. It follows from Eq. (8.12) that

$$S_1 = \frac{\delta}{\alpha} - \frac{\beta}{\alpha} S_2, \tag{8.15}$$

and Eq. (8.13) yields

$$R = \frac{\delta}{\theta} I. \tag{8.16}$$

Substituting (8.15) and (8.16) into (8.9), we express S_2 in terms of I :

$$S_2 = \frac{\alpha}{\alpha - \beta} - \frac{\alpha\delta}{\alpha^2 - \alpha\beta} - \frac{\alpha\theta + \alpha\delta}{\alpha\theta - \beta\theta} I. \tag{8.17}$$

Finally, using (8.15) and (8.17), we also obtain the expression for S_1 in terms of I :

$$S_1 = \frac{\delta}{\alpha} - \frac{\beta}{\alpha - \beta} + \frac{\beta\delta}{\alpha^2 - \alpha\beta} + \frac{\beta\theta + \beta\delta}{\alpha\theta - \beta\theta} I. \tag{8.18}$$

Substituting (8.15)–(8.17) into (8.11), collecting like terms, and performing simple algebraic manipulations, we obtain a quadratic equation for I:

$$A_1 I^2 + A_2 I + A_3 = 0, \tag{8.19}$$

where the coefficients A_1, A_2, and A_3 are defined, respectively, by

$$A_1 = \alpha\beta\Big(\theta + \delta\Big),$$
$$A_2 = \beta(\delta\theta - \delta\theta\xi + \gamma\theta + \gamma\delta) + \alpha(\omega\theta + \omega\delta + \delta\theta\xi - \beta\theta), \tag{8.20}$$
$$A_3 = \theta\Big(\omega(\delta - \alpha) + \gamma(\delta - \beta)\Big).$$

If $\alpha - \beta \neq 0$, $\alpha \neq 0$, $\beta \neq 0$, and $\theta \neq 0$, solutions to Eq. (8.19) are given by

$$I_{1,2} = \frac{-A_2 \pm \sqrt{\Delta}}{2A_1} \tag{8.21}$$

where $\Delta = A_2^2 - 4A_1A_3$. Substituting (8.21) into (8.15)–(8.17), we get

$$S_1^{1,2} = \frac{\beta - \delta}{\beta - \alpha} + \frac{-A_2 \pm \sqrt{\Delta}}{2\alpha\theta(\alpha - \beta)},$$
$$S_2^{1,2} = \frac{\delta - \alpha}{\beta - \alpha} - \frac{-A_2 \pm \sqrt{\Delta}}{2\beta\theta(\alpha - \beta)}, \tag{8.22}$$
$$R^{1,2} = \frac{-A_2 \pm \sqrt{\Delta}}{2\delta A_1}\theta.$$

In what follows, we use the notation:

$$\kappa = \sqrt{\Delta} - A_2. \tag{8.23}$$

Lemma 1 *Assume that $\alpha < \beta$. Then, the following assertions hold:*

(i) If $\delta \leq \alpha < \beta$, Eq. (8.19) has one positive and one negative solution.
(ii) If $\alpha < \beta \leq \delta$, Eq. (8.19) has either two real negative solutions or two complex conjugate solutions with a negative real part.
(iii) There exists a δ_0, $\alpha < \delta_0 < \beta$ satisfying the equation:

$$\frac{\gamma(\beta - \delta_0)}{\omega(\delta_0 - \alpha)} = 1 \tag{8.24}$$

such that Eq. (8.19) has the zero solution and a negative solution.

Proof

Case (i). Note that A_1 is always positive. Since the first factor θ in the expression for A_3 is always positive, the sign of A_3 is determined by the sign of the second factor $\omega(\delta - \alpha) + \gamma(\delta - \beta)$. Observe that if $\delta \leq \alpha < \beta$, then A_3 is negative:

$$A_3 = \omega\underbrace{(\delta - \alpha)}_{\leq 0} + \gamma\underbrace{(\delta - \beta)}_{<0} < 0.$$

By Viète theorem [22], quadratic equation (8.19) should have one positive and one negative root.

Case (ii). Assume now that $\alpha < \beta = \delta$. One can see that A_3 is positive now:

$$A_3 = \omega \underbrace{(\delta - \alpha)}_{>0} + \gamma \underbrace{(\delta - \beta)}_{=0} > 0,$$

and we will prove that in this case, A_2 is also positive. To this end, first, let $\delta = \beta$ in the expression (8.20) for A_2 and collect like terms to get

$$A_2 = \beta^2\theta \underbrace{(1 - \xi)}_{>0} + \alpha\beta\theta \underbrace{(\xi - 1)}_{<0} + \beta\gamma(\theta + \beta) + \alpha\omega(\theta + \beta). \tag{8.25}$$

Substituting β for α in the second term in (8.25) and taking into account that $\alpha < \beta$, we obtain the estimate:

$$A_2 > \underbrace{\beta^2\theta(1 - \xi) + \beta^2\theta(\xi - 1)}_{=0} + (\alpha\omega + \beta\gamma)(\theta + \beta) = (\alpha\omega + \beta\gamma)(\theta + \beta) > 0.$$

Given that A_1, A_2, and A_3 are positive and using the formula (8.21) for the roots of the quadratic equation (8.19), we conclude that the roots may be either complex with a negative real part or real negative.

Case (iii). It follows from Cases (i) and (ii) that if $\Delta > 0$, for $\delta \leq \alpha < \beta$, we have two roots of different sign, and for $\alpha < \beta = \delta$, we have two negative roots. Since the roots of the quadratic equation given by (8.21) depend continuously on α, β, and δ, decreasing the value of δ from $\delta = \beta$ where we have both roots negative, there should exist a $\delta_0 \in (\alpha, \beta)$ such that (8.19) has the zero root and a negative root. Decreasing further the value of δ, for $\delta < \delta_0$, we obtain two roots of different signs, as illustrated in Fig. 8.2. Since the roots of Eq. (8.19) are given by (8.21), quadratic equation has the zero root provided that $\kappa = 0$, which implies that $A_3 = 0$. In terms of the coefficients of the system (8.2)–(8.5), this means that condition (8.24) holds; it can be used for finding δ_0 when all other parameters are known.

□

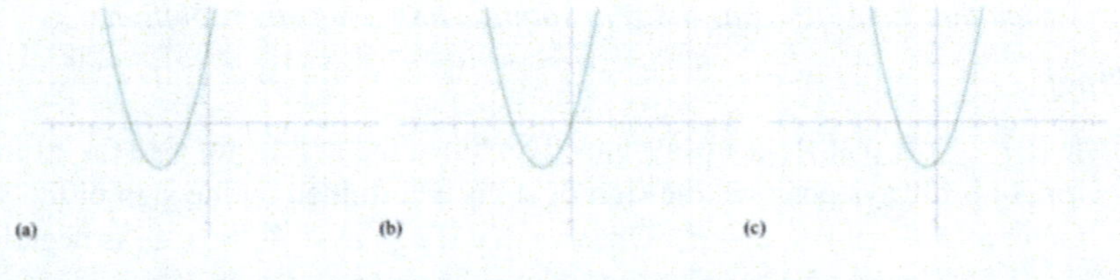

Fig. 8.2 Location of the roots of quadratic equation (8.19) for different values of δ: **(a)** $\alpha < \beta \leq \delta$; **(b)** $\delta = \delta_0$; **(c)** $\delta \leq \alpha < \beta$

It follows now from Eqs. (8.21), (8.22), and Lemma 1 that the endemic equilibrium $E_* = \left(S_1^*, S_2^*, I^*, R^*\right)$ of the system (8.2)–(8.5) is given by

$$E_* = \left(\frac{2\theta\alpha\,(\beta-\delta)-\kappa}{2\theta\alpha\,(\beta-\alpha)}, \frac{2\theta\beta\,(\delta-\alpha)-\kappa}{2\theta\beta\,(\beta-\alpha)}, \frac{\kappa}{2\alpha\beta(\theta+\delta)}, \frac{\theta\kappa}{2\alpha\beta\delta(\theta+\delta)}\right) \tag{8.26}$$

where κ is defined by (8.23).

8.3.3 Stability Analysis

Using condition (8.9) relating all four variables, we rewrite Eqs. (8.2)–(8.5) as a system of three differential equations by eliminating the last, fourth equation for R which becomes redundant in view of (8.9):

$$\begin{aligned}
\dot{I} &= \alpha I(t)S_1(t) + \beta I(t)S_2(t) - \delta I(t),\\
\dot{S}_1 &= -\alpha I(t)S_1(t) - \gamma S_1(t) + \omega S_2(t) + \theta(1-\xi)(1-I-S_1-S_2),\\
\dot{S}_2 &= -\beta I(t)S_2(t) + \gamma S_1(t) - \omega S_2(t) + \theta\xi(1-I-S_1-S_2).
\end{aligned} \tag{8.27}$$

To determine stability properties of the equilibria, we calculate the Jacobian matrix J:

$$J = \begin{pmatrix} \alpha S_1 + \beta S_2 - \delta & \alpha I & \beta I \\ -\alpha S_1 - \theta\,(1-\xi) & -\alpha I - \gamma - \theta\,(1-\xi) & \omega - \theta\,(1-\xi) \\ -\beta S_2 - \theta\xi & \gamma - \theta\xi & -\beta I - \omega - \theta\xi \end{pmatrix}.$$

We analyze first stability properties of the disease-free equilibrium E_0.

Theorem 2 *The equilibrium E_0 is asymptotically stable if and only if*

$$\frac{\alpha\omega + \beta\gamma}{\delta(\omega+\gamma)} < 1. \tag{8.28}$$

Proof The Jacobian matrix J at E_0 assumes the form:

$$J_{E_0} = \begin{pmatrix} \alpha\frac{\omega}{\gamma+\omega} + \beta\frac{\gamma}{\gamma+\omega} - \delta & 0 & 0 \\ -\alpha\frac{\omega}{\gamma+\omega} - \theta\,(1-\xi) & -\gamma - \theta\,(1-\xi) & \omega - \theta\,(1-\xi) \\ -\beta\frac{\gamma}{\gamma+\omega} - \theta\xi & \gamma - \theta\xi & -\omega - \theta\xi \end{pmatrix},$$

and its eigenvalues are

$$\lambda_1 = -\theta, \qquad \lambda_2 = -\gamma - \omega, \qquad \lambda_3 = \frac{\alpha\omega + \beta\gamma}{\omega + \gamma} - \delta.$$

The disease-free equilibrium is asymptotically stable if and only if the real parts of all three eigenvalues are negative. Observe that λ_1 and λ_2 are always negative, while λ_3 is negative if and only if condition (8.28) is satisfied. □

Remark 1 It follows from the proof of Theorem 2 that (8.28) is equivalent to

$$\frac{\gamma(\beta - \delta)}{\omega(\delta - \alpha)} < 1. \tag{8.29}$$

Note that the left-hand side of the inequality (8.29) has the same form as the left-hand side of Eq. (8.24) derived when we studied the existence of the endemic equilibrium.

Since at the endemic equilibrium E_*, we have $\alpha S_1 + \beta S_2 - \delta = 0$, the Jacobian matrix assumes the form:

$$J_{E_*} = \begin{pmatrix} 0 & \alpha I^* & \beta I^* \\ -\alpha S_1^* - \theta(1-\xi) & -\alpha I^* - \gamma - \theta(1-\xi) & \omega - \theta(1-\xi) \\ -\beta S_2^* - \theta\xi & \gamma - \theta\xi & -\beta I^* - \omega - \theta\xi \end{pmatrix}.$$

The characteristic polynomial for J_{E_*} is defined by

$$\zeta(\lambda) = \lambda^3 + B_2\lambda^2 + B_1\lambda + B_0$$

with the coefficients

$$\begin{aligned} B_2 &= \gamma + \omega + \theta\xi + \alpha I^* + \beta I^* - \theta(\xi - 1), \\ B_1 &= \left(\omega + \theta\xi + \beta I^*\right)\left(\gamma + \alpha I^* - \theta(\xi - 1)\right) - (\gamma - \theta\xi)(\omega + \theta(\xi - 1)) \\ &\quad + \beta I^*\left(\theta\xi + \beta S_2^*\right) - \alpha I^*\left(\theta(\xi - 1) - \alpha S_1^*\right), \\ B_0 &= (\gamma - \theta\xi)\left((\omega + \theta(\xi - 1))\left(\gamma + \alpha I^* - \theta(\xi - 1)\right) - \beta I^*\left(\theta(\xi - 1) - \alpha S_1^*\right)\right) \\ &\quad - \left((\gamma - \theta\xi)\omega + \theta(\xi - 1)) - \beta I^*\left(\theta\xi + \beta S_2^*\right)\right)\left(\gamma + \alpha I^* - \theta(\xi - 1)\right) \\ &\quad + \alpha I^*\left(\theta\xi + \beta S_2^*\right)(\omega + \theta(\xi - 1)) - \alpha I^*\left(\theta(\xi - 1) - \alpha S_1^*\right)\left(\omega + \theta\xi + \beta I^*\right). \end{aligned}$$

Although eigenvalues of J_{E_*} can be computed explicitly with the help of computer algebra, this leads to complex expressions which can be simplified only when the values of coefficients are given. Therefore, we use a necessary and sufficient condition for the negativity of all roots of a cubic polynomial, a particular case of the general Routh-Hurwitz criterion [11, p. 180], to derive the conditions for the stability of the endemic equilibrium.

Theorem 3 *The endemic equilibrium E_* of the system (8.2)–(8.5) is asymptotically stable if and only if all three coefficients B_2, B_1, and B_0 of the cubic polynomial $\zeta(\lambda)$ are positive and $B_2 B_1 > B_0$.*

8.3.4 Basic Reproduction Number

Epidemiology defines the basic reproduction number $\mathcal{R}_0$ as the expected number of cases generated by one case in a population. This number provides a useful insight into the dynamics of the infectious disease. A tendency of the basic reproduction number to increase might indicate an outbreak of infection, thus requiring fast and efficient decisions by the local authorities and the central government aimed at the prevention of the spread of disease.

One can obtain the value of $\mathcal{R}_0$ for specific models by using the next generation method developed by van den Driessche and Watmough's [21]. For the convenience of the reader, we include the details necessary for the understanding of the method. Consider the general disease transmission model

$$\dot{x}_i = f_i(x) = \mathcal{F}_i(x) - \mathcal{V}_i(x), \qquad i = 1, \ldots, n \tag{8.30}$$

where n is the number of states in the system, m is the number of infected states, $\mathcal{F}_i(x)$, $i = 1, \ldots, n$ is the rate of appearance of new infections in a compartment i, $\mathcal{V}_i^+(x)$, $i = 1, \ldots, n$ is the rate of transfer of individuals into a compartment i by all other means, $\mathcal{V}_i^-(x)$, $i = 1, \ldots, n$ is the rate of transfer of individuals out of compartment i, and $\mathcal{V}_i(x) = \mathcal{V}_i^-(x) - \mathcal{V}_i^+(x)$. Furthermore, let $\mathbf{X}_s \overset{\text{def}}{=} \{x \geq 0 \mid x_i = 0,\ i = 1, \ldots, m\}$ be the set of all disease-free states and let the matrices F and V be defined by

$$F = \left[\frac{\partial \mathcal{F}_i}{\partial x_j}(x_0)\right] \quad \text{and} \quad V = \left[\frac{\partial \mathcal{V}_i}{\partial x_j}(x_0)\right] \quad \text{with } 1 \leq i,\ j \leq m. \tag{8.31}$$

Theorem 4 ([21, Theorem 2]) *Let the functions $\mathcal{F}_i$, $\mathcal{V}_i^-$, and $\mathcal{V}_i^+$ in the general model (8.30) satisfy the following conditions:*

(A1) *If $x \geq 0$, then $\mathcal{F}_i, \mathcal{V}_i^+, \mathcal{V}_i^- \geq 0$ for $i = 1, \ldots, n$.*
(A2) *If $x_i = 0$, then $\mathcal{V}_i^-(x) = 0$. In particular, if $x \in \mathbf{X}_s$, then $\mathcal{V}_i^-(x) = 0$ for $i = 1, \ldots, m$.*
(A3) *$\mathcal{F}_i(x) = 0$ if $i > m$.*
(A4) *If $x \in \mathbf{X}_s$, then $\mathcal{F}_i(x) = 0$ and $\mathcal{V}_i^+(x) = 0$ for $i = 1, \ldots, m$.*
(A5) *If $\mathcal{F}(x)$ is set to zero, then all eigenvalues of $Df(x_0)$ have negative real parts.*

If x_0 is a disease-free equilibrium of the model, then x_0 is locally asymptotically stable if $\mathcal{R}_0 < 1$ but unstable if $\mathcal{R}_0 > 1$, where the basic reproduction number $\mathcal{R}_0$ is defined in terms of the spectral radius $\rho(A)$ of a matrix A:

$$\mathcal{R}_0 = \rho\left(FV^{-1}\right). \tag{8.32}$$

Let $x = (I, S_1, S_2)^T$ be the vector solution to the system (8.27).

Lemma 2 *The system* (8.27) *satisfies all assumptions of Theorem* 4.

Proof Defining the vectors describing the generation of new infected individuals and the flow between the states by

$$\mathcal{F}(x) = \left[\alpha I S_1 + \beta I S_2,\ 0,\ 0\right]^{\mathrm{T}},$$

$$\mathcal{V}(x) = \begin{bmatrix} \delta I \\ \alpha I(t)S_1(t) + \gamma S_1 - \omega S_2 - \theta(1-\xi)(1 - I - S_1 - S_2) \\ \beta I(t)S_2(t) - \gamma S_1 + \omega S_2 - \theta\xi(1 - I - S_1 - S_2) \end{bmatrix},$$

$$\mathcal{V}^{+}(x) = \begin{bmatrix} 0 \\ \omega S_2 + \theta + \theta\xi(I + S_1 + S_2) \\ \gamma S_1 + \theta\xi \end{bmatrix},$$

$$\mathcal{V}^{-}(x) = \begin{bmatrix} \delta I \\ \alpha I S_1 + \gamma S_1 + \theta(I + S_1 + S_2) + \theta\xi \\ \beta I S_2 + \omega S_2 + \theta\xi(I + S_1 + S_2) \end{bmatrix},$$

we can write the system (8.27) in the form (8.30). Since all variables in (8.27) take on nonnegative values, for $x \geq 0$, the functions $\mathcal{V}_i^{+}$, $\mathcal{V}_i^{-}$, and $\mathcal{F}_i$ defined above also take on nonnegative values for $i = 1, 2, 3$. Therefore, condition (A1) holds. Condition (A2) is verified immediately since for $I = 0$ one has $\mathcal{V}_1^{-}(x) = \delta I = 0$. Furthermore, for $i > 1$, $\mathcal{F}_i = 0$, and thus, (A3) is also satisfied. For the disease-free state we have $I = 0$. Thus, $\mathcal{F}_1 = 0$, and (A4) holds.

Finally, to prove (A5), we calculate the Jacobian matrix of the system when $\mathcal{F}(x) = 0$ and $Df(x) = D(\mathcal{F}(x) - \mathcal{V}(x)) = D(-\mathcal{V}(x))$:

$$Df(x) = \begin{bmatrix} -\delta & 0 & 0 \\ -\alpha S_1 - \theta(1-\xi) & -\alpha I - \gamma - \theta(1-\xi) & \omega - \theta(1-\xi) \\ -\beta S_2 - \xi\theta & \gamma - \xi\theta & -\omega - \omega - \xi\theta) \end{bmatrix}.$$

At the disease-free equilibrium E_0, we have

$$Df(x_0) = \begin{bmatrix} -\delta & 0 & 0 \\ -\alpha\frac{\omega}{\gamma+\omega} - \theta(1-\xi) & -\gamma - \theta(1-\xi) & \omega - \theta(1-\xi) \\ -\beta\frac{\gamma}{\gamma+\omega} - \xi\theta & \gamma - \xi\theta & -\omega - \xi\theta \end{bmatrix}.$$

A straightforward computation yields that all three eigenvalues

$$\lambda_1 = -\delta, \quad \lambda_2 = -\theta, \text{ and } \quad \lambda_3 = -\omega - \gamma$$

are negative. Thus, all conditions of [21, Theorem 2] are verified. □

We apply now Theorem 4 to find the basic reproduction number and state conditions for asymptotic stability of the disease-free equilibrium.

Corollary 2 *The conclusion of Theorem 4 about stability of the disease-free equilibrium holds for the systems* (8.27) *and* (8.2)–(8.5)*, and the basic reproduction number* $\mathcal{R}_0$ *is given by*

$$\mathcal{R}_0 = \frac{\alpha\omega + \beta\gamma}{\delta(\gamma + \omega)}. \tag{8.33}$$

Proof To calculate $\mathcal{R}_0$, we evaluate $\mathcal{F}(x)$ and $\mathcal{V}(x)$ at the disease-free equilibrium E_0 defined by (8.14). Using

$$x_0 = \left(\frac{\omega}{\gamma + \omega}, \frac{\gamma}{\gamma + \omega}, 0\right),$$

we compute $\mathcal{F}(x) = \left[(\alpha I S_1 + \beta I S_2),\ 0,\ 0\right]^{\mathrm{T}}$ and

$$\mathcal{V}(x) = \begin{bmatrix} \delta I \\ \alpha I(t)S_1(t) + \gamma S_1 - \omega S_2 - \theta(1-\xi)(1 - I - S_1 - S_2) \\ \beta I(t)S_2(t) - \gamma S_1 + \omega S_2 - \theta\xi(1 - I - S_1 - S_2) \end{bmatrix}.$$

In our case, two Jacobian matrices (8.31) simplify to

$$F = \frac{\alpha\omega}{\gamma + \omega} + \frac{\beta\gamma}{\gamma + \omega} = \frac{\alpha\omega + \beta\gamma}{\gamma + \omega} \quad \text{and} \quad V = \delta.$$

Thus, we find the basic reproduction number computing the spectral radius for a 1×1 matrix, which yields

$$\mathcal{R}_0 = \rho(FV^{-1}) = \rho\left(\frac{\alpha\omega + \beta\gamma}{\delta(\gamma + \omega)}\right) = \frac{\alpha\omega + \beta\gamma}{\delta(\gamma + \omega)}.$$

The proof is complete now. □

Remark 2 Note that the expression for the basic reproduction number $\mathcal{R}_0$ of the system (8.2)–(8.5) appears on the left-hand side of the inequality (8.28) which ensured asymptotic stability of the disease-free equilibrium E_0 through the negativity of all three eigenvalues of the Jacobian matrix J evaluated at E_0.

8.3.5 Transition Between Two Susceptible Sub-cohorts

One of the difficulties in the development of mathematical models formulated in terms of differential equations is related to the need to "convert" proportions of certain quantities into the rate of change of these quantities. The situation is simpler in the one-dimensional case. To provide an example from mathematical epidemiology, assume that the average time to recover from the disease is three days. Although complete recovery is achieved after three days, every day a step toward recovery is made to avoid "one-in-three-days jumps." Mathematically, this means that in a continuous model, we want to remove daily one third of individuals from the infected state $x(t)$. Think now of a simple linear differential equation where the rate of change $\delta = -0.33$ is negative because individuals are "continuously" removed from the infected compartment:

$$\dot{x}(t) = -0.33x(t). \tag{8.34}$$

The solution to Eq. (8.34) satisfying the initial condition $x(0) = 1$ is given by $x(t) = e^{-0.33t}$, and its value at $t = 1$ is $x(1) = e^{-0.33} = 0.7189$. Since we expect that after one day two thirds should remain infected, the estimate provided by the solution of (8.34) differs significantly from the target value 0.66. The discrepancy becomes even more visible when we look at $x(10) = 0.03688$ in comparison to a target value of $0.66^{10} = 0.0156833688$. To reduce the discrepancy between the expected target value $x(1) = 0.66$ and that provided by the solution of a linear first-order differential equation, we find δ by solving the transcendental equation $0.66 = e^{-\delta}$, which yields $\delta = -\ln(0.66) = 0.41551544396$.

Solving a new differential equation with the value $\delta = 0.41551544396$, we get $x(1) = 0.66$ and $x(10) = 0.0156833688$, a very good fit to our target values of $0.66^1 = 0.66$ and $0.66^{10} = 0.0156833688$. Therefore, one adjusts the coefficient in a one-dimensional linear differential equation using the relation:

$$\delta = -\ln(1 - p), \tag{8.35}$$

where p is the target proportion of the quantity x that should be removed within one unit of time (day, month, year, etc.).

However, this approach does not solve the problem for higher-dimensional systems. For instance, consider a bidirectional transfer between two quantities [10], in our case, sub-cohorts of susceptible S_1 and S_2. To find the values of transition parameters γ and ω in the SSIR model (8.2)–(8.5), we start with a discrete system:

$$s_1(n+1) = as_1(n) + bs_2(n), \tag{8.36}$$

$$s_2(n+1) = (1-a)s_1(n) + (1-b)s_2(n) \tag{8.37}$$

which describes a day by day evolution of a population split into two groups $s_1(n)$ and $s_2(n)$, $(n = 0, 1, \ldots)$. Here, a is the proportion of the members of s_1 that remain in s_1, and b is the proportion of members of s_2 that move from s_2 to s_1 within one unit of time. We assume that $0 \leq b \leq a \leq 1$, which corresponds to the case when group members move between the two groups without drastic changes to the demographic situation, a kind of "inertia" property. We also assume that the variables $s_1(n)$ and $s_2(n)$ are normalized to add up to unity, $s_1(n) + s_2(n) = 1$. The equilibrium $[s_1^*, s_2^*]^T$ of the system (8.36), (8.37) is found by assuming that $s_1(n) = s_1(n+1)$ and $s_2(n) = s_2(n+1)$. Then, Eqs. (8.36) and (8.37) yield, respectively,

$$s_1^* = \frac{b}{1-a+b} \quad \text{and} \quad s_2^* = \frac{1-a}{1-a+b}. \tag{8.38}$$

If the values of s_1 and b are given, a is found by using Eq. (8.38):

$$a = 1 + b - \frac{b}{s_1}.$$

We can write now the discrete system (8.36), (8.37) in the matrix form:

$$\mathbf{X_{n+1}} = A\mathbf{X_n}, \tag{8.39}$$

where $\mathbf{X_n} = [s_1(n), s_2(n)]^T$ and

$$A = \begin{bmatrix} a & b \\ 1-a & 1-b \end{bmatrix}.$$

Condition $a \geq b$ ensures that both eigenvalues of A are positive. The characteristic equation is

$$\lambda^2 - (1 + (a-b))\lambda + (a-b) = 0,$$

so by Viète theorem [22] the eigenvalues are $\lambda_1 = 1$ and $\lambda_2 = a - b$. The corresponding eigenvectors $\mathbf{v_1}$ and $\mathbf{v_2}$ are

$$\mathbf{v_1} = \left[1, \frac{1-a}{b}\right]^T \quad \text{and} \quad \mathbf{v_2} = [-1, 1]^T.$$

One can write the general solution to the system (8.39) in terms of the eigenvalues and eigenvectors as

$$\mathbf{X_n} = \begin{bmatrix} s_1(s) \\ s_2(s) \end{bmatrix} = c_1\mathbf{v_1}\lambda_1^n + c_2\mathbf{v_2}\lambda_2^n.$$

The vector $\mathbf{X}_n$ is now used for constructing the fundamental matrix Φ of a continuous system that generates the same general solution. Our aim is to find the matrix B of a continuous system:

$$\dot{\mathbf{x}}(s) = B\mathbf{x}(s) \tag{8.40}$$

where $\mathbf{x}(s) = [x_1(s), x_2(s)]^{\mathrm{T}}$ is a continuous vector function. First, we express the general solution $\mathbf{x}(s)$ to the continuous system (8.40) in terms of the general solution $\mathbf{X}_n$ to the discrete system (8.39):

$$\mathbf{x}(s) = \begin{bmatrix} x_1(s) \\ x_2(s) \end{bmatrix} = c_1\mathbf{v}_1\lambda_1^s + c_2\mathbf{v}_2\lambda_2^s.$$

Since $\mathbf{x}(s)$ is the general solution, the fundamental matrix of (8.40) is given by

$$\Phi(s) = \begin{bmatrix} v_{11}\lambda_1^s & v_{12}\lambda_2^s \\ v_{21}\lambda_1^s & v_{22}\lambda_2^s \end{bmatrix} = \begin{bmatrix} v_{11} & v_{12}\lambda_2^s \\ v_{21} & v_{22}\lambda_2^s \end{bmatrix}.$$

Recall that λ_1=1 and $0 \le \lambda_2 \le 1$ because $\lambda_2 = a - b \ge 0$ and $0 \le b \le a \le 1$. The fundamental matrix $\Phi(s)$ satisfies the differential equation (8.40):

$$\dot{\Phi}(s) = B\Phi(s). \tag{8.41}$$

Therefore, $B = \dot{\Phi}(s)\Phi^{-1}(s)$ where the derivative of a non-singular matrix $\Phi(s)$ is given by

$$\dot{\Phi}(s) = \begin{bmatrix} 0 & v_{12}\lambda_2^s \ln(\lambda_2) \\ 0 & v_{22}\lambda_2^s \ln(\lambda_2) \end{bmatrix}.$$

Using this approach, we find the values of parameters γ and ω which determine the transition rates between two groups of susceptible S_1 and S_2. Application of a similar idea for finding the values of parameters α, β, δ, and θ is not possible because transition from S_1 and S_2 to I is expressed by using nonlinear terms. Therefore, for finding parameters α, β, δ, and θ, we employ the approach based on relation (8.35).

8.4 Numerical Simulations

8.4.1 Selection of Parameters

Selection of realistic parameter values in a model is important for making simulations as accurate and reliable as possible. Difficulties with finding "right" values for the parameters vary a lot and depend on the availability of the real data and the need to simulate some. In our SSIR model, we have a mix of both. Since the impact of travel on the spread of Covid-19 was widely discussed in European countries and influenced decisions taken by local and national health authorities in Norway, we use coronavirus infection as a prototype for some parameters in our model. According to [17], the average duration of the infectious period of Covid-19 is 15.2 days. Following the procedure described in Sect. 8.3.5, we conclude that

$$\delta = -\ln\left(1 - \frac{1}{15.2}\right) = 0.06805346324.$$

Since the standard deviation of the average duration of the infectious period is 10.3 days, the value of δ in real circumstances can vary a lot.

To estimate the number of regular travelers, we first collected the data from the Norwegian statistics bureau (SSB) [19] choosing 2019, a year prior to Covid-19 outbreak. SSB reports data quarterly accounting for four types of travel: short trips (1–3 days), long trips (4+ days), domestic business trips, and outbound business trips. In our SSIR model, we assume that the average duration of a short trip is 2 days and 9.5 days for a long trip. Using the information provided by the Travel Leaders Corporate [20], we assume that the average duration of a business trip is 3.05 days for a domestic trip and 5.82 days for an outbound one. We calculated the number of all days of travel per year given that the population of Norway was 5.348 million in 2019. In total, this gave 115.3796 million days of travel in a year. With the maximum number of $365 \cdot 5.348 = 1952.02$ million days the Norwegian population could travel in 2019, we deduce that

$$\sigma = 115.3796/1952.02 = 0.0591,$$

that is, on average, 5.91% of Norwegians were traveling at a given time in 2019. This estimate is based on several assumptions including those on an average amount of travel days and on a uniform distribution of travel throughout the year.

It follows from the above estimate that we expect the quota of non-travelers $\eta = 1 - \sigma$ in our model to be equal to $1 - 0.0591 = 0.9409$, and 25% of all travelers should move daily from a mobile cohort to a nonmobile one since the average duration of a trip is assumed to be 4 days. The likelihood a for an individual to stay in S_1 is $a = 1 + b - b/\eta = 0.98429695$. We can set up now a discrete system describing the transition between the two sub-cohorts of susceptible:

$$\mathbf{X}_{n+1} = \begin{bmatrix} 0.98429695 & 0.25 \\ 0.01570305 & 0.75 \end{bmatrix} \mathbf{X}_n.$$

By using the method explained in Sect. 8.3.5, this discrete system is converted into a continuous one with the matrix A defined by

$$A = \begin{bmatrix} 0.0591\ln(1.36185) & -0.9409\ln(1.36185) \\ -0.0591\ln(1.36185) & 0.9409\ln(1.36185) \end{bmatrix}.$$

This matrix contains the values of two important parameters used in the simulations:

$$\gamma = 0.0591 \ln(1.36185) = 0.0182526845$$

and

$$\omega = 0.9409 \ln(1.36185) = 0.29059138499.$$

If the value of the basic reproduction number $\mathcal{R}_0$ given for the SSIR model by (8.33) is known, along with the values of δ, ω, and γ, we can calculate α if β is given and vice versa. Unfortunately, $\mathcal{R}_0$ varies substantially during an outbreak depending also on the measures taken by the government and health authorities. To date, we do not possess reliable real data describing the spread of infection in two sub-cohorts of susceptible needed to estimate $\mathcal{R}_0$ in our model. The use of the basic reproduction number from the empirical research is not the most reliable way to calculate α or β, but it still can be used to model specific outbreaks. The same remark applies to the determination of the rate at which individuals move from recovered to susceptible. Firstly, there is a significant variation in data, and secondly, arguments and data to support any firm conclusion are still lacking. For our purposes, we assume that recovery takes on average 14 days and we set:

$$\theta = -\ln\left(1 - \frac{1}{14}\right) = 0.07410797215.$$

8.4.2 *Simulations for SIR and SSIR Models*

In this section, we use eight different numerical simulations referred to as "simulation No. 1 (2, 3, ..., 8)" to compare the dynamics of SSIR and SIR models. We use the same notation for the variables and associated parameters in both models for as long as no confusion arises and assign subscripts to parameters identifying their relation to SIR or SSIR models as, for instance, in α_{SSIR} and α_{SIR} whenever needed. We recast an SIR model with reinfections in the form that can be easily compared to the SSIR model:

$$\dot{S} = -\alpha I(t)S(t) + \theta R(t), \tag{8.42}$$

$$\dot{I} = \alpha I(t)S(t) - \delta I(t), \tag{8.43}$$

$$\dot{R} = \delta I(t) - \theta R(t). \tag{8.44}$$

We assume that $N(t) = S(t) + I(t) + R(t)$ is normalized so that

$$N(0) = S(0) + I(0) + R(0) = 1 \tag{8.45}$$

Table 8.1 Reference values for parameters $\delta, \theta, \gamma, \omega$, and $\sigma = \xi = q$

Parameters	Transfer rate between compartments	Value
δ	From I to R (recovery rate)	0.06805346324
θ	From R to S_1 and S_2	0.07410797215
γ	From S_1 to S_2	0.01825268450
ω	From S_2 to S_1	0.29059138499
$\sigma = \xi = q$	Distribution between S_1 and S_2	0.0591

and consider the system (8.42)–(8.44) with the initial condition

$$\begin{bmatrix} S(0) \\ I(0) \\ R(0) \end{bmatrix} = \begin{bmatrix} 0.9999 \\ 0.0001 \\ 0 \end{bmatrix}. \tag{8.46}$$

For the SSIR model, we use the same initial condition, but susceptible are split into two sub-cohorts:

$$\begin{bmatrix} S_1(0) \\ S_2(0) \\ I(0) \\ R(0) \end{bmatrix} = \begin{bmatrix} (1-q)\cdot 0.9999 \\ q \cdot 0.9999 \\ 0.0001 \\ 0 \end{bmatrix}. \tag{8.47}$$

To compare two models on similar terms, we first pick values for α_{SSIR} and β_{SSIR} in the SSIR model and choose the infection rate $\alpha_{\text{SIR}_{\text{adj}}}$ for nonmobile individuals in the SIR model that reflects the impact of a higher infection rate β for mobile sub-cohort in SSIR model.

In our simulations, we use the relation between the transmission rates $\beta_{\text{SSIR}} = 4\alpha_{\text{SSIR}}$; see Sect. 8.5 for more details. In this case, the weighted average is

$$\alpha_{\text{SIR}_{\text{adj}}} = \alpha_{\text{SSIR}}(1-\sigma) + \beta_{\text{SSIR}} \cdot \sigma = \alpha_{\text{SSIR}}(1+3\sigma), \tag{8.48}$$

where σ is the proportion of traveling individuals in the SSIR model. Every time the travel pattern changes, one needs to adjust the value of σ and recalculate the values of four parameters: γ, the transfer rate from S_2 to S_1; ω, the transfer rate from S_1 to S_2; q, the initial distribution between S_1 and S_2; and ξ, the distribution between S_1 and S_2 after the recovery and transfer from recovered to susceptible. The values of $\delta, \theta, \gamma, \omega$, and $\sigma = \xi = q$ for the first four simulations are shown in Table 8.1.

The results of Simulation No. 1 are shown in Fig. 8.3. Numerical test was designed to explore the effect of including travel in an SIR model. The first two plots in Fig. 8.3 compare the dynamics of a classical SIR (a) to that of an SSIR model (b). The values of α and $\mathcal{R}_0$ are given in Table 8.2. Note that the values of infection rate α_{SIR} and α_{SSIR} are equal and quite high which yields very high basic reproduction number $\mathcal{R}_{0_{\text{SIR}}}$ and even higher value for $\mathcal{R}_{0_{\text{SSIR}}}$. Very high reproductive

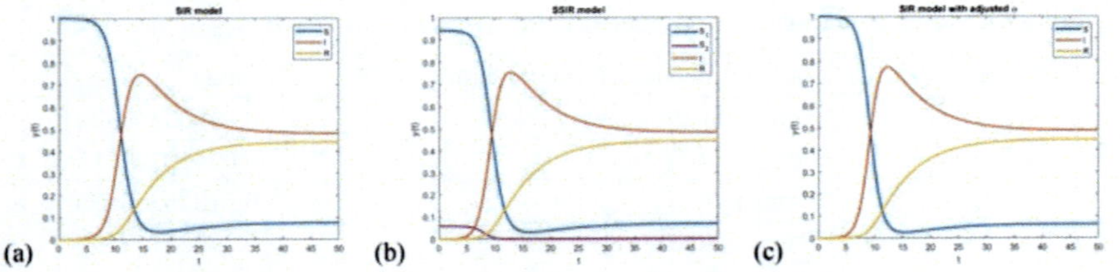

Fig. 8.3 Plots for Simulation No. 1 with parameter values given in Table 8.2

Table 8.2 Values of parameters α_{SIR}, α_{SSIR}, $\alpha_{\text{SIR}_{\text{adj}}}$, $\mathcal{R}_{0_{\text{SIR}}}$, and $\mathcal{R}_{0_{\text{SSIR}}}$ used in Simulation No. 1

Simulation No.	α_{SIR} (a)	α_{SSIR} (b)	$\alpha_{\text{SIR}_{\text{adj}}}$ (c)	$\mathcal{R}_{0_{\text{SIR}}}$	$\mathcal{R}_{0_{\text{SSIR}}}$
1	0.9	0.9	1.05957	13.2249	15.5697

numbers contribute to the fast outbreak of epidemic which happens earlier in case (b). To find the value of α which generates for the SIR model the same dynamics and the same basic reproduction number $\mathcal{R}_{0_{\text{SSIR}}}$ as in the SSIR model, we use the relation (8.48):

$$\alpha_{\text{SIR}_{\text{adj}}} = 0.9 \cdot 0.0591 + 4 \cdot 0.9 \cdot 0.0591 = 1.05957.$$

The adjusted value $\alpha_{\text{SIR}_{\text{adj}}}$ can be found in the Table 8.2; it is noticeably higher than the initial value of α_{SIR}. The dynamics of the "adjusted" SIR model is shown in Fig. 8.3c. One can observe that it matches well that for an SSIR model shown in Fig. 8.3b.

In the next simulation, we use in our SSIR model the basic reproduction number $\mathcal{R}_0 = 4.18$ suggested for Covid-19 by Spencer et al. [17] and test how well does it match the dynamics of the new model. Recall that the value of $\mathcal{R}_{0_{\text{SSIR}}}$ for an SSIR model is given by (8.33). Surprisingly, the latter expression does not contain the coefficient θ related to the return of recovered individuals to one of the two susceptible sub-cohorts and their possible reinfection. This is the consequence of the fact that the eigenvalues of the Jacobian matrix J at the disease-free equilibrium E_0 do not depend on ξ. It is not difficult to deduce from (8.33) that the value of $\mathcal{R}_{0_{\text{SSIR}}}$ can be used to calculate α_{SSIR}, provided that we know δ, γ, and ω:

$$\alpha_{\text{SSIR}} = \mathcal{R}_{0_{\text{SSIR}}} \frac{\delta(\gamma + \omega)}{\omega + 4\gamma}. \tag{8.49}$$

In fact, the value $\alpha_{\text{SSIR}} = 0.2416$ in Table 8.3 was computed using (8.49) with the values δ, γ, and ω in Table 8.1 and $\mathcal{R}_{0_{\text{SSIR}}} = 4.18$. The value of $\mathcal{R}_{0_{\text{SIR}}}$ was then calculated using the formula for the basic reproduction number for an SIR model in [5, p. 115]

$$\mathcal{R}_{0_{\text{SIR}}} = \frac{\alpha}{\delta}, \tag{8.50}$$

Table 8.3 Values of parameters α_{SIR}, α_{SSIR}, $\mathcal{R}_{0_{\mathrm{SIR}}}$, and $\mathcal{R}_{0_{\mathrm{SSIR}}}$ used in Simulation No. 2

Simulation No.	α_{SIR}	α_{SSIR}	$\mathcal{R}_{0_{\mathrm{SIR}}}$	$\mathcal{R}_{0_{\mathrm{SSIR}}}$
2	0.2416	0.2416	3.55	4.18

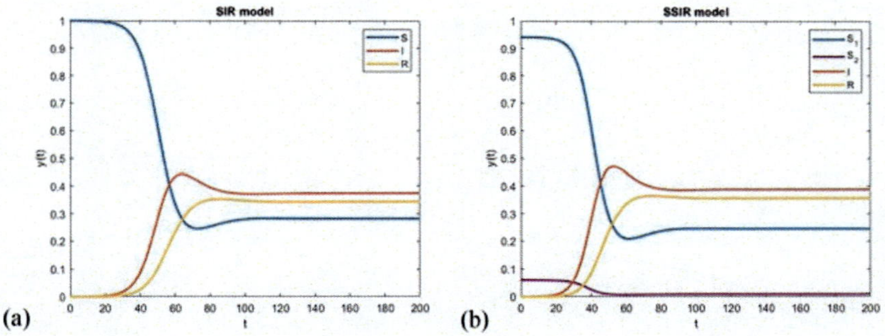

Fig. 8.4 Plots for Simulation No. 2 with parameter values values given in Table 8.3

Table 8.4 Values of parameters α_{SIR}, α_{SSIR}, $\mathcal{R}_{0_{\mathrm{SIR}}}$, and $\mathcal{R}_{0_{\mathrm{SSIR}}}$ used in Simulations No. 3 and No. 4

Simulation No.	α_{SIR}	α_{SSIR}	$\mathcal{R}_{0_{\mathrm{SIR}}}$	$\mathcal{R}_{0_{\mathrm{SSIR}}}$
3	0.0520	0.0520	0.764106	0.9
4	0.0636	0.0636	0.93456	1.1

with the value of δ in Table 8.1 and $\alpha_{\mathrm{SIR}} = 0.2416$. Note that with lower values of α_{SIR} and α_{SSIR} the basic reproduction numbers $\mathcal{R}_{0_{\mathrm{SIR}}}$ and $\mathcal{R}_{0_{\mathrm{SSIR}}}$ are also lower and solution graphs in Fig. 8.4 flatten faster compared to those in Fig. 8.3. Figure 8.4 illustrates the dynamics of SIR (Fig. 8.4a) and SSIR (Fig. 8.4b) models for the same value of α. Observe that again the outbreak starts earlier for the latter model which is associated with a higher value of $\mathcal{R}_{0_{\mathrm{SSIR}}}$ in Table 8.3.

The next test simulates epidemic situation at a border line when the value of the basic reproduction number $\mathcal{R}_{0_{\mathrm{SSIR}}}$ is close to unity. To this end, we change our initial condition for an SIR model to

$$\begin{bmatrix} S(0) \\ I(0) \\ R(0) \end{bmatrix} = \begin{bmatrix} 0.7 \\ 0.2 \\ 0.1 \end{bmatrix}. \tag{8.51}$$

To illustrate qualitative difference between an SIR model and our SSIR model, we run two simulations, one with the value of $\mathcal{R}_{0_{\mathrm{SSIR}}}$ smaller than 1 (Simulation No. 3) and another with $\mathcal{R}_{0_{\mathrm{SSIR}}}$ greater than 1 (Simulation No. 4); see the last column in Table 8.4. The values of α and $\mathcal{R}_{0_{\mathrm{SIR}}}$ are calculated by using formulas (8.49) and (8.50), respectively.

Simulation No. 3 illustrates the dynamics of the SIR model (Fig. 8.5a) and SSIR model (Fig. 8.5b) when the values of $\mathcal{R}_{0_{\mathrm{SIR}}}$ and $\mathcal{R}_{0_{\mathrm{SSIR}}}$ are lower than unity. As expected, in both cases, we observe fast decay of the number of infected to zero. The results of Simulation No. 4 shown in Fig. 8.5c and d are different. Although

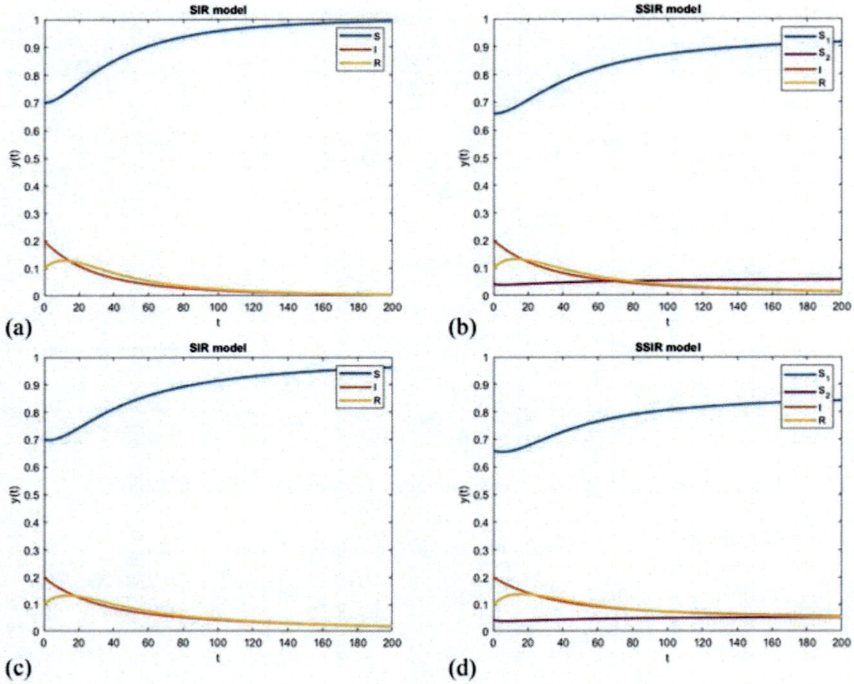

Fig. 8.5 Plots for Simulations No. 3 and 4 with parameter values given in Table 8.4

the number of infected individuals in an SIR model still decays fast to zero as in Fig. 8.5a, the SSIR model exhibits qualitatively different long-term behavior, and the values of all variables approach the endemic equilibrium E_*. The explanation to this situation can be found in Table 8.4. The values of α_{SIR} and α_{SSIR} are higher than in Simulation No. 3 which affects the values of $\mathcal{R}_{0_{\mathrm{SIR}}}$ and $\mathcal{R}_{0_{\mathrm{SSIR}}}$. The value of $\mathcal{R}_{0_{\mathrm{SIR}}}$ is higher than in Simulation No. 3 but still remains smaller than 1, while the value of $\mathcal{R}_{0_{\mathrm{SSIR}}}$ now exceeds 1.

We ran Simulations No. 5–8 for our SSIR model with four sets of parameters given in Table 8.5 to illustrate the impact of a traveling sub-cohort on the spread of disease. Keeping the value of α fixed, we gradually increased the value of σ from 0.001 to 0.15. This corresponds to the increase of the proportion of active travelers from 0.1 to 15 percent of all susceptible. One can observe the changes in the dynamics of the SSIR model in Fig. 8.6. While the plots in Fig. 8.6a and b show a very similar behavior, increasing σ further leads to a significantly earlier outbreak in Fig. 8.6c. The effect of active traveling is even more pronounced in Fig. 8.6d corresponding to a higher value of σ.

It is enlightening to reflect about the differences in the dynamics of the SSIR model in relation to the corresponding values of the basic reproduction number $\mathcal{R}_0$. Although the value of σ in Simulation No. 6 is ten times higher than in Simulation No. 5, the difference between the corresponding values of $\mathcal{R}_0$ is relatively small,

Table 8.5 Values of parameters α_{SIR}, σ, γ, ω, and $\mathcal{R}_0$ used in Simulations No. 5–8

Simulation No.	α_{SSIR}	$\sigma = \xi = q$	γ	ω	$\mathcal{R}_0$
5	0.15	0.001	0.00028801701	0.28772899933	2.210762
6	0.15	0.01	0.00291056373	0.28814580936	2.270274
7	0.15	0.0591	0.01825268450	0.29059138499	2.594945
8	0.15	0.15	0.05224635708	0.29606269012	3.196017

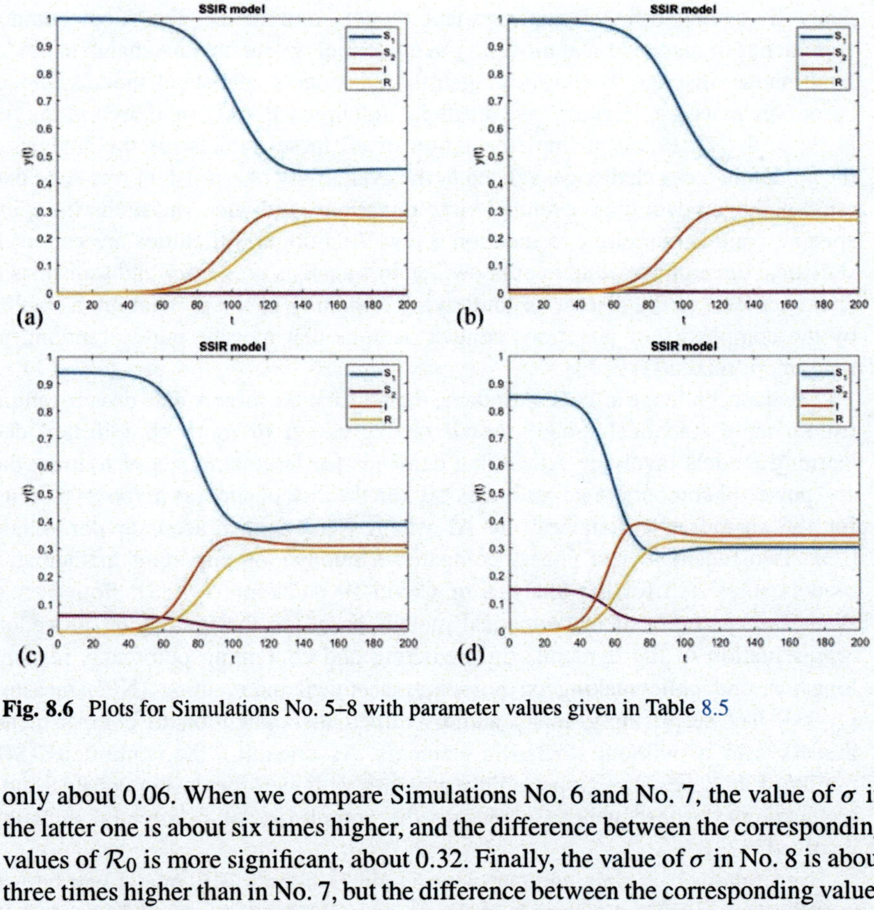

Fig. 8.6 Plots for Simulations No. 5–8 with parameter values given in Table 8.5

only about 0.06. When we compare Simulations No. 6 and No. 7, the value of σ in the latter one is about six times higher, and the difference between the corresponding values of $\mathcal{R}_0$ is more significant, about 0.32. Finally, the value of σ in No. 8 is about three times higher than in No. 7, but the difference between the corresponding values of $\mathcal{R}_0$ is quite large, 0.6. This analysis suggests that the higher is the proportion of traveling individuals among the susceptible, the larger impact travel has on the spread of infection. This agrees with the relevant research reported in the literature; see Sect. 8.5.

8.5 Discussion

Sharp increase of the interest to mathematical epidemiology during the last decade has been fuelled by the recent outbreaks of contagious diseases such as severe acute respiratory syndrome (SARS) in 2002, the Ebola virus in the West Africa in 2014 to 2016, avian flu in 2005, and Covid-19 in 2019. These outbreaks caused numerous concerns, affected millions of people in different countries, took away many lives, and brought mathematical models into focus. The most common approaches to mathematical modeling in epidemiology include mechanistic models (continuous, discrete, stochastic), agent-based models, statistical models, models based on machine learning and artificial intelligence (AI), and hybrid models [1, 2, 5, 6, 24]. Practical implementation of all these models for the analysis of the epidemic faces challenges related to the availability of consistent real-time data, rapid pathogen evolution, eventual virus mutations, and wide variations in region-specific control measures, to mention a few. Additional difficulties are caused by statistical uncertainties due to unknown factors such as properties and transmission characteristics of the disease or underlying immunity of the population, as well as by the complexity of infectious contact patterns that may be purely random, not random, or mixed [1, 2, 23].

The data challenge affects most those models that are largely data driven, ranging from simple statistical models based on regression to more complicated deep learning models involving AI. Strong demands for large amounts of training data and powerful computational resources explain the lack of success in recent forecasts for the spread of coronavirus by AI which were neither accurate nor reliable [23]. Two recent review papers compared a number of important mathematical models suggested for the analysis of Covid-19 pandemic [1, 23]. Both surveys stress the centrality of mathematical models based on the underlying procedural representation of the dynamics in predicting and containing outbreaks, response logistics, and policymaking on non-pharmaceutical intervention (NPI) measures [1, 23]. Not surprisingly, many authors turned their attention to compartmental models with or without stochastic elements. As a result, "the continuous (SIR) family of models seems to be rapidly growing with ever-increasing considerations and variables, subsequently increasing the complexity of differential equations involved" [23].

Even relatively simple compartmental models like an SIR model can be used to efficiently predict the spread of the disease, the total number of infected, the duration of the epidemics, etc. They provide useful information for governmental and medical authorities and policymakers at different levels. However, questions asked at different stages of the pandemic vary, starting with the basic understanding of the pathogen and evaluation of risks of its spread locally and internationally in the beginning and proceeding with the forecast of the dynamics of epidemic and control strategies including the allocation of medical resources and NPI at later stages. When pandemic slows down, the focus shifts to the recovery issues and long-term impact of the pandemic [1]. NPI measures can be taken by individuals and communities to slow down the spread of infectious diseases. Most popular NPI measures include (i) public communication and information campaigns, (ii) personal protective measures (social distancing, hand hygiene, use of facial masks), (iii) adherence to public policies for individual and collective behavior (shift to home offices and remote work mode, restrictions on gatherings, stay at home requirements), and (iv) travel-related measures including post-travel quarantine and border closure [23, 24].

The primary goal of our extension of the classic SIR model was the analysis of the impact of travel restrictions, one of the important NPI measures, on the dynamics of the infectious disease. Given that journal articles and reference books related to travel medicine cite that between 22 and 64% of international travelers become ill during or after travel [3], it is natural to expect that frequent travelers often facilitate the spread of infection. However, the research on this topic based on the use of mathematical models is scarce; we found some relevant information only in two pre-Covid-19 and two post-Covid-19 papers [7, 9, 13, 24]. This agrees with the claim that although travel restrictions are often suggested as an efficient way to reduce the spread of a contagious disease, only a few papers address this issue [7]. Furthermore, the conclusions regarding the impact of travel in the cited papers differ significantly.

Camitz and Liljeros [7] used survey data about travel patterns between municipalities in Sweden collected over 3 years in a stochastic model simulating an outbreak of a disease similar to SARS with three different levels of travel restrictions. Nine scenarios, all starting with a single infected individual in Stockholm and with 1000 realizations, were tested. The simulation period of sixty days was chosen to allow sufficient time for a possible extinction of the disease and to reduce the effect of stochasticity. The authors concluded that a ban on journeys over 50 km would drastically reduce the speed and geographical spread of outbreak and travel restrictions might be the most expedient form of intervention during the pandemic [7].

Hollingsworth et al. [13] employed a synchronous stochastic susceptible-exposed-infectious-recovered (SEIR) model to simulate outbreaks of a SARS-like and an influenza-like airborne respiratory infection in a population in which only 1% of the population travels twenty times more frequently than the rest of the population. The simulations show that frequent travelers accelerate international spread of epidemics only if they are infected early in an outbreak and the outbreak does not expand rapidly. If the epidemic growth rate is high, differences in travel patterns are frequently less noticeable due to the large number of infected persons in the population and increasing likelihood that some of them will travel [13].

Recently, Chinazzi et al. [9] analyzed travel limitations in China during Covid-19 outbreak using an individual-based, stochastic, and spatial global epidemic and mobility model (GLEAM) integrating a meta-population network approach with real data. The world is divided into subpopulations centered around major transportation hubs, and the Covid-19 transmission within each subpopulation is simulated with a compartmental model accounting for four states, susceptible, latent, infectious, and removed. Contrary to the optimistic predictions by Camitz and Liljeros [7], Chinazzi et al. [9] argue that additional travel limitations (up to 90% of traffic) have only a modest effect and should be paired with public health interventions and behavioral changes that can facilitate a considerable reduction in disease transmissibility. Furthermore, their simulations suggest that the Wuhan travel ban in early February 2020 would have reduced the number of cases imported from mainland China to other countries by 77%, but this would have delayed the pandemic only by two weeks [9].

In order to analyze the impact of government policies on the COVID-19 pandemic, Wilk et al. [24] constructed a set of individualized relatively simple SEIR models not allowing repeated infections. They treated a subset of parameters as common for twenty one European countries and the rest as country specific. Thirteen different pandemic control measures were incorporated into models, and their parameters were estimated without prior assumptions. The analysis suggests that the most effective measures are public information campaigns followed closely by income support, school and workplace closures, cancellation of public events, and open testing policy, whereas intuitively powerful measures such as facial coverings, restrictions on gatherings, and stay at home requirements ranked relatively low and would require frequent and strict controls by the authorities. Neither restrictions on internal movement nor international travel controls came out as efficient NPI measures [24].

The literature search revealed lack of reliable information for the evaluation of the impact of travel on the spread of infectious diseases. This concern became the starting point for the design of our SSIR model. As mentioned above, acquisition of realistic data constitutes one of the biggest problems, and little data are currently available for the relative frequency of travel [13]. For our model, we needed the information regarding the difference in infection rates for mobile and nonmobile susceptible sub-cohorts. Our literature search provided only hints rather than answers. The data from Statistics Canada show that, as of April 7, 2020, about 26% of COVID-19 cases in Canada have been related to travel exposure [18]. On a similar note, Lunney et al. [15] reported that 21.5 per 1000 international travelers tested positive for COVID-19.

Hollingsworth et al. [13] modeled a population of 10 million persons with a 0.005 probability of flying per day in large cities in China. Based on a study on domestic flying in Norway from 2000, they simulated the spread of infection in a population in which 1% of the population travel 20 times more frequently than the rest of the population. This, however, might not be the best estimate given that the air travel in Norway is very frequent for large parts of the country where other public transportation means are not available.

The most useful for us information was found in a paper by Bradley et al. on the difference between traveling and non-traveling children with regard to the spread of malaria in Equatorial Guinea [4]. They reported that children who had traveled to the mainland in the previous eight weeks were at greater risk of infection than those who had not traveled (56 vs. 26% in 2013; 42 vs. 18% in 2014). Furthermore, children who had not traveled but were living in localities with the highest proportion of travelers were significantly more likely to be infected compared to those in localities with the smallest proportion of travelers. These data agree with the conclusions by Hollingsworth et al. [13]. Reflecting about the speed of spread of malaria and viral infections, we finally decided to relate infection rates α and β by $\beta = 4\alpha$.

To provide more accurate advice to policymakers, reliable quantitative results are needed [5, p. 117]. As discussed earlier, this is challenging for many reasons, we name just a few. The modeling results in [9] suggest that in mainland China only one in four infections is detected and confirmed. Wilk et al. [24] point out considerable inconsistencies and gaps in reporting numbers of tests. In addition, statistical uncertainties distort the data and require more complex models [1, 23]. Further improvements in the predictions provided by compartmental models could be achieved when more reliable data are available.

8.6 Conclusions

In this chapter, a new model for the analysis of the spread of infectious disease is suggested. The distinctive features are the differentiation between two susceptible sub-cohorts with respect to their travel patterns and the possibility for reinfection. The suggested SSIR model complements existing models designed to explore the impact of travel on the spread of pandemic. Contrary to the models discussed in Sect. 8.5, it does not depend on statistical or survey data and is purely deterministic which is one of its advantages. On the other hand, our SSIR model has not been tested yet for prediction since we lack important real data.

In fact, due to the lack of reliable data regarding the dependence of the infection rate on the intensity of individuals' travel, we ran numerical simulations under the basic assumption that the infection rate is four times higher in the sub-cohort of frequently traveling individuals than in that of non-traveling, cf. [4, 13]. Several scenarios were explored including the comparison of the dynamics of an SIR model with reinfection and our SSIR model and the exploration of the impact of the gradually increasing from 0.1% to 15% travel activity on the spread of infection. We observed that the higher is the proportion of traveling people, the sooner the epidemic outbreak starts, and the larger proportion of the population is affected. The increase of transmission rate and travel frequency may have different impact on the basic reproduction number and thus on the progress of a pandemic. Our simulations confirm that active travel significantly impacts the dynamics of the spread of the disease, and timely travel restrictions help to slow down epidemic. As soon as reliable data are available for the SSIR model regarding the infection transmission rates for two susceptible sub-cohorts are available, the model can be used to predict the spread of infection and provide advice on the control of disease by introducing temporary restrictions on travel.

Regarding future work, we are interested in the acquisition of real data and thorough testing of the model. We would like to explore the reasons for why possibility of reinfection does not affect equilibria and their stability and thus does not influence the basic reproduction number. We also reflect about possibilities to expand the SSIR model by adding new states such as quarantine state, exposed state [8], or hospitalization state. It would be also interesting to model travel to the areas with very bad epidemic situation or to explore the impact of contact network of active travelers on the spread of disease. If, for instance, frequent travelers have almost exclusively contact among themselves, the spread of infection will be significantly restricted. Finally, to account for latent periods, the introduction of the delay into the model would be useful.

Acknowledgments K.H. and Y.R. thank the Department of Mathematical Sciences, Faculty of Engineering and Science, and MatRIC, The Centre for Research, Innovation, and Coordination of Mathematics Teaching at the University of Agder for financial support to attend the International Conference on Mathematical Analysis and Applications in Science and Engineering in Porto, Portugal during June 27–29, 2022 where parts of this research were reported. J.R. and Y.R. gratefully acknowledge the support from the project MeMoV, No. CZ.02.2.69/0.0/0.0/18_053/0016962 funded by the European Union, the Ministry of Education, Youth and Sports of Czech Republic, and Brno University of Technology. The authors also wish to express sincere gratitude to the anonymous reviewer for careful reading of the first draft and useful suggestions for improving the presentation of our results.

References

1. A. Adiga, D. Dubhashi, B. Lewis, M. Marathe, S. Venkatramanan, A. Vullikanti, Mathematical models for COVID-19 pandemic: a comparative analysis. J. Indian Inst. Sci. **100**, 793–807 (2020). https://doi.org/10.1007/s41745-020-00200-6
2. L.J.S. Allen, An introduction to stochastic epidemic models, in: *Mathematical Epidemiology*, ed. by F. Brauer, P. van den Driessche, J. Wu. Lecture Notes in Mathematics, vol 1945 (Springer, Berlin, 2008). https://doi.org/10.1007/978-3-540-78911-6_3
3. K.M. Angelo, P.E. Kozarsky, E.T. Ryan, L.H. Chen, M.J. Sotir, What proportion of international travellers acquire a travel-related illness? A review of the literature. J. Travel Med. **24**, tax046 (2017). https://doi.org/10.1093/jtm/tax046
4. J. Bradley, F. Monti, A.M. Rehman, C. Schwabe, D. Vargas, G. Garcia, D. Hergott, M. Riloha, I. Kleinschmidt, Infection importation: a key challenge to malaria elimination on Bioko Island, Equatorial Guinea. Malar. J. **14**, 46 (2015). https://doi.org/10.1186/s12936-015-0579-5
5. F. Brauer, Mathematical epidemiology: past, present, and future. Infect. Dis. Model. **2**(2), 113–127 (2017). https://dx.doi.org/10.1016/j.idm.2017.02.001
6. F. Brauer, Compartmental models in epidemiology, in *Mathematical Epidemiology*, ed. by F. Brauer, P. van den Driessche, J. Wu. Lecture Notes in Mathematics, vol 1945 (Springer, Berlin, 2008). https://doi.org/10.1007/978-3-540-78911-6_2
7. M. Camitz, F. Liljeros, The effect of travel restrictions on the spread of a moderately contagious disease. BMC Med. **4**, 32 (2006). https://doi.org/10.1186/1741-7015-4-32
8. J.M. Carcione, J.E. Santos, C. Bagaini, J. Ba, A simulation of a COVID-19 epidemic based on a deterministic SEIR model. Front. Public Health **8** (2020). https://www.frontiersin.org/article/10.3389/fpubh.2020.00230
9. M. Chinazzi, J.T. Davis, M. Ajelli, C. Gioannini, M. Litvinova, S. Merler, A. Pastore y Piontti, K. Mu, L. Rossi, K. Sun, C. Viboud, X. Xiong, H. Yu, M.E. Halloran, I.M. Longini Jr., A. Vespignani, The effect of travel restrictions on the spread of the 2019 novel coronavirus (COVID-19) outbreak. Science **368**(6489), 395–400 (2020). https://doi.org/10.1126/science.aba9757

10. T.Z. Fahidy, Solving chemical kinetics problems by the Markov-chain approach. Chem. Eng. Educ. **27**(1), 42–43 (1993). https://journals.flvc.org/cee/article/view/123821
11. F.R. Gantmacher, *The Theory of Matrices*, vol. 2 (Chelsea Publishing Company, New York, 1959). https://archive.org/details/gantmacher-the-theory-of-matrices-vol-2-1959
12. W.H. Hamer, Epidemic disease in England—the evidence of variability and of persistence. Lancet **16**, 733–738 (1906). https://doi.org/10.1016/S0140-6736(01)80187-2
13. T.D. Hollingsworth, N.M. Ferguson, R.M. Anderson, Frequent travelers and rate of spread of epidemics. Emerg. Infect. Dis. **13**, 1288–1294 (2007). https://doi.org/10.3201/eid1309.070081
14. W.O. Kermack, A.G. McKendrick, A contribution to the mathematical theory of epidemics. Proc. R. Soc. Lond. A: Math. Phys. Eng. Sci. **115**, 700–721 (1927). https://doi.org/10.1098/rspa.1927.0118
15. M. Lunney, P.E. Ronksley, R.G. Weaver, L. Barnieh, N. Blue, M.T. Avey, E. Rolland-Harris, F.M. Khan, J.X.Q. Pang, E. Rafferty, T.D. Scory, L.W. Svenson, R. Rodin, M. Tonelli, COVID-19 infection among international travellers: a prospective analysis. BMJ Open. **11**, e050667 (2021). https://doi.org/10.1136/bmjopen-2021-050667
16. R. Ross, *The Prevention of Malaria*, 2nd edn. (John Murray, London, 1910). http://www.archive.org/details/pr00eventionofmalarossrich
17. J.A. Spencer, D.P. Shutt, S.K. Moser, H. Clegg, H.J. Wearing, H. Mukundan, C.A. Manore, Distinguishing viruses responsible for influenza-like illness. J. Theor. Biol. **545**, 111145 (2022). https://doi.org/10.1016/j.jtbi.2022.111145
18. Statistics Canada, *Travel-Related Exposure to COVID-19*. https://www150.statcan.gc.ca/n1/pub/89-28-0001/2018001/article/00018-eng.htm#shr-pg0
19. Statistisk sentralbyrå, 11268: *Trips, by Type of Accomodation and Type of Trip (Million Trips) 2016K1–2021K4*. https://www.ssb.no/en/statbank/table/11268/
20. Travel Leaders Corporate, *Travel Leaders Corporate Releases Q3 Business Travel Trends Data* (2016). https://www.bit.ly/37tgi17
21. P. van den Driessche, J. Watmough, Reproduction numbers and sub-threshold endemic equilibria for compartmental models of disease transmission. Math. Biosci. **180**(1), 29–48 (2002). https://doi.org/10.1016/S0025-5564(02)00108-6
22. F. Viète, *Opera Mathematica* (Londini: Apud Franciscum Bouuier). https://doi.org/10.5479/sil.379493.39088006177687
23. V. Vytla, S.K. Ramakuri, A. Peddi, K. Srinivas, N.N. Ragav, Mathematical models for predicting Covid-19 pandemic: a review. J. Phys.: Conf. Ser. **1797**, 012009 (2021). https://doi.org/10.1088/1742-6596/1797/1/012009
24. A.M. Wilk, K. Łakomiec, K. Psiuk-Maksymowicz, K. Fujarewicz, Impact of government policies on the COVID-19 pandemic unraveled by mathematical modelling. Sci. Rep. **12**, 16987 (2022). https://doi.org/10.1038/s41598-022-21126-2

Chapter 9
Mathematical Oncology: Tumor Evolution Models

Paula Nagy, Eva H. Dulf, and Levente Kovacs

9.1 Introduction

Cancer is the second leading cause of death globally, accounting for an estimated 9.6 million deaths, or one in six deaths, in 2018. Lung, prostate, colorectal, stomach, and liver cancers are the most common types of cancer in men, while breast, colorectal, lung, cervical, and thyroid cancers are the most common among women [1]. The mathematical oncology, the development of mathematical models for tumor evolution, began in the 1060s [2], and today there are different approaches and interpretations in the specialized literature that consider both the parameters that cause the increase in tumor volume and the parameters corresponding to inhibitory or chemotherapeutic drugs.

The tumor represents the exaggerated and abnormal multiplication of cells in the body, which make up a certain type of tissue, having the ability to affect other tissues in the body. Normally, the formation of new cells in the body, starting from the existing ones, is a beneficial process for the body, in the conditions where the

P. Nagy
Faculty of Automation and Computer Science, Automation Department, Technical University of Cluj-Napoca, Cluj-Napoca, Romania

E. H. Dulf (✉)
Automation Department Technical University of Cluj-Napoca, Cluj-Napoca, Romania
e-mail: Eva.Dulf@aut.utcluj.ro

L. Kovacs
Physiological Controls Research Center, Obuda University, Budapest, Hungary
e-mail: kovacs@uni-obuda.hu

C. M. A. Pinto, C. M. Ionescu (eds.), *Computational and Mathematical Models in Biology*, Nonlinear Systems and Complexity 38,
https://doi.org/10.1007/978-3-031-42689-6_9

tissue needs young cells to maintain its proper functioning (e.g., if at the level of the epithelial tissue an injury, it is necessary to form new epithelial cells to replace the destroyed ones). This process is preceded by the death of aged or damaged cells, a stage controlled by certain genes with a role in tissue destruction and multiplication.

Sometimes, due to several causes, errors occur in the cell division process, and mutations occur at the level of these genes, which facilitate the uncontrolled multiplication of the tissue, allowing the survival of abnormal or old cells, forming the primary tumor.

The first mathematical models that describe the evolution of a tumor are simple, based on differential equations, of the first order, which follow the growth of the cancerous tissue, in relation to the physiological processes in the human body, but also the destructive or constructive effect of the administered treatment. One of the first models to appear in the literature is the Hahnfeldt model, describing an antiangiogenic therapy [3]. Among the drugs that block angiogenesis are angiogenesis inhibitors, called anti-angiogenetic agents, which are an important part of the treatment applied to the patient. They are named Endostatin, Angiostatin, and TNP-470 [4]. Another known model is the minimal and extended bilinear model, which is also based on the phenomenon of angiogenesis. The difference from the previous model is that the inhibitor applied is bevacizumab [5]. Over time, models have become more complex and have incorporated additional variables such as the effect of obesity [6], the effect of the drug delivery method [7], and the effect of periodic pulsed drug injection [8] included in the model. In addition, there are various models, which include various therapies such as radiotherapy, chemotherapy, immunotherapy, and a combination of these methods.

To be able to develop a model that describes the dynamics of the evolution of tumors in the body, the main factors that influence the formation and propagation of this process must first be known. Angiogenesis [9] represents the formation of new blood vessels, starting from existing ones. A tumor needs nutrients and oxygen to grow and develop, and blood contains these basic ingredients. Even if the degree of functioning of the angiogenesis process in adults is low, as mentioned in the paper [10], there are several factors that contribute to the prosperity of this process, described, and discussed in the paper [11]:

(a) Vascular endothelial growth factor (VEGF)
(b) Platelet-derived growth factor (PDGF)
(c) Tumor necrosis factor (TNF)

Obesity and excess weight are two causes that intervene in the development of malignant tumors that form at the level of different tissues in the body. Several experimental studies [12–15] have demonstrated that the obesity-cancer relationship is a close one. In paper [16], a mathematical model with differential equations is analyzed that considers the risk factors caused by obesity, but also the connection between the cell's excess fat and the decrease of the patient's immunity.

Fractional mathematics is a branch of mathematics that applies non-integer calculus to address physiological problems. Fractional calculus relevance has constantly grown in recent years, being widely used in very different applications.

The fractional operators (differential or integral) have proved to be suitable to model non-local behavior of materials in time or space, which is crucial in physiological phenomena, but these cannot be described properly by standard mathematical tools [17–20].

There exist several definitions of fractional-order derivative (or integral). One of these is the Riemann–Liouville definition, which states that a fractional-order integral of order $\Re(\alpha) > 0$ is a natural consequence of Cauchy's formula for repeated integrals, expressed as [21]:

$$D_c^{-n} f(t) = \frac{1}{(n-1)!} \int_c^t (t-\tau)^{n-1} f(\tau)\, d\tau, \quad t > c, n \in Z^+ \tag{9.1}$$

This equation can be also expressed as [18]:

$$D_c^{-\alpha} f(t) = \frac{1}{\Gamma(\alpha)} \int_c^t (t-\tau)^{\alpha-1} f(\tau)\, d\tau, \quad t > c, \alpha \in R^+ \tag{9.2}$$

where

$$\Gamma(n) = \int_0^\infty t^{n-1} e^{-t} dt \tag{9.3}$$

is the Euler's *Gamma* function, which is a generalization of a factorial and $n \in R^+$ is an extension of the fractional integral.

This definition became [21]:

$$D^{-\alpha} f(t) = \frac{1}{\Gamma(\alpha)} \int_0^t (t-\tau)^{\alpha-1} f(\tau)\, d\tau, \quad t > 0, \alpha \in R^+ \tag{9.4}$$

for dynamic systems, where $f(t)$ is a causal function of t.

The fractional-order derivative of order $\alpha \in$ R+ can be defined using the Riemann–Liouville formula [21]:

$${}_R D^\alpha f(t) = \frac{d^m}{dt^m} \left[\frac{1}{\Gamma(m-\alpha)} \int_0^t \frac{f(\tau)}{(t-\tau)^{\alpha-m+1}} d\tau \right], \quad m-1 < \alpha < m, \quad m \in N \tag{9.5}$$

or by the alternative definition introduced by Caputo [21]:

$${}_C D^\alpha f(t) = \frac{1}{\Gamma(m-\alpha)} \int_0^t \frac{f^{(m)}(\tau)}{(t-\tau)^{\alpha-m+1}} d\tau, \quad m-1 < \alpha < m, \quad m \in N \tag{9.6}$$

Due to its importance in applications, it must be also enumerated the Grünwald–Letnikov's definition of the fractional-order derivative, based on the generalization of the backward difference [21]:

$$_{GL}D^{\alpha} f(t)=\sum_{k=0}^{m} \frac{f^{(k)}\left(0^{+}\right) t^{k-\alpha}}{\Gamma(m+1-\alpha)}+\frac{1}{\Gamma(m+1-\alpha)} \int_{0}^{t}(t-\tau)^{m-\alpha} f^{(m+1}(\tau)\, d\tau$$

$$m > \alpha-1 \tag{9.7}$$

The purpose of this chapter is to analyze different existing mathematical models for tumor evolution in specialized literature and the generalization of these models using fractional calculus. The goal is to reduce the modeling errors by the use of fractional-order equations in the dynamics description of tumor evolution. Thus, the mean square errors of the existing integer order models are analyzed and calculated, and then a fractional-order α and suitable parameters are chosen so that the value of the mean square error is reduced, closely following the trajectory of the experimental data.

This chapter is structured in five sections. After this first introductory part, Sect. 9.2 presents the analyzed mathematical models with the corresponding fractional-order generalization. In Sect. 9.3, the obtained results are detailed and discussed in Sect. 9.4. This chapter ends with concluding section.

9.2 Methods

In this section, the most used mathematical models existing in the specific literature will be presented.

9.2.1 Hahnfeldt Model

The methods proposed by Hahnfeldt describe the increase of tumor volume under the influence of the vascular network. This is called by authors carrying capacity of the vasculature and it is a variable directly proportional with the extension of blood vessels and which changes over time.

The tumor growth pattern depends on three biological, stimulating factors: the spontaneous loss of functional vascularity, the ability to stimulate the tumor, and the endogenous inhibition of the vascular system (death of endothelial cells or their disaggregation), as well as on an external factor, the inhibition of tumor vascularity due to administered inhibitors. Thus, the model is strictly based on the vascular system of the body, and the control will be bidirectional, whereby a tumor regulates the growth or vascular suppression associated with the tissue, and the tumor vascularization in turn controls the growth of the tumor.

$$V' = -\lambda_1 V \ln\left(\frac{V}{K}\right) \tag{9.8}$$

$$K' = -\lambda_2 K + bV + dKV^{\frac{2}{3}} - eKg(t) \tag{9.9}$$

where K represents carrying capacity of the vasculature, V is tumor volume, $g(t)$ is inhibitor concentration, and $\lambda_1, \lambda_2, b, d$ depend on the inhibitor applied (Endostatin, Angiostatin, or TNP-470). For example, in the paper [22], the growth parameters λ_2, b, and d were identified using a Monte Carlo algorithm (random algorithm for which the outputs are considered random variables, used for numerical simulation, most of the algorithms are based on Hamilton's equations, as in the work [23]), and the parameters e and a, by Gompertz fitting, as in [24]. Gompertz's law express that a model can be classified as a Gompertz model, if in that model there is an initially exponential growth for the elements in the model that are small compared to the elements in the external environment, and which then stabilizes around a point of balance. The differential equation that describes the Gompertz model is:

$$\frac{dN}{dt} = -rN \ln\left(\frac{N}{k}\right) \tag{9.10}$$

where N represents the elements in the model, r is growth rate, and k represents the loading capacity.

The experiments were carried out with the help of laboratory mice, which were implanted with a malignant tumor similar to the characteristics of a tumor in the human body. At the end of the treatment with these inhibitors, the mice were separated into four groups, following the identification of two situations:

(a) The mice responded positively to the treatment, the tumor volume changing depending on the applied inhibitor, resulting in the first three groups of mice, which were administered TNP-470, Endostatin, and Angiostatin.
(b) The mice did not respond to the treatment, resulting in the fourth group of mice (untreated mice).

Figure 9.1 illustrates how the tumor volume changes using the Hahnfeldt model, both in the case of treated and untreated mice, following the administration of these inhibitors. It can be observed that in the case of TNP-470 therapy, the volume tends to grow exponentially, reaching a rather large value, which does not satisfy the patient's expectations. In the case of the Angiostatin inhibitor, the tumor volume does not show considerable variations; it remains around the value of 200 mm^3 and in the case of the Endostatin inhibitor, the volume decreases during the first 2 days of administration, and then remains at the same value. In the case of mice from the fourth group, those for which the treatment with these inhibitors did not work, it can be observed that, in the first days of treatment, the tumor volume stagnates (a longer period of time in the case of the TNP-470 inhibitor and less in the case of Endostatin and Angiostatin), and then the volume increases uncontrollably, reaching a very high upper limit at which it stops. It can also be said that the evolution of the response of untreated mice is similar to the evolution of the response of a first-order system with dead time.

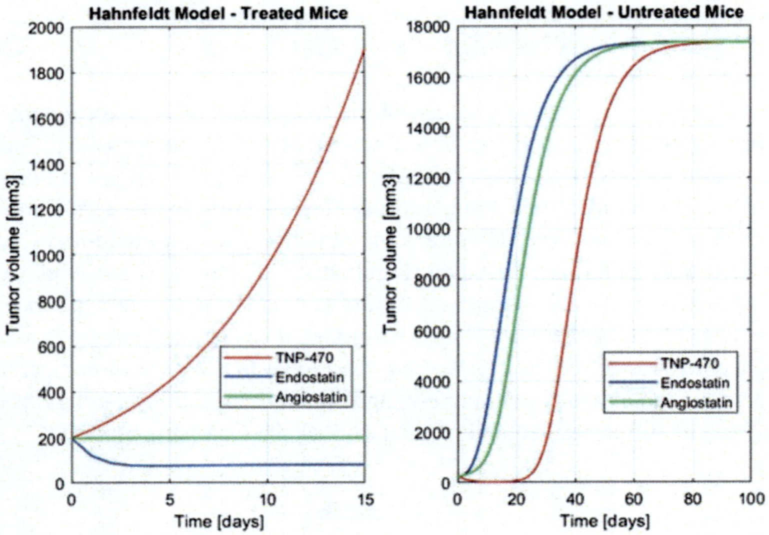

Fig. 9.1 Hahnfeldt model tumor volume evolution

9.2.2 Minimal Bilinear Model

It is a second-order model that contains linear terms and a bilinear term that models the dynamics of the tumor volume and the administered inhibitor level. The difference from the Hahnfeldt model is that the treatment used for identification is based on bevacizumab, a drug that belongs to the class of anti-angiogenic inhibitors [25]. The amount of administered inhibitor changes over time, according to equation:

$$\dot{x} = ax - bxy \tag{9.11}$$

$$\dot{y} = -cy \tag{9.12}$$

where x represents the evolution over time of the tumor volume, y is time variation of the amount of inhibitor, a represents the tumor growth rate, b is tumor inhibition rate, and c is speed of spreading the inhibitor (clearance).

Let $f(x, y) = ax - bxy$ and $h(y) = -cy$.

$$f'(x, y) = \begin{bmatrix} a - by \\ -bx \end{bmatrix} = 0 => x = 0; y = 0$$

The point (0,0) is a stationary point for $f(x, y)$.

$$h'(y) = \begin{bmatrix} 0 \\ -c \end{bmatrix} = 0.$$

The point (0,0) is a stationary point for the model.

$$\begin{bmatrix} a - by & -bx \\ 0 & -c \end{bmatrix}\Bigg|_{x=0, y=0} = \begin{bmatrix} a & 0 \\ 0 & -c \end{bmatrix} \tag{9.13}$$

From Eq. (9.13), it can be said that the stability of the model depends only on the parameters a and c, not on the parameter b. Assuming that it is desired that the administered inhibitor be the same, bevacizumab, the parameter c will be constant, and the stability of the model will be influenced only of the values that the parameter a can have, influencing the eigenvalues of the model.

Let $H = \begin{bmatrix} a & 0 \\ 0 & -c \end{bmatrix}$ => $\det(H - sI) = 0$

$$\begin{vmatrix} a - s & 0 \\ 0 & -c - s \end{vmatrix} = 0 => s^2 + s\,(c - a) - ac = 0$$

$$\Delta = a^2 + c^2 + 2ac$$

$$\begin{cases} \lambda_1 = \frac{a-c-\sqrt{\Delta}}{2} \\ \lambda_2 = \frac{a-c+\sqrt{\Delta}}{2} \end{cases} \tag{9.14}$$

Since $\sqrt{\Delta} > 0$, for any a and c, the model can have an eigenvalue with a positive real part if $a > 0$. Therefore, a parameter can be modified as follows:

$a > 0$ – unstable model with growth rate a
$a \leq 0$ – stable model

Figure 9.2 illustrates the variation of the tumor volume depending on the parameter a. In the case when $a > 0$, the tumor volume increases exponentially, the model being unstable. When $a < 0$, the volume decreases and stabilizes near 0, and when $a = 0$, the volume decreases, stabilizes around the value of 140 mm^3, but fails to reach 0 mm^3, this being the value of interest for the patient.

The second variable of interest is the amount of inhibitor administered, which is no longer constant, as in the case of the Hahnfeldt model. The variation of the amount of bevacizumab is present in Fig. 9.3. It can be observed that, with the passage of time, the amount of bevacizumab required decreases, approaching 0 in the last days of administration of the inhibitor.

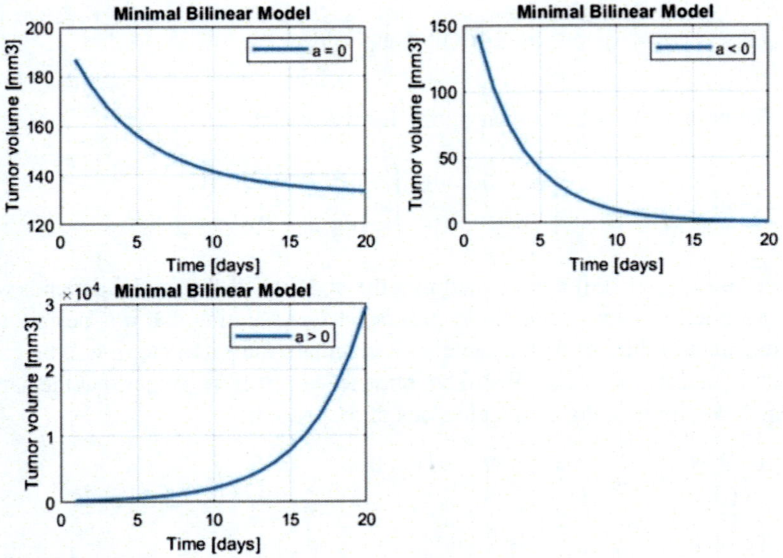

Fig. 9.2 Tumor volume depending on parameter a

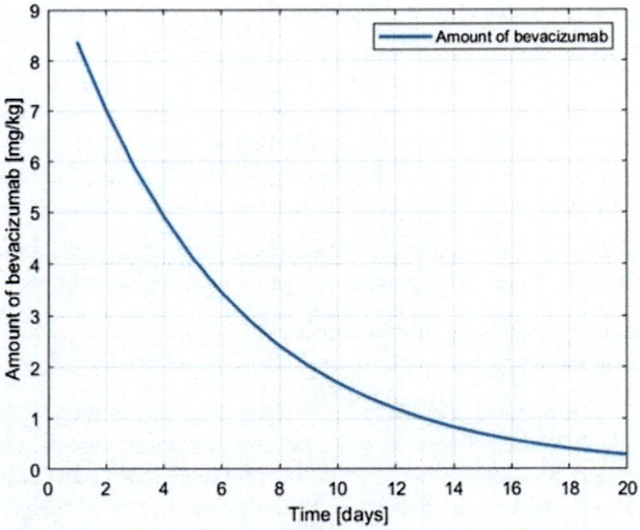

Fig. 9.3 Variation over time of the amount of inhibitor

9.2.3 Extended Bilinear Model

This model incorporates the phenomenon of necrosis [26], which is a stimulator of angiogenesis. Necrosis represents the death of a cell in a living tissue, following an injury. The total volume of the tumor consists of the volume of healthy cells and the volume of degraded cells, their death being an important factor in the formation of the size of the tumor, which must then be treated or removed. The model discussed in the work [27] is of the third order, which describes the most important physiological phenomena, such as the spread of the tumor and the effect of the inhibitor, but less so the dynamics of blood vessels.

$$\dot{x}_1 = (a - n)\,x_1 - b\frac{x_1 x_3}{\mathrm{ED}_{50} + x_3} \tag{9.15}$$

$$\dot{x}_2 = n x_1 + b\frac{x_1 x_3}{\mathrm{ED}_{50} + x_3} \tag{9.16}$$

$$\dot{x}_3 = -c\frac{x_3}{K_{\mathrm{B}} + x_3} - b_{k\frac{x_1 x_3}{\mathrm{ED}_{50}+x_3}+u} \tag{9.17}$$

$$\dot{y} = a x_l \tag{9.18}$$

$$y = x_1 + x_2 \tag{9.19}$$

where the meaning of variables and constants and variables is:

x_1 – time evolution of the proliferation tumor volume
x_2 – time evolution of the necrotic tumor volume
x_3 – time variation of the level inhibitor
u – inhibitor injection speed
a – tumor growth rate
b – tumor inhibition rate
c – clearance of the inhibitor
n – necrosis rate
b_{k} – rate of change of level inhibitor
K_{B} – Michaelis–Menten constant of the inhibitor
ED_{50} – average dose of inhibitor
y – time evolution of tumor volume

A particularity of this model is the existence of equilibrium points, also analyzed in the paper [28]. From Eqs. (9.15), (9.16), and (9.17), the following equilibrium point results:

$$\begin{cases} x_{10} = 0 \\ x_{20} = f\,(x1, x3) \\ x_{30} = \frac{\mathrm{ED}_{50}(n-a)}{-b-n+a} \\ u_0 = c\frac{x_3}{K_B+x_3} + b_{k\frac{x_1 x_3}{\mathrm{ED}_{50}+x_3}} \end{cases} \tag{9.20}$$

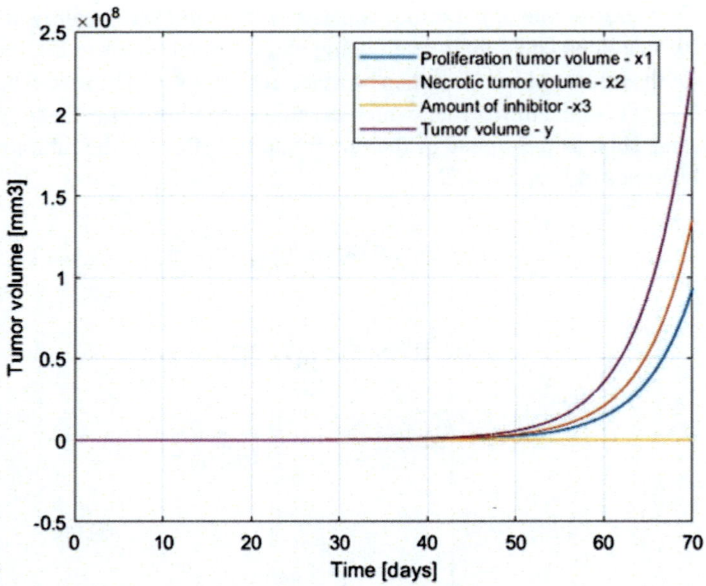

Fig. 9.4 Tumor volume – extended bilinear model, $x_3 < 0$

According to the system of Eq. (9.13), it can be said that the model reaches equilibrium when the tumor volume is 0; in other words, there is no more proliferating tumor in the body. The volume of the necrotic tumor (x_2) is a function that depends on the volume of the proliferating tumor (x_1) and the amount of inhibitor (x_3), and the input applied to the model will be the injection speed of the inhibitor, also depending on x_1 and x_3. The amount of inhibitor applied (x_3) cannot be negative, thus distinguishing three cases:

(a) $(n - a) < 0 => x_3 < 0$
(b) $(n - a) > 0 => x_3 > 0$
(c) $n = a => x_3 = 0$

Figure 9.4 illustrates the variation of the tumor volume, according to the extended bilinear model, when it is assumed that $x_3 < 0$. Although this situation is not a realistic one, the study of this model can be a starting point for the following directions. Thus, in this case, both the volume of the proliferating tumor and the volume of the necrotic tumor are increasing, and the amount of inhibitor also increases during administration. The volume of the necrotic tumor shows a greater increase, favoring the spread of tumor cells to other tissues as well.

The other situation is when $x_3 \geq 0$, as in Fig. 9.5. Following the application of a considerable amount of inhibitor, the volume of the proliferating tumor decreases, but the volume of the necrotic tumor increases exponentially, the final volume being the sum of the two, according to Eq. 9.19. The volume of the necrotic tumor shows too much growth rate for 70 days of simulation, compared to the real situation, when

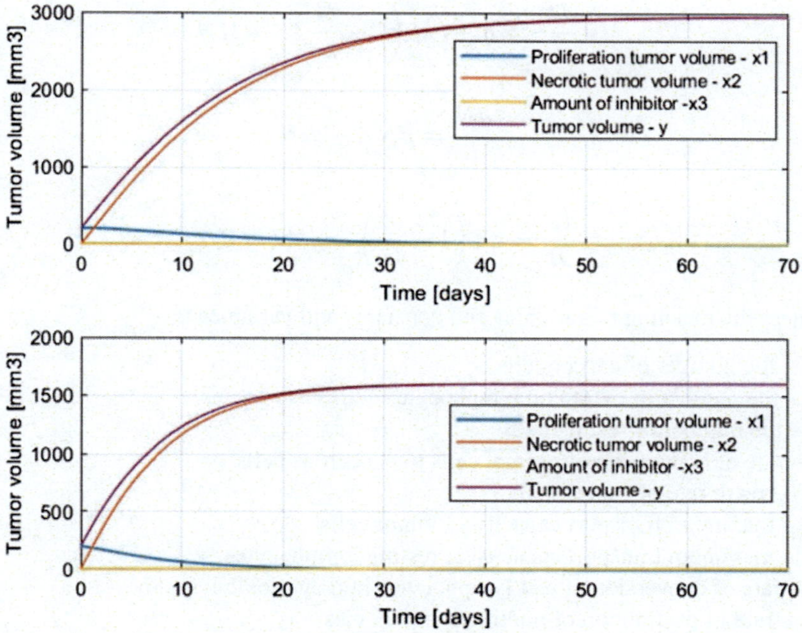

Fig. 9.5 Tumor volume – extended bilinear model, $x_3 \geq 0$

the human body quickly eliminates dead cells through the inflammatory/immune system. Thus, the first limitation of this model takes into account the fact that it is not possible to decrease the volume. A control law is needed to obtain the desired result.

9.2.4 Predatory-Predator Model Without Delay in Conversion of Resting Cells of Hunting Cells

In the work [29], a third-order nonlinear model is analyzed, which describes the dynamics of the evolution of malignant tumor cells, together with cytotoxic and resting T cells. It is assumed that the transformation of resting cells into hunting cells is done almost instantaneously, without considerable delays, and once activated, by eliminating cytokines, cytotoxic cells cannot return to the resting state. Both cytotoxic and resting T lymphocytes have the possibility to act on tumor cells, or they can die, being destroyed by certain external factors. This process is part of the system of self-defense mechanisms of the human body, and thus no tumor inhibitor or chemotherapeutic drug is administered, with tumor regression taking place following the internal destruction of tumor cells, as described by the equations:

$$\frac{dM}{dt} = q + rM\left(1 - \frac{M}{k_1}\right) - \alpha MN \tag{9.21}$$

$$\frac{dN}{dt} = \beta NZ - d_1 N \tag{9.22}$$

$$\frac{dZ}{dt} = sZ\left(1 - \frac{Z}{k_2}\right) - \beta NZ - d_2 Z \tag{9.23}$$

where the meaning of variables and constants and variables is:

M – the number of cancer cells
N – the number of cytotoxic lymphocytes cells
Z – the number of resting cells
q – rate of conversion of normal cells to cancerous cells
r – growth rate of cancer cells
k_1 – maximum transport capacity of tumor cells
k_2 – maximum transport capacity of resting lymphocytes
β – rate of conversion of rest lymphocytes into cytotoxic lymphocytes
d_1 – rate of destruction of hunting lymphocytes
d_2 – rate of destruction of resting lymphocytes
s – resting rate of hunting lymphocytes

9.2.5 Predatory-Predator Model with Delay in Conversion of Resting Cells of Hunting Cells

This model is based on the action of cytotoxic or hunting T lymphocytes. Hunting cells can induce programmed cell death (apoptosis) of malignant tumor cells, or instant death by destroying or ingesting them. To increase the number of cytotoxic cells, the resting cells are activated and transformed into hunting cells by secreting substances called cytokines.

The model is based on the equations of the model in paper [30], a nonlinear system of the third order, to which is added the delay produced by the conversion of resting cells into hunting cells, expressed by the delay factor τ.

$$\frac{dM}{dt} = r_1 M\left(1 - \frac{M}{k_1}\right) - \alpha_1 MN \tag{9.24}$$

$$\frac{dN}{dt} = \beta NZ(t - \tau) - d_1 N - \alpha_2 \tag{9.25}$$

$$\frac{dZ}{dt} = r_2 Z\left(1 - \frac{Z}{k_2}\right) - \beta NZ(t - \tau) \tag{9.26}$$

where the meaning of variables and constants and variables is:

M – the number of cancer cells
N – the number of cytotoxic lymphocytes cells
Z – the number of resting cells
r_1 – growth rate of cancer cells
r_2 – growth rate of resting T lymphocytes
α_1 – rate of distruction of tumor cells by cytotoxic lymphocytes
α_2 – rate of degradation of cytotoxic lymphocytes by tumor cells
β – rate of conversion of rest lymphocytes into cytotoxic lymphocytes
k_1 – maximum transport capacity of tumor cells
k_2 – maximum transport capacity of resting lymphocytes
d_1 – rate of destruction of hunting lymphocytes

9.2.6 Fractional-Order Models

In order to try to reduce the modeling errors, fractional-order equivalents of the classical models are proposed, which can describe more precisely the evolution dynamics [31].

Replacing the first-order derivative with a non-integer order derivative $\alpha > 0$, the Hahnfeldt model will have the following form:

$$\frac{dV(t)^{\alpha}}{dt} = -\lambda_1 V(t) \ln\left(\frac{V(t)}{K(t)}\right) \tag{9.27}$$

$$\frac{dK(t)^{\alpha}}{dt} = -\lambda_2 K(t) + bV(t) + dK(t)V(t)^{\frac{2}{3}} - eK(t)g(t) \tag{9.28}$$

where $\alpha \in (0;1)$ and K represents carrying capacity of the vasculature, V is tumor volume, $g(t)$ is inhibitor concentration, and λ_1, λ_2, b, d depend on the inhibitor applied (Endostatin, Angiostatin, or TNP-470).

The fractional-order derivative of the minimal bilinear model will have the following form:

$$\frac{dx(t)^{\alpha}}{dt} = ax(t) - bx(t)y(t) \tag{9.29}$$

$$\frac{dy(t)^{\alpha}}{dt} = -cy(t) \tag{9.30}$$

where $\alpha \in (0;1)$, x represents the evolution over time of the tumor volume, y is time variation of the amount of inhibitor, a represents the tumor growth rate, b is tumor inhibition rate, and c is speed of spreading the inhibitor (clearance).

The form of the fractional-order extended bilinear model will be:

$$\frac{dx_1(t)^\alpha}{dt} = (a - n)\,x_1 - b\frac{x_1 x_3}{\mathrm{ED}_{50} + x_3} \tag{9.31}$$

$$\frac{dx_2(t)^\alpha}{dt} = n x_1 + b\frac{x_1 x_3}{\mathrm{ED}_{50} + x_3} \tag{9.32}$$

$$\frac{dx_3(t)^\alpha}{dt} = -c\frac{x_3}{K_\mathrm{B} + x_3} - b_{k\frac{x_1 x_3}{\mathrm{ED}_{50}+x_3}+u} \tag{9.33}$$

$$y = x_1 + x_2 \tag{9.34}$$

with $\alpha \in (0;1)$ and the meaning of variables and constants are as follows:

x_1 – time evolution of the proliferation tumor volume
x_2 – time evolution of the necrotic tumor volume
x_3 – time variation of the level inhibitor
u – inhibitor injection speed
a – tumor growth rate
b – tumor inhibition rate
c – clearance of the inhibitor
n – necrosis rate
b_k – rate of change of level inhibitor
K_B – Michaelis–Menten constant of the inhibitor
ED_{50} – average dose of inhibitor
y – time evolution of tumor volume

9.3 Results

To analyze the effectiveness of the models, it will be analyzed if the simulated model follows the evolution of the experimental data from laboratory mice, by calculating the mean squared error:

$$\mathrm{MSE} = \frac{1}{n}\sum_{i=1}^{n}\left(y_i - \hat{y}_i\right)^2 \tag{9.35}$$

Table 9.1 Mean squared error for the Hahnfeldt model

Treated mice	TNP - 470	Endostatin	Angiostatin
MSE	204.892	24.14	0.0014

Table 9.2 Mean squared error for the bilinear model

Bilinear model	Minimal	Extended
MSE	38.88	35.23

where n represents the number of samples for each model, y_i are the experimental data values, and $\hat{y}_i$ the values estimated by the models.

The used experimental data are from [27], where the tumor evolution was monitored on laboratory mice, aged 6–8 weeks, which were implanted with a tumor with a volume of 200 mm^3. These mice were divided into three groups and administered tumor inhibitory substances such as TNP-470, Endostatin, and Angiostatin. The volume of the tumor was measured on days 4, 7, 10, and 13 of treatment. The best results obtained with the Hahnfeldt model are presented in Table 9.1. It can be seen that the highest value of the error is in the case of the administration of TNP-470, followed by the situation in which Endostatin was administered. The lowest value of the mean square error is obtained in the case of the administration of Angiostatin, 0.0014.

In the case of the bilinear model, the experiment that monitored the evolution of the tumor took place on laboratory mice was divided into two groups: control and treated case. Mice in the control group were given a single dose of bevacizumab at the beginning of the treatment, and those in the case group were given 0.05 mg/kg of bevacizumab every day. The volume of the tumor was measured on days 4, 7, 10, 13, 16, and 19 of treatment. It can be seen in Table 9.2 that in the case of this model, the mean squared errors have quite high values, which suggests that the simulated model does not accurately follow the experimental data.

The same simulation was carried out for the fractional-order equivalents of these models. The simulation are realized for a period of 20 days. The initial volume is considered $V = 200$ mm^3. Are applied the same inhibitor concentrations as in the real experiments: 10 mg/ml bevacizumab for the minimal fractional bilinear model (MFBM) and 0.171 mg/ml bevacizumab in the first day for the extended fractional bilinear model (EFBM). The used model parameters are in Table 9.4. The evolution of fractional-order bilinear models are shown in Fig. 9.7.

In the case of TNP-470 administrations, the smallest mean squared error is obtained for $\alpha = 0.6$. As can be seen in Fig. 9.6a), for $\alpha < 0.6$, the average error increases, for $\alpha = 0.6$ it decreases, and when $\alpha > 0.6$, the average error starts to increase again.

In the case of Endostatin inhibitor administrations, the smallest mean squared error is obtained for $\alpha = 0.5$. As can be seen in Fig. 9.6b), for $\alpha < 0.5$, the average error increases, for $\alpha = 0.5$ it decreases, and when $\alpha > 0.5$, the average error starts to increase again.

In the case of Angiostatin administration, the threshold of α is 0.7, as illustrated in Fig. 9.6c). Therefore, for $\alpha < 0.7$, the average error increases, for $\alpha = 0.7$, it

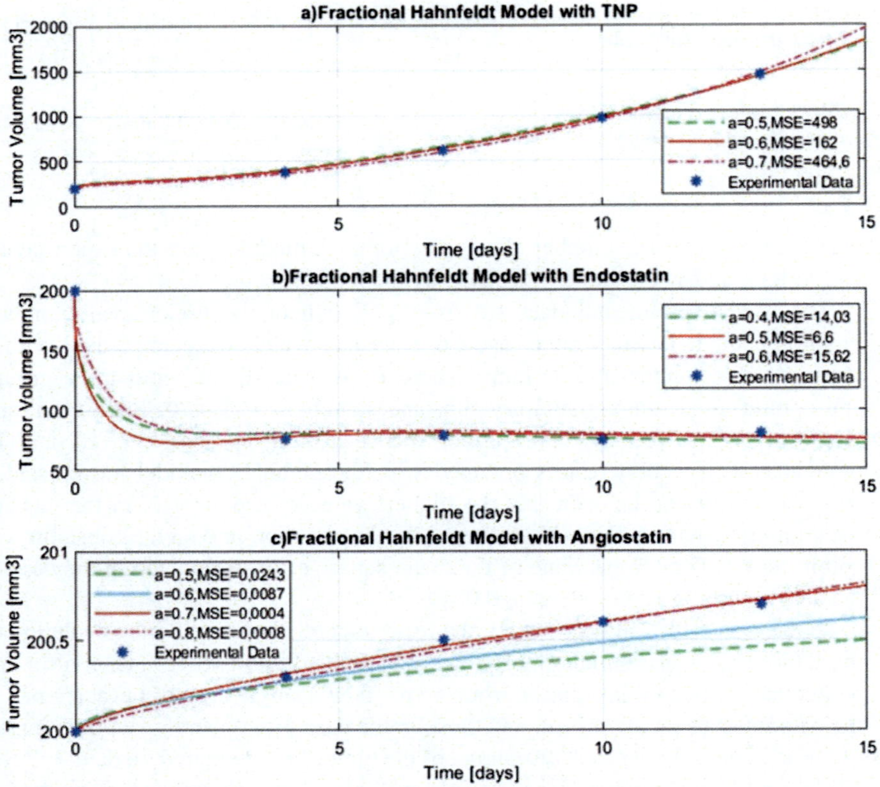

Fig. 9.6 Comparison of different α values for fractional Hahnfeldt model

Table 9.3 The parameters used for the simulation of the fractional-order Hahnfeldt model

	e(day^{-1})	clr (day^{-1})	λ_1 (day^{-1})	λ_2 (day^{-1})	b (day^{-1})	d(day^{-1} vol$^{-2/3}$)	K(vol)	a	α
TNP-470	1.3	1.1	0.25[a]	1.5[a]	7.1[a]	0.01[a]	625[a]	1.5	0.6
Endostatin	1.3	0.66	0.4[a]	2.5[a]	0.08[a]	0.008[a]	625[a]	0.45	0.5
Angiostatin	0.15	0.38	0.192[a]	0.5[a]	5.85[a]	0.00873[a]	625[a]	7	0.7

[a]tumor growth rate

decreases and reaches the lowest value, and when $\alpha > 0.7$, the average error starts to increase again.

For each previously determined α, starting from the idea of reducing the mean squared error between the experimental data and the model, the best results were obtained for the parameters in Table 9.3.

The simulation are realized for a period of 20 days. The initial volume is considered $V = 200$ mm^3. Are applied the same inhibitor concentrations as in the real experiments: 10 mg/ml bevacizumab for the minimal fractional bilinear model

Table 9.4 The parameters used for the simulation of the fractional-order bilinear model

	$a(\text{day}^{-1})$	$b(\text{day}^{-1})$	$c(\text{day}^{-1})$	n (day^{-1})	b_k mg/(ml × day)	K_B (mg/ml)	ED_{50} (mg/ml)	xo (mm^3)	α
MFBM	0.58	1.25	2.18	0.103	$1.0839 \cdot 10^{-6}$	0.4409	$50 \cdot 10^{-6}$	200	0.7
EFBM	2.2	0.1685	0.1825	0.103	$1.0839 \cdot 10^{-6}$	0.4409	$50 \cdot 10^{-6}$	200	0.4

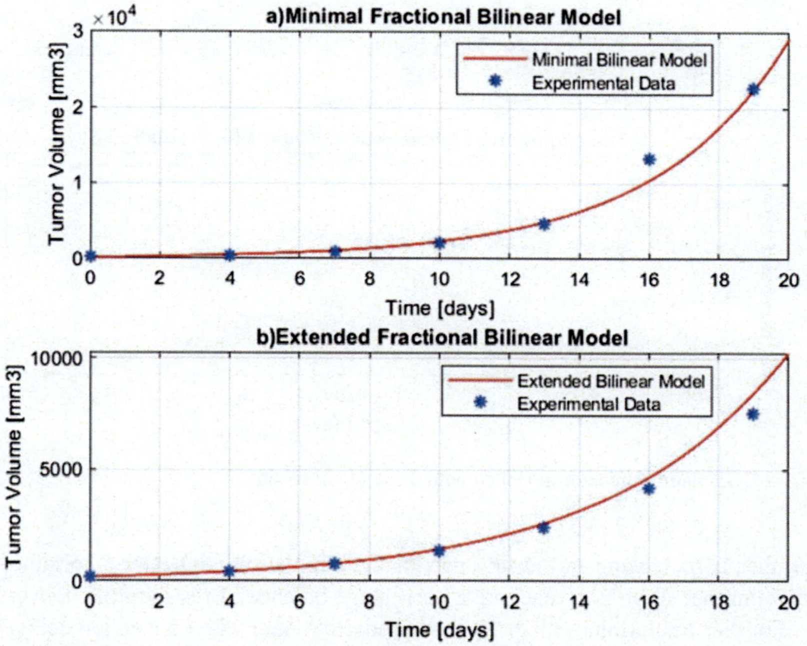

Fig. 9.7 Minimal and extended fractional-order bilinear model simulation results

Table 9.5 Mean squared error for fractional Hahnfeldt model and fractional bilinear model

Model	Hahnfeldt model			Bilinear model	
	TNP	Endostatin	Angiostatin	Minimal	Extended
MSE	162	6.6	0.0004	464	>500

(MFBM) and 0.171 mg/ml bevacizumab in the first day for the extended fractional bilinear model (EFBM). The used model parameters are in Table 9.4. The evolution of fractional-order bilinear models are shown in Fig. 9.7.

The obtained mean squared errors for these models are presented in Table 9.5.

The results obtained for the predatory-predator model are presented in Fig. 9.8, showing the evolution of the number of the three types of cells, following the simulation for a period of 25,000 days. It can be observed that first the number of tumor cells increases, then decreases, and finally it shows constant oscillations, while the number of resting lymphocytes decreases, turning into hunting lymphocytes, whose

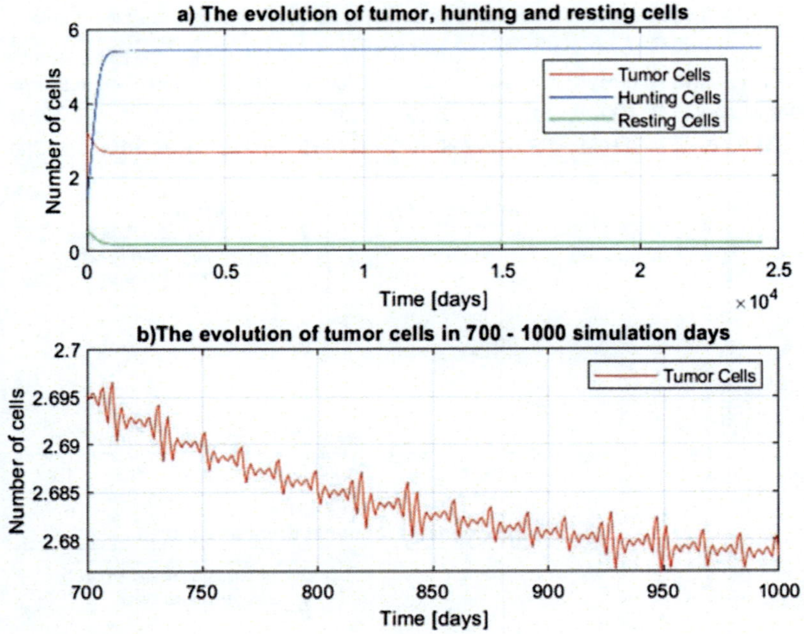

Fig. 9.8 Predatory-predator model without delay of conversion

number is increasing in the first part of the simulation, and after a certain period, their number stabilizes, reaching a percentage balance of rest-hunting conversion.

For the simulation of the predatory-predator model with conversion delay, a time of 5000 days was used, and the delay factor is 46 days. The results of this simulation are presented in Fig. 9.9. In this situation, the number of cytotoxic lymphocytes decreases strongly, and the number of those at rest increases, due to the delay that occurs in the rest-hunting conversion (Fig. 9.10). As in the case of the predatory-predator model without a delay factor, the number of tumor cells fluctuates strongly, but first their number increases and does not show any decrease, a situation that is not desired to occur.

9.4 Discussion

In Fig. 9.6, it can be seen that the shape in which the tumor volume evolves following the simulation of the fractional-order Hahnfeldt model, applying the inhibitors TNP-470, Endostatin, and Angiostatin, does not differ from that of the full order Hahnfeldt model. The corresponding fractional-order models have reduced mean squared errors, as are presented in Table 9.5. This indicates that the introduction of the fractional order and the estimation of the new parameters lead

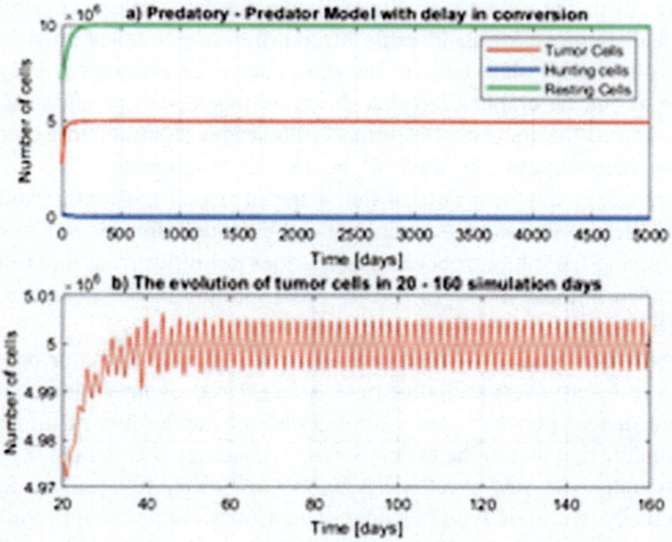

Fig. 9.9 Predatory-predator model with delay of conversion

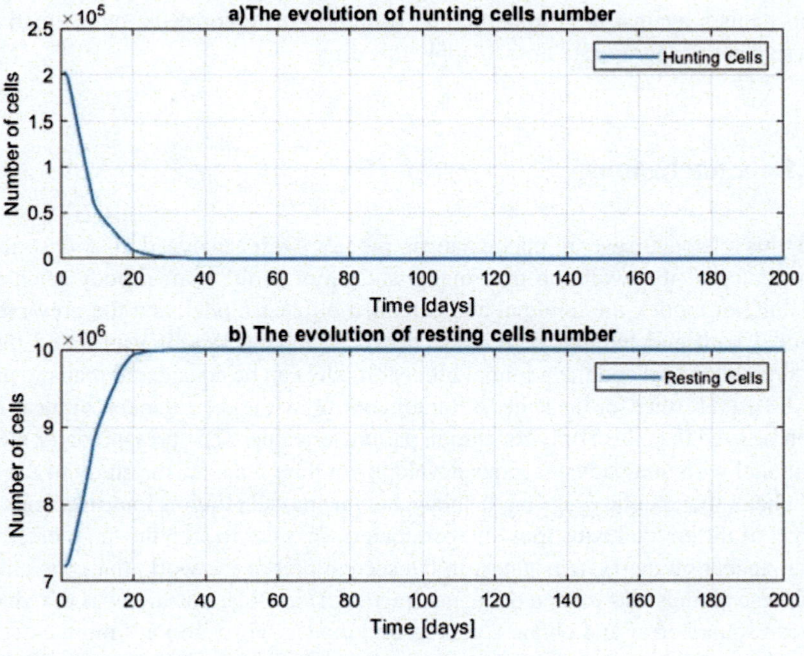

Fig. 9.10 Evolution of the number of lymphocytes with delay factor

to a better fit of the model with the experimental data, reflecting better the tumor evolution dynamics. A considerable error difference is noted in the case of the simulation with TNP-470, but also the simulation with Endostatin and Angiostatin. However, it can be emphasized that the inhibitors TNP-470 and Angiostatin fail to contribute to the reduction of tumor volume, but contrary to expectations, the volume increases, quite a lot for TNP, but less for Angiostatin.

In Fig. 9.7, it can be seen that the shape in which the tumor volume evolves following the simulation of the minimal and extended bilinear model of fractional order, applying the inhibitor bevacizumab, does not differ from that of the minimal and extended bilinear model of integer order, but the mean squared errors can be reduced by proper choice of this fractional order.

In the case of predatory-predator models, the number of tumor cells constantly oscillates in a very short period of time, a fact that does not reflect reality, as such a frequent increase or decrease is not possible of the number of tumor cells. This problem may arise due to the estimation of parameters β (the rate of conversion of resting lymphocytes into cytotoxic lymphocytes) or α_1 (the rate of destruction of tumor cells by cytotoxic lymphocytes), their values not being appropriate, because they fail to control the oscillations in the system. However, this model could be used further if trying to estimate the activation/conversion rate as realistically as possible. The advantage of this model is that it could represent a starting point for the application of a chemotherapeutic drug, which complements the body's self-defense mechanism against the formation and evolution of tumor cells, by estimating the dose and the appropriate time to apply this drug.

9.5 Conclusions

In this chapter, several mathematical models were analyzed that describe the dynamics of the evolution of a malignant tumor in the human body, such as the Hahnfeldt model, the minimal and extended bilinear model, and the prey-predator model with and without conversion delay. Following this, it was found that the Hahnfeldt model and the minimal bilinear model can be considered realistic models to be used further for the study of tumor control, while the extended bilinear model can be used in control only for certain parameter values. The prey-predator model is a model with quite advantageous development directions for the study of the effect of chemotherapeutic drugs on the body, but this model requires improvements at the level of differential equations and parameters, in order to stop the large oscillations that appear on the system states. In the second part of the work, the generalization of these models was carried out using fractional calculus. The aim was to reduce the mean square error and obtain a generalized model, which follows more closely the experimental data evolutions. It is shown that choosing the right fractional order, the tumor dynamics can be captured more realistic with these models.

On the other hand, it is concluded that in some cases, the inhibitors fail to reduce the volume of the tumor. This leads to the future goal of designing a control law on the model that will compute the proper dose and amount of inhibitor or cocktail of inhibitors to be administered, customized to the requirements and needs of each patient, as in the work [32].

Acknowledgment This work was supported by a grant of the Ministry of Research, Innovation and Digitization, CNCS-UEFISCDI, project number PN-III-P4-PCE-2021-0750, within PNCDI III.

References

1. WHO, Cancer website (2014). http://www.who.int/cancer/en/. Last accessed 28 Nov 2022
2. M. Schwartz, A biomathematical approach to clinical tumor growth. Cancer **14**, 1272–1294 (1961)
3. A.L. Harris, Angiogenesis as a new target for cancer control. EJC Suppl., 1–12 (2003)
4. Z. Lin, Q. Zhang, W. Luo, Angiogenesis inhibitors as therapeutic agents in cancer: Challenges and future directions. Eur. J. Pharmacol., 76–81 (2016)
5. J. Sápi, L. Kovács, D.A. Drexler, P. Kocsis, D. Gajári, Z. Sápi, Tumor volume estimation and quasi-continuous administration for most effective bevacizumab therapy. PLoS One, 1–20 (2015)
6. R.A. Ku-Carrillo, S.E. Delgadillo, B. Chen-Charpentier, A mathematical model for the effect of obesity on cancer growth and on the immune system response. Appl. Math. Model., 7–8 (2016)
7. L. Cattaneo, P. Zunino, A computational model of drug delivery through microcirculation to compare different tumor treatments. Int. J. Numeric. Methods Biomed. Eng. **30**(11), 1347–1371 (2014)
8. J.C. Panetta, A mathematical model of periodically pulsed chemotherapy: Tumor recurrence and metastasis in a competitive environment. Bull. Math. Biol. **58**(3), 425–447 (1996)
9. J.H. Distler, A. Hirth, M. Kurowska-Stolarska, R.E. Gay, S. Gay, O. Distler, Angiogenic and angiostatic factors in the molecular control of angiogenesis. Q. J. Nucl. Med. **47**(3), 149–161 (2003)
10. J. Drevs, M. Medinger, C. Schmidt-Gersbach, R. Weber, C. Unger, Receptor tyrosine kinases: The main targets for new anticancer therapy. Curr. Drug Targets, 113–121 (2003)
11. A.L. Harris, Angiogenesis as a new target for cancer control. Eur. J. Cancer Suppl. **1**, 1–12 (2003)
12. H. Rubin, Promotion and selection by serum growth factors drive field cancerization, which is anticipated in vivo by type 2 diabetes and obesity. Proc. Natl. Acad., 13927–13931 (2013)
13. E.A. Ehsanipout, X. Sheng, Y.W. Behan, X. Wang, A. Butturini, V.I. Avramis, S.P. Mittelman, Adipocytes cause leukemia cell resistence to l-asparagines via release of glutamin. Cancer Res., 2998–3006 (2013)
14. G. Taubes, Unravelling the obesity-cancer connection. Science **335**, 28–32 (2012)
15. S.D. Hursting, Minireview: The year in obesity and cancer. Mol. Endocrinol., 1961–1966 (2012)
16. R.A. Ku-Carrilloa, S.E. Delgadillo, B.M. Chen-Charpentier, A mathematical model for the effect of obesity on cancer growth and on the immune system response. Appl. Math. Model. **40**, 4908–4920 (2016)
17. T. Alinei-Poiana, E.H. Dulf, L. Kovacs, Fractional calculus in mathematical oncology. Sci. Rep. **13**, 10083 (2023). https://doi.org/10.1038/s41598-023-37196-9
18. E.H. Dulf, D.C. Vodnar, A. Danku, C.I. Muresan, O. Crisan, Fractional-order models for biochemical processes. Fractal Fract **4**(2), 12 (2020)

19. E.H. Dulf, F.V. Dulf, C.I. Muresan, Fractional model of the cryogenic (13C) isotope separation column. Chem. Eng. Commun. **202**(12), 1600–1606 (2015)
20. E.H. Dulf, C.I. Pop, F.V. Dulf, Fractional calculus in 13 C separation column control. SIViP **6**, 479–485 (2012)
21. C.A. Monje, Y.Q. Chen, B.M. Vinagre, D. Xue, V. Feliu, *Fractional-Order Systems and Controls* (Springer, 2010). https://doi.org/10.1007/978-1-84996-335-0
22. P. Hahnfeldt, D. Panigrahy, J. Folkman, L. Hlatky, Tumor development under angiogenic signaling: A dynamical theory of tumor growth, treatment response, and postvascular dormancy. Cancer Res., 4770–4775 (1999)
23. A.M. Horowitz, A generalized guided Monte Carlo algorithm. Theory Division., Switzerland, 247–252 (1991)
24. A. Ali, M. Hussain, A. Ghaffar, Z. Ali, K. Sooppy Nisar, M.R. Alharthi, W. Jamshed, Numerical simulations and analysis for mathematical model of avascular tumor growth using Gompertz growth rate function. Alex. Eng. J., 3731–3740 (2021)
25. L. Kovács, B. Czakó, D.A. Drexler, G. Eigner, T. Ferenci: Integrative cybermedical systems for computer-based drug delivery, Research results of the physiological controls Research Center of Óbuda University, Cap 9, (2020)
26. G.I. Ghiorghita, Programmed cell death and its mechanism (in Romanian), in *Editura Academiei Oamenilor de Ştiinţă din România*, (Bucureşti, 2012), pp. 22–29
27. D.A. Drexler, J. Sápi, L. Kovác, Modeling of tumor growth incorporating the effects of necrosis and the effect of bevacizumab. Complexity, 1–11 (2017)
28. D.A. Drexler, I. Nagy, V. Romanovski, J. Tóth, L. Kovács, Qualitative analysis of a closed-loop model of tumor growth control, in *18th IEEE International Symposium on Computational Intelligence and Informatics*, (2018), pp. 21–22
29. R.R. Sarkar, S. Banerjee, Cancer self remission and tumor stability—A stochastic approach. Math. Biosci., 65–81 (2005)
30. S. Banerjee, R.R. Sarkar, Delay-induced model for tumor-immune interaction and control of malignant tumor growth. Biosystems **91**(1), 268–288 (2008)
31. H. Dulf Eva, L. Kovacs, L. Mocan, Fractional order tumor growth model, in *International Conference on Mathematical Analysis and Applications in Science and Engineering*, (Porto-Portugal, 2022, June 27–July 1)
32. B. Czako, J. Sapi, L. Kovacs, Model-based optimal control method for cancer treatment using model predictive control and robust fixed point method, in *21st International Conference on Intelligent Engineering Systems*, (2017, October 20–23)

Chapter 10
A Model-Based Optimal Distributed Predictive Management of Multidrug Infusion in Lung Cancer Patient Therapy

Anca Maxim and Clara Mihaela Ionescu

10.1 Introduction

Optimizing therapy profiles in terms of frequency and dose rates for treating any form of cancer requires availability of a model, next to other physiological variables of interest describing the state of the patient and treatment evolution. Monitoring pre- and posttreatment can often require alterations to the initially planned therapy profiles. In particular, for lung cancer-diagnosed patients, it has been recently suggested that a cocktail of multidrug therapy has better outcomes in terms of clinical effect and patient recovery period [1, 2]. This counteracts elder studies of randomized trials where cocktails of multidrug weekly chemotherapy regimen combining various drugs did not necessarily lead to significant improvements [3]. Other recent investigations in mouse models indicated that such multidrug therapy with adaptive protocols achieves better outcomes in terms of cell turnover and therapies which used as little drug as possible worked best [4]. These adaptive protocol strategies used agent-based models where more competition delivered better outcomes at the cost of higher drug resistance. The amount of dose did not affect much the outcome but more so the intervals of administration.

A. Maxim
"Gheorghe Asachi" Technical University of Iasi, Iasi, Romania
e-mail: anca.maxim@ac.tuiasi.ro

C. M. Ionescu (✉)
Ghent University, Ghent, Belgium
e-mail: claramihaela.ionescu@ugent.be

C. M. A. Pinto, C. M. Ionescu (eds.), *Computational and Mathematical Models in Biology*, Nonlinear Systems and Complexity 38,
https://doi.org/10.1007/978-3-031-42689-6_10

However, models are scarce and main advances are at molecular scale of tumor dynamic profiles and dose-effect kinetics [5, 6]. As part of the breathing dynamic system, the tumor volume and consistency affect respiratory mechanics, which in turn affect pathophysiology and quality of life in patient. Other models are based on time-frequency approximations of linear dynamics and limited to analysis purposes [7]. Consequently, one may define also a macroscale description of the lung tumor volume dynamics and use lung function test information to complement models [8]. Recent advances in modelling lung tumor growth dynamics have proposed simplistic yet descriptive compartmental models for characterizing the pharmacokinetics (PK) and pharmacodynamics (PD) of lung cancer therapy in non-small cell lung cancer (NSCLC). The most prevalent therapy is antiangiogenesis and a model has been proposed in [9, 10]. In clinical practice, this is enhanced with periods of stereotactic body radiotherapy (SBRT) [11], and in severe cases, additional immunotherapy [12] is applied. We proposed a PKPD model encompassing all three therapies [13, 14] and validated it on a set of lung cancer patients [15]. Having this model available at hand, it is possible to evaluate control algorithms for optimizing multiple scenarios of therapy profiles.

Distributed agent optimal control strategies often involve the existence of an equilibrium point, whereas all agents converge toward a common objective or a self-centered objective. As such, one may project a theory that multiagent systems seeking optimality can well approximate the multidrug therapy protocols seeking optimal tumor volume response in patients. In this paper, we propose to use distributed agent-based cooperative predictive control as to mimic clinical practice which uses the information from patient and medical expertise to provide a tailored therapy profile in a personalized medicine context. Several scenarios are analyzed in terms of clinical outcome, ranging from a conventional single therapy profile with additional fixed inputs to a full multidrug dosing optimal control strategy.

The next section describes the multidrug PKPD model and its features characterizing drug interactions and dose-effect relationship. The third section introduces briefly the predictive control strategy employed here, and the results are given with a discussion onto their relevance for tumor volume control objectives.

10.2 Model Description

A tumor volume growth model has been proposed in [13], calibrated in [16] and validated with real data from patients in [17]. It consists of first-order compartmental PK models with linear dynamics, followed by dose-effect response nonlinear gains. Although this is a well-known Wiener type model, identification is difficult due to lack of modalities for persistent excitation of drug dosing profiles [18, 19]. When combining three therapies in a single compartmental PKPD model, we obtain [13]:

$$\begin{aligned}
\dot{x}_1 &= (a-n)x_1 & -E\cdot x_1\\
\dot{x}_2 &= x_2(0)+nx_1 & +E\cdot x_1\\
\dot{x}_3 &= -c_a x_3 & +u_a\\
\dot{xe}_3 &= -c_a xe_3 & +Et_a\cdot x_3\\
\dot{x}_4 &= -c_i x_4 & +u_i\\
\dot{xe}_4 &= -c_i xe_4 & +Et_i\cdot x_4\\
\dot{x}_5 &= -c_r x_5 & +u_r\\
\dot{xe}_5 &= -c_r xe_5 & +Et_r\cdot x_5
\end{aligned} \tag{10.1}$$

where the model states are described as follows: x_1 in mm^3 is the proliferating tumor volume, x_2 in mm^3 is the necrotic tumor volume, x_3 in mg/mL is the concentration of the antiangiogenesis drug profile, xe_3 in mg/mL is the effect drug concentration for the antiangiogenesis drug profile, x_4 in mg/mL is the concentration of the immunotherapy drug profile, xe_4 in mg/mL is the effect drug concentration for the immunotherapy drug profile, x_5 in mg/(mL·day) is the concentration of the radiotherapy drug profile, and xe_5 in mg/mL is the effect drug concentration for the radiotherapy drug profile. As the dead cells are eliminated from the body in a natural way, the necrotic volume does not depend on the past value; hence, $x_2(0) = 0$.

The inputs are selected as the three therapies' drug profiles, where u_a, u_i, and u_r in mg/(mL·day) are the dose rates for the antiangiogenesis, the immunotherapy, and the radiotherapy, respectively.

Note that from the control point of view, one can choose the states x_3, x_4, and x_5 as system's outputs, to be controlled directly through the manipulated inputs u_a, u_i, and u_r.

The model parameters are denoted as follows: a is the tumor growth rate; n is the necrosis rate; c_r is the clearance rate on the Michaelis-Menten kinetics $\frac{x_1 x_3}{ED50_r + x_3}$ (mm^3/day); c_a is the clearance rate bevacizumab; c_i is the clearance rate nivolumab; and Et_a, Et_i, and Et_r are the synergic effects between tumor cells and the chosen antiangiogenesis, immunotherapy, and radiotherapy therapy, respectively. The dose-effect pharmacodynamic model is in fact a nonlinear gain (for single drug) or a nonlinear surface (for multiple drugs) and can be considered as averaged effects of all therapies (in absence of real data at molecular level to confirm this hypothesis) for the tumor Et_{all} and drug Ed_{all} interactions:

$$\begin{aligned}
Et_{all} &= \frac{Et_a+Et_i+Et_r}{3}\\
Ed_{all} &= \frac{E_{ai}+E_{ar}+E_{ir}}{3}\\
E &= \frac{Et_{all}+Ed_{all}}{2}
\end{aligned} \tag{10.2}$$

colorred with Et_x denoting the interaction between tumor cells and each drug, while E_{xy} denotes interaction among drugs. When surface models are used to characterize synergic effects among drugs, the effect drug concentrations xe are normalized to their potency, i.e., to their corresponding half effect concentration C_{50}. The combined effects of two drugs U_A and U_B are considered as a new drug and expressed as a Hill curve dose-response relationship 3D surface:

$$Effect = \frac{I^{\gamma}}{1 + I^{\gamma}} \tag{10.3}$$

with I denoting the interaction term:

$$I = Un_A + Un_B + \sigma Un_A \cdot Un_B \tag{10.4}$$

with $Un_A = \frac{U_A}{C_{50A}}$ and $Un_B = \frac{U_B}{C_{50B}}$ the normalized drug effect concentrations and C_{50} the concentrations at half effect 50%. The term γ denotes the nonlinearity of the surface, which represents how a patient responds to the drug (effectiveness or resistance of therapy). The term σ denotes the degree of synergy present between the drugs. Values for the model coefficients are those reported in [13] (open access).

10.3 Predictive Control Strategy

Predictive control has been used in SBRT for compensating the breathing pattern effects in 3D volume changes in lung tumor tissue, to minimize toward zero the radiation of healthy tissue around the tumor. The breathing pattern was identified online, and the robot arm used for therapy was guided with feedforward compensation using a specially designed disturbance model in model predictive control (MPC) [20]. When multiple objectives are envisaged within changing context of execution, a prioritized optimizing scheme can be used as that proposed in [21], reducing both computational and numerical complexities involved when Pareto front is used instead.

In this paper, a centralized MPC strategy with state-space formulation is proposed. The algorithm is derived starting from the velocity-form methodology from [22], extended to a multivariable system.

Let us consider the PKPD model from (10.1) described by

$$\begin{aligned} x_p(k+1) &= A_p x_p(k) + B_p u(k) \\ y(k) &= C_p x_p(k) \end{aligned} \tag{10.5}$$

where k is the discrete time instant and $u \in \mathbb{R}^{n_u}$, $y \in \mathbb{R}^{n_y}$, and $x_p \in \mathbb{R}^{n_{xp}}$ are the input, output, and state variables, respectively.

Aiming at implementing an offset-free constant reference tracking algorithm, an integral action is applied to the control increments obtained at the controller output. To get the control increments at the controller output for predictive control design, a velocity-form model will be used.

Using the methodology described in [22], the difference operation is applied on both sides of (10.5) resulting:

$$\begin{aligned} \Delta x_p(k+1) &= A_p \Delta x_p(k) + B_p \Delta u(k), \\ \Delta y(k+1) &= C_p A_p \Delta x_p(k) + C_p B_p \Delta u(k) \end{aligned} \tag{10.6}$$

Note that (10.6) introduces the increments of the variables x_p, u, and y, with $\Delta y(k+1) = y(k+1) - y(k)$.

A new state variable $x(k) = \left[\Delta x_p(k)^T y(k)\right]^T$ is introduced, resulting the augmented model:

$$\underbrace{\begin{bmatrix} \Delta x_p(k+1) \\ y(k+1) \end{bmatrix}}_{x(k+1)} = \underbrace{\begin{bmatrix} A_p & O^T_{n_y \times n_{x_p}} \\ C_p A_p & 1 \end{bmatrix}}_{A} \underbrace{\begin{bmatrix} \Delta x_p(k) \\ y(k) \end{bmatrix}}_{x(k)} + \underbrace{\begin{bmatrix} B_p \\ C_p B_p \end{bmatrix}}_{B} \Delta u(k)$$

$$y(k) = \underbrace{\begin{bmatrix} O_{n_y \times n_{x_p}} & I_{n_y} \end{bmatrix}}_{C} \begin{bmatrix} \Delta x_p(k) \\ y(k) \end{bmatrix} \tag{10.7}$$

which will be used to design the predictive controller. Note that the new input of the state-space model in velocity form is $\Delta u(k)$.

The A, B, and C are matrices with adequate dimensions.

The model (10.7) can be written in a compressed form as

$$\begin{cases} x(k+1) = Ax(k) + B\Delta u(k) \\ y(k) = Cx(k) \end{cases} \tag{10.8}$$

The centralized MPC cost function is defined as

$$J(x(k), \Delta U(k)) = (R_s - Y)^T (R_s - Y) + \Delta U(k)^T R \Delta U(k) \tag{10.9}$$

depending on the output future predictor

$$Y = \left[y(k+1|k) \ \ldots y(k+N_p|k)\right]^T$$

and the future input sequence

$$\Delta U(k) = [\Delta u(k|k) \ \ldots \ \Delta u(k+N_c-1|k)]^T$$

with N_p the prediction horizon and N_c the control horizon ($N_c \leq N_p$). In this paper, for simplicity, we consider $N_c = N_p$. The predicted reference trajectory $R_s \in \mathbb{R}^{N_p}$ assumed constant and equal with the set point at time instant k and the input weight matrix has the form $R = \alpha I_{N_c}$, $\alpha \geqslant 0$.

Using the velocity-form model (10.8), the prediction of state and input variables can be computed. Based on model (10.8), the state predictor is obtained as

$$x(k+l|k) = A^l x(k) + A^{l-1} B\Delta u(k|k) + \ldots + A^{Np-l} B\Delta u(k+l-1|k)$$
$$l = \overline{1, N_p} \tag{10.10}$$

and using the output from (10.8), the output predictor is achieved:

$$y(k+l|k) = CA^l x(k) + CA^{l-1} B\Delta u(k|k) + \ldots + CA^{N_P - l} B\Delta u(k+l-1|k)$$
$$l = \overline{1, N_p} \tag{10.11}$$

The output predictor Y is computed in the following matrix form using (10.11):

$$Y = \tilde{A}x(k) + \tilde{B}\Delta U(k) \tag{10.12}$$

where

$$\tilde{A} = \quad \left[CA\; CA^2 \ldots CA^{N_p}\right]^T \tag{10.13}$$

$$\tilde{B} = \begin{bmatrix} CB & 0 & \ldots\; 0 \\ CAB & CB & \ldots\; 0 \\ \ldots & \ldots & \ldots\;\ldots \\ CA^{N_p-1}B & CA^{N_p-2}B & \ldots\; CB \end{bmatrix}.$$

After the substitution of Eq. (10.12) into the cost function (10.9) and minimizing it, the centralized $\Delta U^*(k)$ solution is obtained:

$$\Delta U^* = (\tilde{B}^T\tilde{B} + R)^{-1}\tilde{B}^T[R_s - \tilde{A}x(k)]. \tag{10.14}$$

Following the receding horizon principle, only the first n_u elements from the optimal solution vector are send to the process.

10.4 Optimality for Distributed Multidrug Predictive Control

In terms of dose-effect relationship versus drug resistance, the problem of optimum seeking is both exploitative and explorative, as in reinforcement learning theory. Such learning schemes are highly relevant in network systems with heterogeneous entities, whereas these entities represent different drug therapy profiles, and the agents in the network are the respective drug selection combinations. Finding optimum implies finding an equilibrium point where the cell turnover outperforms the cell growth in tumor volume. As such, this corresponds broadly to a combination of game theoretic models with learning-based approaches, which we will employ in our analysis. In a decentralized exchange of information, it allows heterogeneous agents to strategically interact with each other, e.g., the choice of drugs affects the degree of synergy effects, and learn to adjust their behaviors, i.e., an adaptive therapy protocol strategy. Information here refers to the structure used to model the knowledge the players in the game acquire and the history of their decision effect when they make the decisions for their next move. We introduce here some concepts used hereafter:

- Players are the participants in a game, in competition against each other. In our context, these are the different multidrug selection and protocols competing for the best patient outcome.
- Actions of a player, denoting here the drug profiles and timeline administered to the patient.
- Information in game theory refers to acquiring knowledge about the game, skills, and forecasting of move effects in finding optimality; in our context, this refers to the knowledge of how the patient responses to the drug profile both past and forecasted in optimum seeking algorithms.
- Strategy refers to the association between a player's move and the information available at that moment; this is fairly similar in our context denoting the controller optimal solution seeking protocol and can be cooperative or noncooperative, static or adaptive, etc.
- Utility (or reward) is part of the optimization cost variable, and for our case, this is the minimal amount of drug which maximizes the patient outcome, i.e., minimizes a relative ratio between volume growth and cell death rate.

It is necessary to explicitly represent the dynamic nature of the game theory parallelism to multidrug decision system, as it evolves over a period of time, i.e., the active treatment period in patient. The current state of the tumor volume specifies the current situation of the dynamic game (dose-effect relationships), including the set of players (choice of drug cocktail), actions available to them (drug profiles expressed in amount and time interval dosage), and their utilities at this time (relative tumor volume reduction). As an example, a subclass of Markov games with multiagent sequential decision-making under uncertainties has been discussed in [23]. The decision-making process is based on a reinforcement learning (RL) principle, where the future choices of the actions are shaped by feedback of a reward function, in our case this being the therapeutic effect in patient. The gradient play is most relevant here as it indicates a convergence of the RL scheme toward a Nash equilibrium in dynamic environments. An asymptotic behavior of such systems has been broadly discussed in [23].

When analyzing systems with limited resources, noncooperative games have been proven to be good candidates in reaching Nash equilibrium (NE). To minimize the risk for drug resistance and side effects of drug therapy in patient and improve quality of life in cancer-treated patients, it is desirable to minimize the amount of drug or intensity of radiation profiles. If we have multiple players denoting multiple therapy profile strategies, reaching NE implies solving the problem of finding the best strategy for one player, given all other players move at optimality solution. That is, it needs to determine the actions that players should take to achieve the best outcome in response to other player's actions. In terms of finding best treatment protocols, finding NE implies having an adaptive protocol strategy. If the drug selection remains constant, it represents the amount of players is constant. If the drug selection also varies, then it represents the number of players in the game

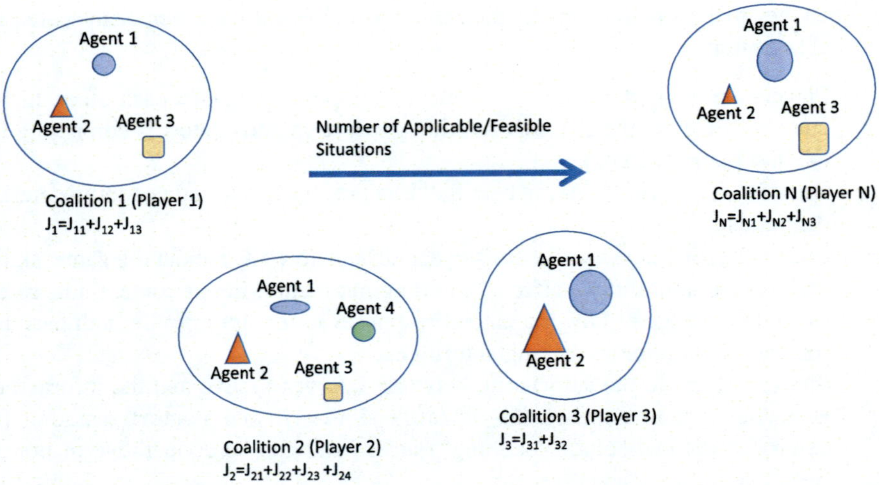

Fig. 10.1 Concept of coalition game NE optimality problem, with players representing various therapy protocols and agents representing different drugs

changes as well. Coalition consists of multiple agents collectively acting as a virtual player to minimize a coalition cost function, defined as the sum of all agent's local cost function, as represented in Fig. 10.1.

To analyze the number of feasible solutions and their convergence, one may employ the already mature theory of Lyapunov analysis, particularly useful in designing continuous time distributed NE seeking algorithms. In this context, Fig. 10.1 depicts a noncooperative game with gradient play for an average consensus collecting the sum of all local cost functions. A convergence analysis for NE seeking in N-coalition games has been discussed in [24].

From a clinical perspective, the selection of drugs to be used in the multidrug combination therapy in cancer patients is crucial, because of their interconnected effects described by drug synergy and drug resistance. This translates to the choice of agents within a coalition set. Aggregative games are a special subclass of noncooperative games where the decision process of each agent depends on the aggregate effect of all agents in the coalition. In this case, convergence is based on monotonicity of convex functions, i.e., a gradient descent in tumor volume effect. Solutions for center-based NE seeking in such population games are presented in [25].

In this paper, we investigate the effect of a distributed NE seeking problem with various coalition profiles, whereas the agents remain the same. In particular, we employ the model from (10.1) for a set of three drugs, with coalitions defined as protocols in the next section.

10.5 Results and Discussion

In this section, we aim to test the MPC algorithm on the PKPD model (10.1) converted to the discrete-time, state-space form (10.5), using the discretization method zero-order hold and the Matlab function `c2dm`, MathWorks(R). Note that the state vector is composed from $x_p = [x_1\ x_3\ x_{e3}\ x_4\ x_{e4}\ x_5\ x_{5e}]$ and does not contain the necrotic tumor volume x_2, since as mentioned before, its value does not depend on past values. To compute it, we use the proliferating tumor volume x_1 and the relationship provided in (10.1). For the optimization purposes, the outputs are selected as the states $y = [x_3\ x_4\ x_5]$, which are directly influenced by the drug dose rates $u = [u_a\ u_i\ u_r]$. Moreover, due to the dynamic nature of PKPD model, the resulting state-space model is time variant. To overcome this difficulty, in the optimization procedure, the system model was updated at each sampling time, according to the interactions from (10.2).

The tuning parameters of the controller are sampling interval of 1-day, a control horizon of $N_c = 5$ samples, and a prediction horizon of $N_p = 5$ samples. The input weight parameter is $\alpha = 1$.

In all our tests, when computing the tumor model, we considered a static environment, i.e., we fixed the patient response/resistance to drug at a value $\gamma = 2.5$, and the effect of drug synergy $\sigma = 4$, i.e., the nominal values taken from [13]. The initial tumor volume was calibrated with clinical real data as reported in [16].

To test the efficiency of our proposed optimization procedure, four coalition protocols were devised, in a 60-day simulation time interval.

Protocol 1 The antiangiogenic and the immunotherapy drug profiles from [13] were used, i.e., with a 0.171 mg/mL and a 0.2 mg/mL dose applied once every week, respectively. The radiotherapy drug profile was optimally computed using MPC strategy. This results in a 1×1 system with one input $u = [u_r]$ and one output $y = [x_5]$. Inputs u_a and u_i are considered known and are introduced in the control system in open loop.

Protocol 2 The antiangiogenic drug profile from Protocol 1 is maintained, whereas the immunotherapy and radiotherapy drug profiles are optimally computed using MPC strategy. This results in a 2×2 centralized system with two inputs $u = [u_i\ u_r]$ and two outputs $y = [x_4\ x_5]$. Input u_a is considered a-priory known and is introduced in the control system in open loop.

Protocol 3 The immunotherapy drug profile from Protocol 1 is maintained, whereas the antiangiogenic and radiotherapy drug profiles are optimally computed using MPC strategy. This results in a 2×2 centralized system with two inputs $u = [u_a\ u_r]$ and two outputs $y = [x_3\ x_5]$. Input u_i is considered known and is introduced in the control system in open loop.

Protocol 4 All three drug profiles are optimally computed using MPC strategy. This results in a full 3×3 centralized system with three inputs $u = [u_a\ u_i\ u_r]$ and three outputs $y = [x_3\ x_4\ x_5]$.

As previously mentioned, our aim is to test different coalitions (i.e., drug selection combinations), in order to minimize the tumor growth. To ensure this outcome, an additional safeguard is employed, by dynamically changing (e.g., online increasing) the drug profiles. When the variation of the total tumor is negative, i.e., $x_1(k-1) - x_1(k) < 0$, it means that the tumor size in day-k has increased its size w.r.t. the previous day (sample time $k-1$). To stop this unwanted, negative trend, the imposed references for the concentration profiles corresponding to the selected drug profiles within the coalition are increased. In our context, the set of players (the drug cocktail) from the employed coalition (protocol) changes their action, thus implicitly changing the drug dosages. As such, each protocol was tested in two case studies (or strategies):

Case 1 Constant reference values imposed for states x_3, x_4, and x_5 at the values $r_3 = 0.4$, $r_4 = 0.2$, and $r_5 = 4$ mg/ml, corresponding to the desired concentrations for antiangiogenesis, immunotherapy, and radiotherapy, respectively.

Case 2 Variable reference values for states x_3, x_4, and x_5. The reference values are modified as follows: while the safeguard is active (i.e., the tumor grew), the references for x_3 and x_4 are incremented with 0.05 mg/ml, whereas the reference for x_5 is incremented with 1 mg/ml; when the safeguard is inactive (i.e., the tumor shrunk w.r.t. the previous day measurement), the reference values are maintained constant at the previous day value.

Similar to the work in [13], the radiotherapy model parameters were recalibrated via a unit transformation $Gy \rightarrow 1/mL$. The reported tumor growth volumes are relative to its initial volume of 1000 mm^3.

The results obtained in Protocol 1 are provided in Figs. 10.2 and 10.3, for Case 1 and Case 2, respectively. The profiles for inputs u_a and u_i were taken from the literature studies [9, 26], respectively. The radiotherapy drug profile u_r was optimally computed using MPC, and the resulting profiles are provided in the lower plots (see Figs. 10.2 and 10.3a). The difference between the two strategies consists in different actions for Agent 3 (changes in SBRT profiles), whereas the actions for Agent 1 and Agent 2 remain constant. Thus, in Case 1, the concentration profile for the radiotherapy, i.e., the imposed reference for x_5, is constant at 4 mg/mL. The effect is that, in this case, by day 23, the total tumor volume starts increasing again, suggesting that the dose effect for u_r reached a plateau. In Case 2, the imposed concentration reference is variable and increases in the first three days (because the growth of the total tumor volume activates the safeguard). After day 4, the imposed reference for x_5 remains constant until day 27, when the safeguard is activated again. The effectiveness of these two strategies is visible in the total tumor volume evolution (ref. Fig. 10.2 and 10.3c).

The results obtained in Protocol 2 are provided in Figs. 10.4 and 10.5, for Case 1 and Case 2, respectively. Here, the profile for u_a is taken from the literature [9], whereas u_i and u_r were resulted from the optimization problem. The effect of the two strategies employed in this coalition is visible when comparing the total tumor volume (see Figs. 10.4 and 10.5c). In Case 2, by day 27 (when the safeguard is

Fig. 10.2 Results for coalition Protocol 1: Case 1. (**a**) The dose rates for the antiangiogenesis (upper plot), immunotherapy (middle plot), and radiotherapy (lower plot) drug profiles, respectively. (**b**) Concentration profiles in body corresponding to the drug profiles. (**c**) Total tumor growth (blue stems) and necrotic tumor volume (red stems) response to therapy

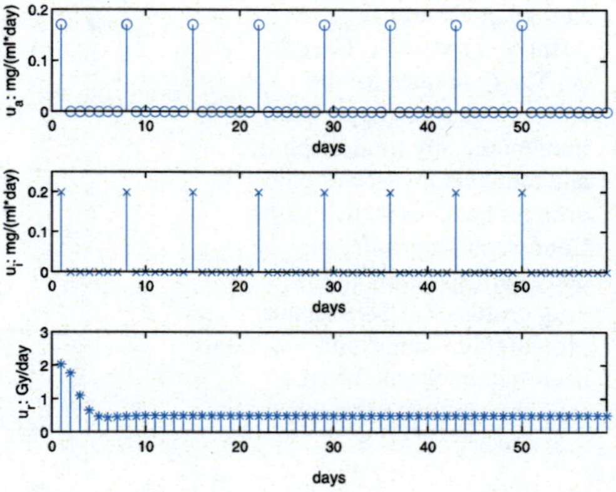

(a)

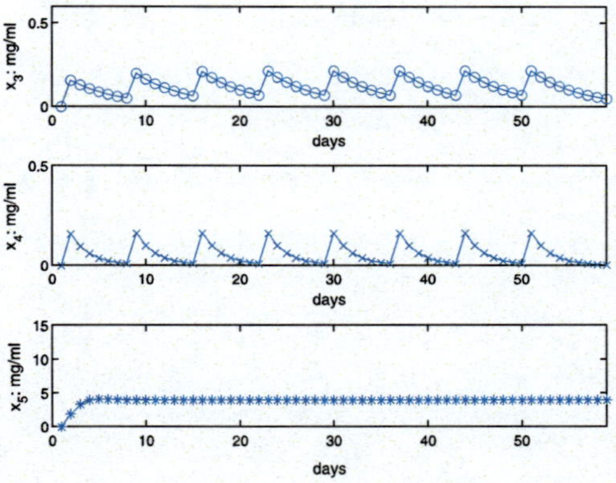

(b)

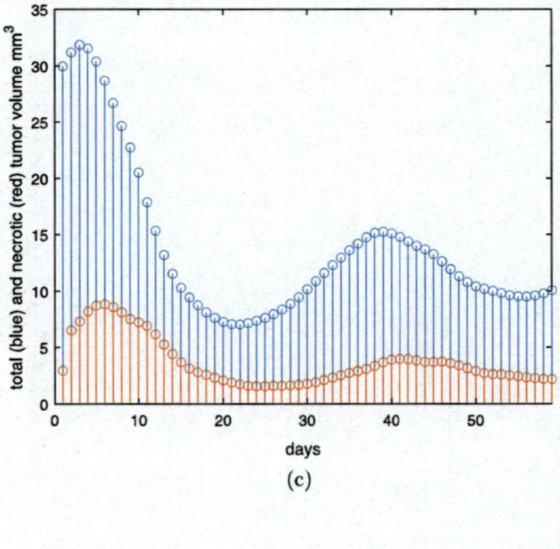

(c)

Fig. 10.3 Results for coalition Protocol 1: Case 2. (**a**) The dose rates for the antiangiogenesis (upper plot), immunotherapy (middle plot), and radiotherapy (lower plot) drug profiles, respectively. (**b**) Concentration profiles in body corresponding to the drug profiles. (**c**) Total tumor growth (blue stems) and necrotic tumor volume (red stems) response to therapy

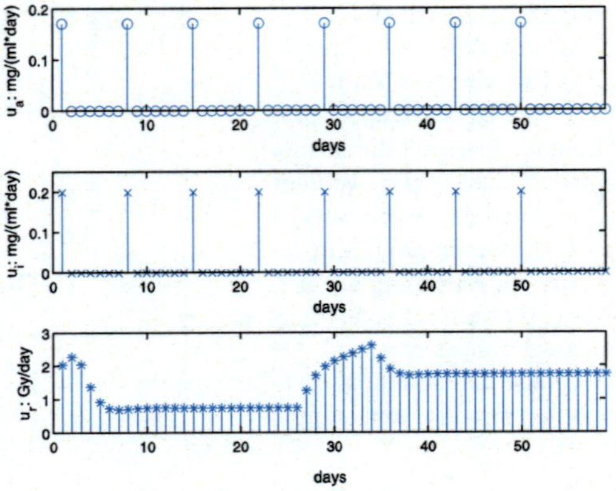

(a)

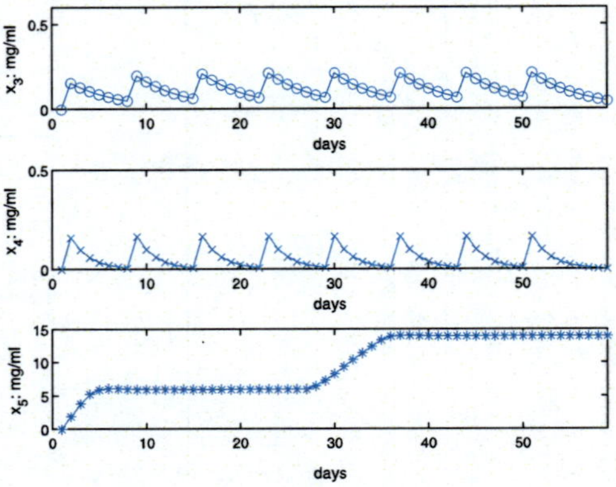

(b)

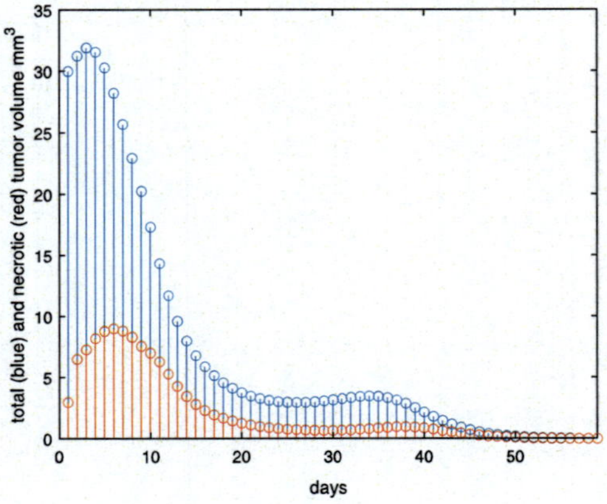

(c)

Fig. 10.4 Results for coalition Protocol 2: Case 1. (**a**) The dose rates for the antiangiogenesis (upper plot), immunotherapy (middle plot), and radiotherapy (lower plot) drug profiles, respectively. (**b**) Concentration profiles in body corresponding to the drug profiles. (**c**) Total tumor growth (blue stems) and necrotic tumor volume (red stems) response to therapy

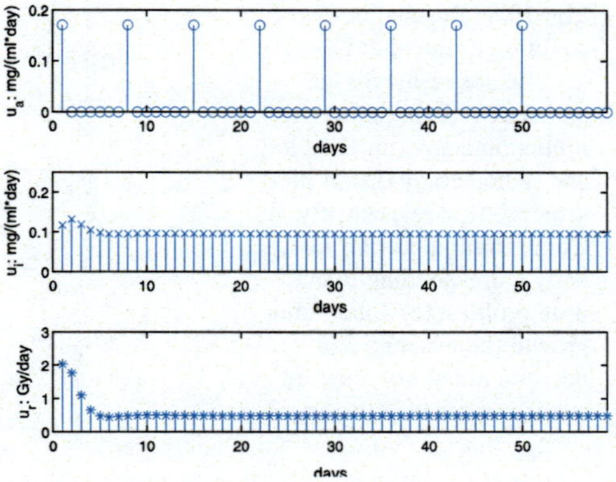

(a)

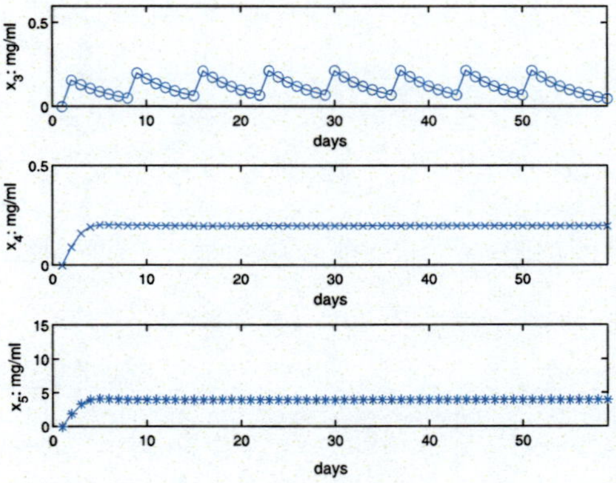

(b)

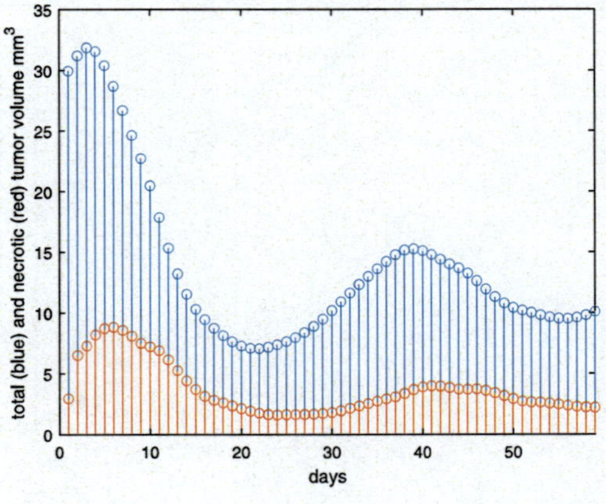

(c)

Fig. 10.5 Results for coalition Protocol 2: Case 2. (**a**) The dose rates for the antiangiogenesis (upper plot), immunotherapy (middle plot), and radiotherapy (lower plot) drug profiles, respectively. (**b**) Concentration profiles in body corresponding to the drug profiles. (**c**) Total tumor growth (blue stems) and necrotic tumor volume (red stems) response to therapy

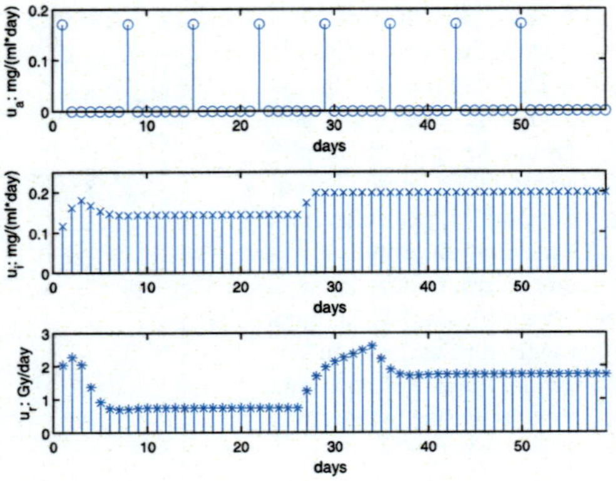

(a)

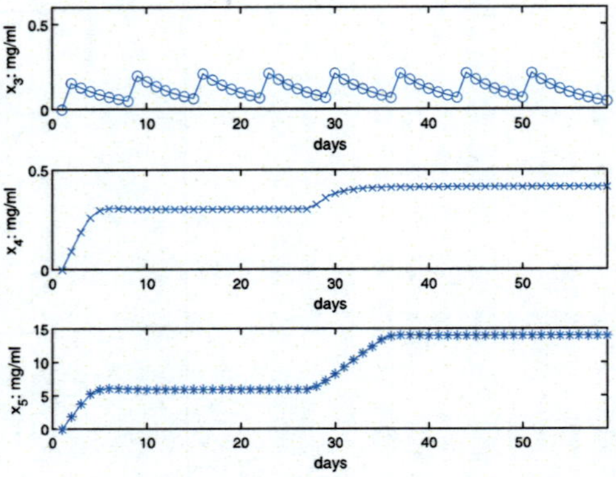

(b)

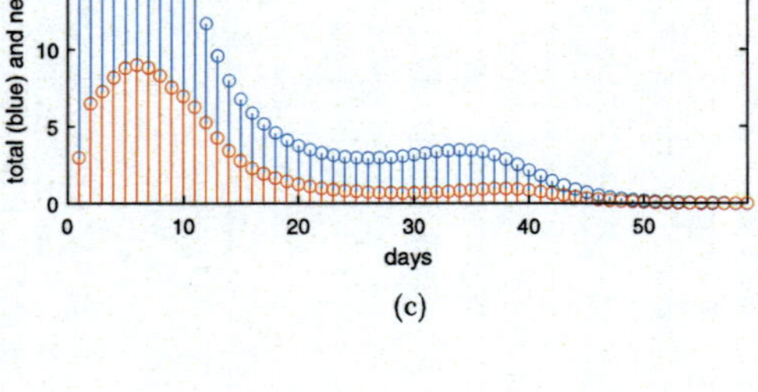

(c)

Fig. 10.6 Results for coalition Protocol 3: Case 1. (**a**) The dose rates for the antiangiogenesis (upper plot), immunotherapy (middle plot), and radiotherapy (lower plot) drug profiles, respectively. (**b**) Concentration profiles in body corresponding to the drug profiles. (**c**) Total tumor growth (blue stems) and necrotic tumor volume (red stems) response to therapy

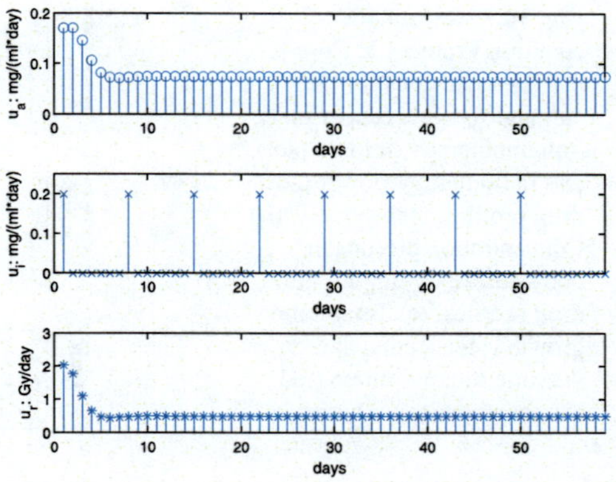

(a)

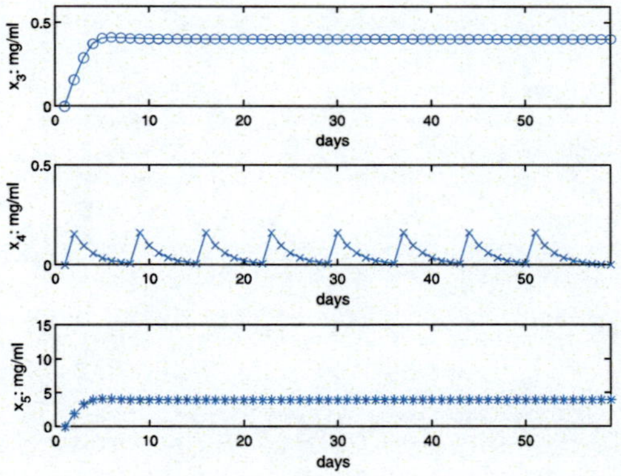

(b)

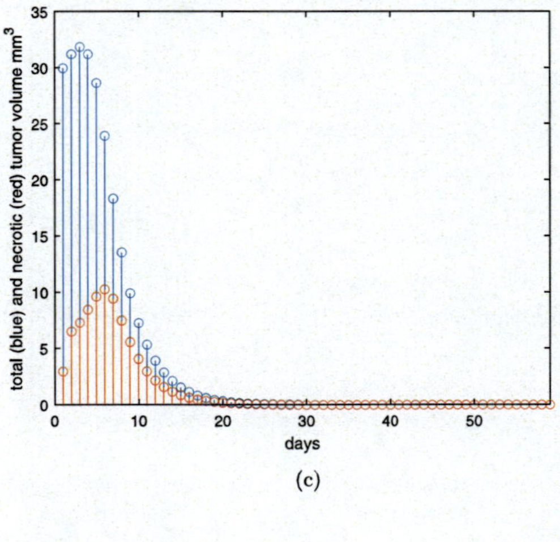

(c)

Fig. 10.7 Results for coalition Protocol 3: Case 2. (**a**) The dose rates for the antiangiogenesis (upper plot), immunotherapy (middle plot), and radiotherapy (lower plot) drug profiles, respectively. (**b**) Concentration profiles in body corresponding to the drug profiles. (**c**) Total tumor growth (blue stems) and necrotic tumor volume (red stems) response to therapy

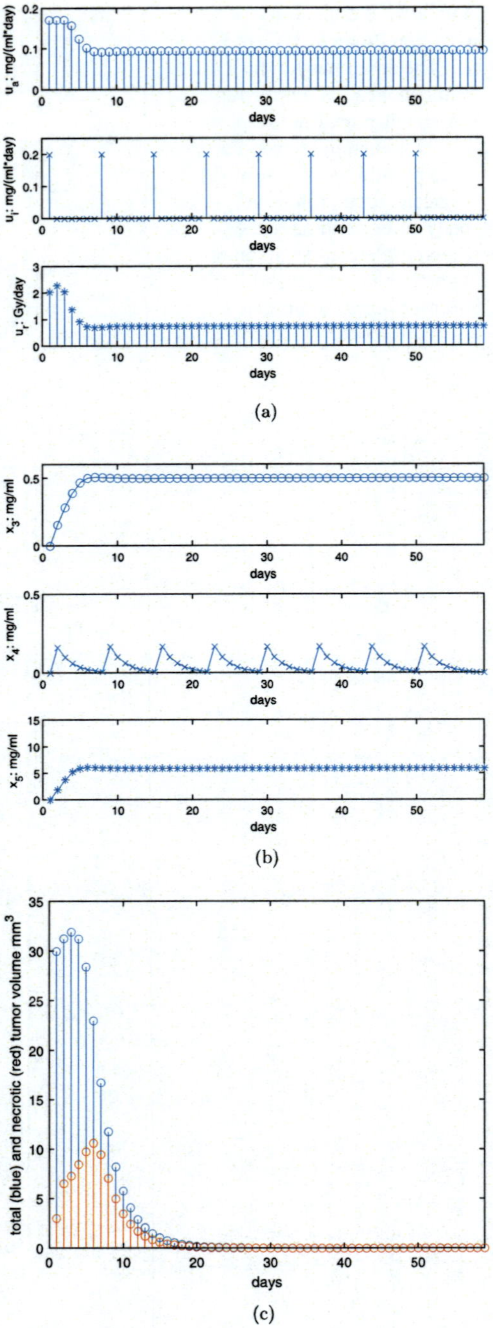

Fig. 10.8 Results for coalition Protocol 4: Case 1. (**a**) The dose rates for the antiangiogenesis (upper plot), immunotherapy (middle plot), and radiotherapy (lower plot) drug profiles, respectively. (**b**) Concentration profiles in body corresponding to the drug profiles. (**c**) Total tumor growth (blue stems) and necrotic tumor volume (red stems) response to therapy

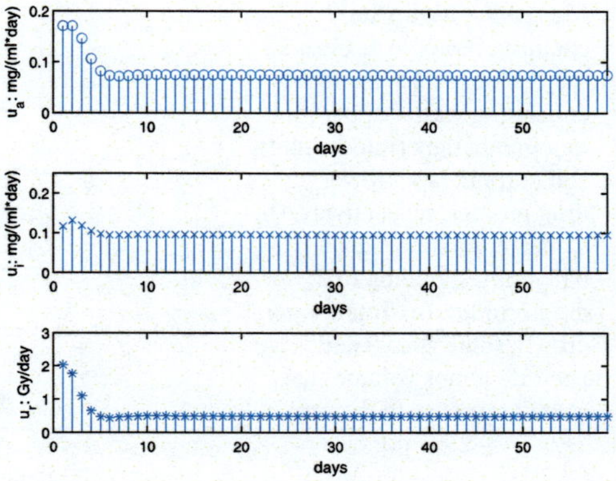

(a)

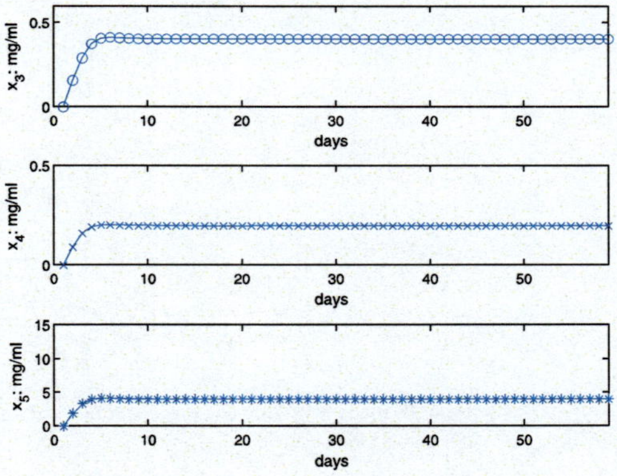

(b)

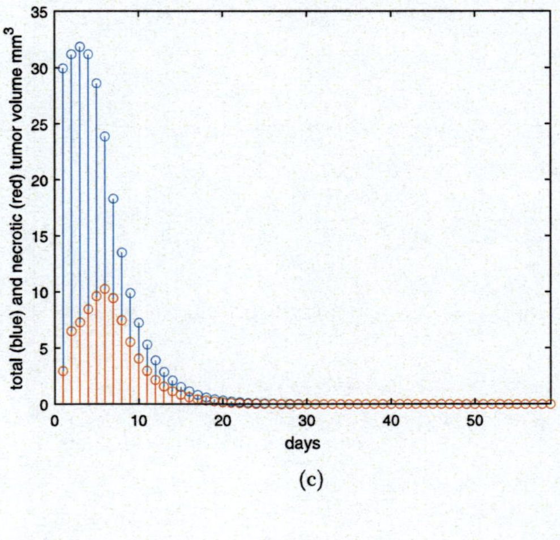

(c)

Fig. 10.9 Results for coalition Protocol 4: Case 2. (**a**) The dose rates for the antiangiogenesis (upper plot), immunotherapy (middle plot), and radiotherapy (lower plot) drug profiles, respectively. (**b**) Concentration profiles in body corresponding to the drug profiles. (**c**) Total tumor growth (blue stems) and necrotic tumor volume (red stems) response to therapy

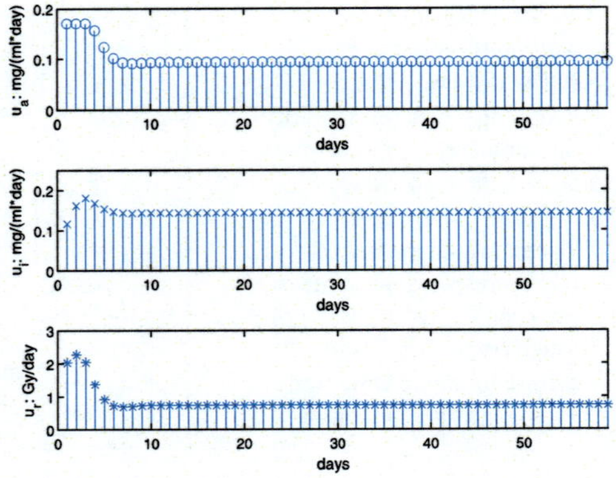

(a)

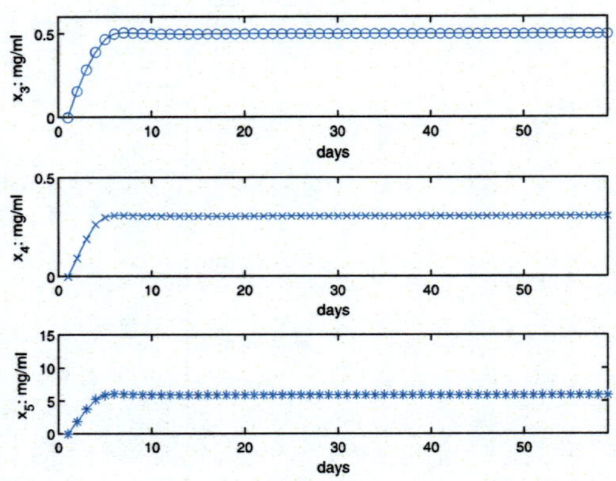

(b)

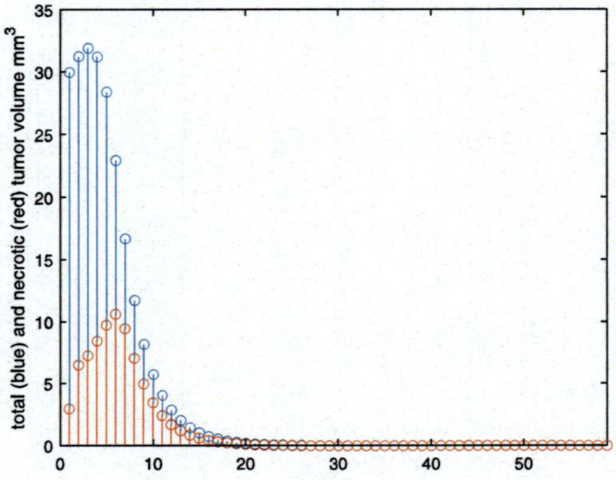

(c)

Table 10.1 Cumulative drug infusion rates

Protocol	u_a	u_i	u_r
Protocol 1—Case 1	1.3680	1.6000	32.9522
Protocol 1—Case 2	1.3680	1.6000	85.4576
Protocol 2—Case 1	1.3680	5.7928	32.9522
Protocol 2—Case 2	1.3680	10.4061	85.4576
Protocol 3—Case 1	4.7313	1.6000	32.9522
Protocol 3—Case 2	5.8723	1.6000	49.0532
Protocol 4—Case 1	4.7313	5.7928	32.9522
Protocol 4—Case 2	5.8723	8.6167	49.0532

activated), both imposed references for concentrations x_4 and x_5 start to increase until day 34 (when the safeguard is disabled) (ref. Fig. 10.5b and c).

The results obtained in Protocol 3 are provided in Figs. 10.6 and 10.7, for Case 1 and Case 2, respectively. Here, the profile for u_i is taken from the literature [9], whereas u_a and u_r were resulted from the optimization problem. In this coalition, the difference between the two strategies employed is given by the variation in drug profiles for u_a and u_r, in the first three days of the simulation. Afterward, the safeguard is disabled, meaning that the concentration references for x_3 and x_5 remain constant at values $r_3 = 0.5$ and $r_5 = 6$ mg/ml, for the remaining of the simulation (see Fig. 10.7b). Using this protocol, by day 20 of the therapy, the tumor volume converges to zero, in both strategies, using slightly different drug profiles.

The results obtained in Protocol 4 are provided in Figs. 10.8 and 10.9, for Case 1 and Case 2, respectively. All drug profiles were optimally computed using MPC. In this coalition, the difference between the two strategies employed (Case 1 and Case 2) is given by the variation in the drug profiles for u_a, u_i, and u_r, in the first three days of the simulation, visible in Case 2. Afterward, because the total tumor growth is exponentially decreasing, the concentration references for x_3, x_4, and x_5 remain constant for the remaining of the simulation time, at the values $r_3 = 0.5$, $r_4 = 0.3$, and $r_5 = 6$ mg/ml, respectively (see Fig. 10.9b).

In Table 10.1, the cumulative drug infusion rates for u_a, u_i, and u_r obtained for each protocol, in both case studies, are given.

The results suggest that the proposed PKPD model is suitable for prediction within a model-based optimization control scheme such as predictive control. It illustrates that significant changes in lung tumor volume can be achieved within the 60 days' timeline, albeit adequate adaptation of the multidrug infusion rates is necessary. This was achieved using a coalitional optimization formulation, with three agents (set of drugs) combined in four coalitions (different therapy protocols in which the drug profiles are changed). Comparing the strategies employed in Protocol 1, i.e., the cumulative drug usage in Case 1 and Case 2, one can observe an increase of radiation therapy by 159.338% (see the cumulative drug infusion rates from Table 10.1). This noticeable difference is expected, if one analyzes the number of active tumor cells from Protocol 1. The low dosage of radiotherapy which is solely manipulated in Protocol 1 is not sufficient to obtain a constant decrease of the tumor volume, in Case 1. In consequence, in the middle of the simulation interval, as the dose effect reached a plateau, the tumor volume starts oscillating. By comparison, in Case 2, although more than double radiotherapy dosage is used, the total tumor volume converges to zero, by day 50. Note that, when comparing Protocol 2 (Case 1) with Protocol 1 (Case 1), the effect of immunotherapy variation in tumor growth is analyzed. Thus, an increase of immunotherapy by 399.138% (see the cumulative drug infusion rates from Table 10.1) has insignificant overall results. This is expected, since the immunotherapy has a supportive role, and does not directly influences the tumor volume growth. Comparing Protocol 3 (Case 1) and Protocol 1 (Case 1) w.r.t. the cumulative drug usage, we observe that an increase of the dosage for antiangiogenic inhibitors by 245.8553% has a meaningful effect in the tumor growth. This is expected, since the antiangiogenic drugs stop the tumor growth by inhibiting the blood supply. Comparing Protocol 4 (Case 1) and Protocol 4 (Case 2) cumulative drug infusion rates, we can observe that increases of 24.116% for u_a, 48.7484% for u_i, and 48.8617% for u_r have minor effects in the total tumor growth. This is expected, since Protocol 4 (Case 1) takes advantage of the full potential of multivariable optimization, by selecting optimal drug profiles. As such, the safeguards inserted in Case 2 are only active at the beginning of the treatment, with negligible effects in the overall clinical outcome for the patient. It is noteworthy to mention that Protocol 3 and Protocol 4 have similar tumor growth effects, since the difference is only in the immunotherapy dosage, which as previously discussed has an ancillary role.

10.6 Conclusions

In this work, a multidrug infusion therapy optimization technique tested on lung tumor growth models was provided. The results show the merit of using centralized MPC as a tool for multivariable control, computing optimal drug profiles for the tumor treatment. Next steps include a stability analysis of Nash equilibrium states when multidrug therapy is constrained within intervals tailored for each patient to mimic personalized profiles; in this case, the patient sensitivity (or alternatively, the resistance) to drug therapy is employed in the analysis. Moreover, using personalized profiles implies the computation of various tumor models, which will be useful in testing the robustness of the proposed control method with respect to model variations.

References

1. W. Huang, L. Chen, L. Kang, M. Jin, P. Sun, X. Xin, Z. Gao, Y. H. Bae, Nanomedicine-based combination anticancer therapy between nucleic acids and small molecular drugs. Adv. Drug Delivery Rev. **115**(1), 82–97 (2017)

2. M. Eisenstein, New lung-cancer drugs extend survival times. Nature **587**, S10–S12 (2020)
3. J.P. Sculier, M. Paesmans, G. Bureau, G. Dabois, P. Libert, G. Vandermoten, O. Van Cutsem, M.C. Berchier, F. Ries et. al, Multiple-drug weekly chemotherapy versus standard combination regimen in small-cell lung cancer: a phase III randomized study conducted by the european lung cancer working party. J. Clin. Oncol. **11**(10), 1858–1865 (1993)
4. D.S. Thomas, L.H. Cisneros, A.R.A. Anderson, C.C. Maley, In silico investigations of multi-drug adaptive therapy protocols. Cancers **14**(11), 2699 (2022)
5. C.M. Rudin, J.T. Poitier, L.A. Byers et al., Molecular subtypes of small cell lung cancer: a synthesis of human and mouse model data. Nat. Rev. Cancer **19**, 289–297 (2019)
6. N. Karachaliou, S. Pilloto, C. Lazzari, E. Bria, F. De Marini, R. Rosell, Cellular and molecular biology of small cell lung cancer; an overview. Trans. Lung Cancer Res. **5**(1), 2–15 (2016)
7. C.M. Ionescu, D. Copot, R. De Keyser, Modelling doxorubicn affect in various cancer therapies by means of fractional calculus, in *American Control Conference* (Boston, 2016), pp. 1283–1288
8. C.M. Ionescu, J.F. Kelly, Fractional calculus for respiratory mechanics: power law impedance, viscoelasticity, and tissue heterogeneity. Chaos Solitons Fractals **102**, 433–440 (2017)
9. D.A. Drexler, J. Sapi, L. Kovacs, Modeling of tumor growth incorporating the effects of necrosis and the effect of bevacizumab. Complexity **2017**, 5985031 (2017)
10. J. Sapi, L. Kovacs, D.A. Drexler, P. Kocsis, D. Gajari, Z. Sapi, Tumor volume estimation and quasi-continuous administration for most effective bevacizumab therapy. PLoS ONE **10**, e0142190 (2015)
11. K.M. Prezzano, S.J. Ma, G.M. Hermann, C. Rivers, J.A. Gomez-Suescun, A.K. Singh, Stereotactic body radiation therapy for non-small cell lung cancer: a review. World J. Clin. Oncol. **10**(1), 14–27 (2019)
12. M.D. Shields, J.A. Marin-Acevedo, B. Pellini, Immunotherapy for advanced non-small cell lung cancer: a decade of progress. Am. Soc. Clin. Oncol. Educ. Book **41**, e105–e127 (2021)
13. C.M. Ionescu, M. Ghita, D. Copot, E. Derom, D. Verellen, A minimal PKPD interaction model for evaluating synergy effects of combined NSCLC therapies. J. Clin. Med. **9**(6), 1832 (2020)
14. M. Ghita, D. Copot, C. Billiet, D. Verellen, C. Ionescu, Lung cancer dynamics using fractional order impedance modeling on a mimicked lung tumor setup. J. Adv. Res. **32**, 61–71 (2021)
15. M. Ghita, D. Copot, C. Billiet, D. Verellen, C.M. Ionescu, Local anomalous drug diffusion at healthy-cancer tissue surface and data-driven tumor growth model prediction, in *American Control Conference* (San Diego, California, USA, 2023)
16. M. Ghita, C. Billiet, D. Copot, D. Verellen, C.M. Ionescu, Model calibration of pharmacokinet-icpharmacodynamic lung tumour dynamics for anticancer therapies. J. Clin. Med. **11**(4), 1006 (2022)
17. M. Ghita, C. Billiet, D. Copot, D. Verellen, C.M. Ionescu, Parameterisation of respiratory impedance in lung cancer patients from forced oscillation lung function test. IEEE Trans. Biomed. Eng. **70**(5), 1587–1598 (2023)
18. A.H. Shaikh, K. Barbe, Study of random forest to identify Wiener-Hammerstein system. IEEE Trans. Instrum. Meas. **70**, 1–12 (2021)
19. A. Haryanto, K.-S. Hong, Maximum likelihood identification of Wiener-Hammerstein models. Mech. Syst. Signal Process. **41**, 57–70 (2003)
20. C.M. Ionescu, C. Copot, D. Verellen, Motion compensation for robotic lung tumour radiotherapy in remote locations: a personalised medicine approach. Acta Astronaut. **132**, 59–66 (2017)
21. C.M. Ionescu, R.A. Cajo-Diaz, S. Zhao, M. Ghita, M. Ghita, D. Copot, A low computational cost, prioritized, multi-objective optimization procedure for predictive control towards cyber physical systems. IEEE Access **8**, 128152–128166 (2020)
22. L. Wang, *Control System Design and Implementation Using MATLAB* (Springer, Berlin, 2009)
23. T. Li, G. Peng, Q. Zhy, T. Basar, The confluence of networks, games and learning a game-theoretic framework for multiagent decision making over network. IEEE Control Syst. Mag. **42**(4), 35–67 (2022)
24. G. Hu, Y. Pang, C. Sun, Y. Hong, Distributed Nash equilibrium seeking. IEEE Control Syst. Mag. **42**(4), 68–86 (2022)

25. G. Belgioioso, P. Yi, S. Grammatico, L. Pavel, Distributed generalized Nash equilibirum seeking. IEEE Control Syst. Mag. **42**(4), 87–102 (2022)
26. A. Capasso, J. Lang, T.M. Pitts, K.R. Jordan, C.H. Lieu, S.L. Davis, J.R. Diamond, S. Kopetz, J. Barbee, J. Peterson et al., Characterization of immune responses to anti-PD-1 mono- and combination therapy in hematopoietic humanized mice implanted with tumor xenographs. J. Immunother. Cancer **7**, 37 (2019)

Chapter 11
Analysis of a Robust Fractional Order Multivariable Controller for Combined Anesthesia and Hemodynamic Stabilization

Cristina I. Muresan, Erwin T. Hegedus, Marcian Mihai, and Isabela R. Birs

11.1 Introduction

Biomedical applications are currently experiencing a major trend toward computer control as a means to assist clinicians in making the optimal decisions. Automated drug delivery systems in diabetes, cancer, or anesthesia, to name just a few [1–5], can help prevent and avoid human errors, under or over dosing, reduce side effects, while minimizing costs related to drugs, medical staff workload, or patient hospitalization period. Recent pandemic has shown that overburdened clinicians combined with reduced access to certain drugs are more likely to lead to doubtful decisions and casualties. A computer-aided control system, which computes optimal drug doses, can potentially minimize the quantity of drugs used. At the same time, it can represent a useful tool in alleviating the workload of clinicians.

General anesthesia is one of these biomedical applications where computer-controlled drug dosing has proven to be more efficient [6, 7] than manual control. It is perhaps one of the most frequently used procedures, both in surgery and emergency care units. General anesthesia requires that three patient states are adequately induced and maintained. These refer to the depth of hypnosis, analgesia, and neuromuscular blockade [8]. Hypnosis in a patient is represented through the

C. I. Muresan (✉) · E. T. Hegedus · M. Mihai
Technical University of Cluj-Napoca, Cluj-Napoca, Romania
e-mail: Cristina.Muresan@aut.utcluj.ro; Erwin.Hegedus@aut.utcluj.ro; Marcian.Mihai@aut.utcluj.ro

I. R. Birs
Technical University of Cluj-Napoca, Cluj-Napoca, Romania

Ghent University, Department of Electromechanical, Systems and Metal Engineering; Flanders Make EEDT Core Lab, Ghent, Belgium
e-mail: Isabela.Birs@aut.utcluj.ro

C. M. A. Pinto, C. M. Ionescu (eds.), *Computational and Mathematical Models in Biology*, Nonlinear Systems and Complexity 38,
https://doi.org/10.1007/978-3-031-42689-6_11

Bispectral Index (BIS) and evaluated based on a measured electroencephalogram (EEG) signal. Traditionally, hypnosis is induced and maintained by the anesthesiologist by modifying propofol doses. Analgesia or lack of pain is evaluated in this chapter through the Ramsay Agitation Score (RASS), while neuromuscular blockade is visible in the electromyogram signal (EMG). The usual drugs used in inducing and maintaining analgesia and neuromuscular blockade are remifentanil and atracurium [9]. From a control engineering perspective, anesthesia is regarded as a multivariable system [10]. But, at the same time, anesthesia is not independent and it is linked to critical hemodynamic parameters, especially in cardiac patients, where mean arterial pressure (MAP) and cardiac output (CO) [11] have to be kept within safe operating ranges by a balanced amount of dopamine (DOP) and sodium nitroprusside (SNP).

Most of the current research papers focus on developing and analyzing various control strategies intended to maintain BIS levels within safe limits. PID controllers, predictive controllers, event-based algorithms, and even fractional order controllers are some of the most popular approaches [12–16]. Control algorithms dealing with the co-administration of propofol and remifentanil for maintaining BIS have been also researched [13, 17]. For regulating the neuromuscular blockade, PID controllers as well as advanced control algorithms have been considered and validated as viable options [18]. To control MAP and CO levels, the antagonistic effect of the two administered drugs is considered, with control strategies ranging from simple PIDs to advanced algorithms, including fractional order control [19, 20].

A multivariable control approach for both anesthesia and hemodynamic variables has not been fully considered, even though research studies show that interactions exist between the subsystems [11, 21–24]. Some recent preliminary control algorithms have been proposed being based on fractional order control [23, 24] or predictive control [22, 25]. Fractional calculus is also employed in this chapter in the design of a multivariable control algorithm for the anesthesia and hemodynamic variables. Fractional order controllers are versatile and flexible, allowing for an increased number of performance specifications to be addressed and met, compared to their integer order counterparts [26]. Such characteristics are preferred since the controllers of the combined anesthesia and hemodynamic subsystems have to account for the synergic effects of remifentanil and propofol on BIS values and for the interactions between the anesthesia and hemodynamic control loops due to the effects of these drugs on MAP and CO. Additional challenges occur since changes in CO lead to changes in the propofol clearance rates, which trigger undesired variations of BIS. Couplings are presented between the CO and MAP control loops, with both dopamine and sodium nitroprusside affecting the cardiac output and the arterial pressure. Maintaining all anesthesia and hemodynamic variables within safe operating ranges is moreover complicated due to the large and varying time delays that affect the hemodynamic and BIS [27] variables. A computer-controlled solution should also be robust to imminent changes due to patient intra- and inter-variability.

In this chapter, fractional order PID (FO-PID) controllers are designed for the anesthesia and hemodynamic system and compared with an existing approach. Fractional order controllers have been chosen due to the increased robustness, flexibility, and reduced control effort compared to traditional PID controllers [26, 28, 29]. A database of 24 patients [11] is used to validate the proposed control solution. An analysis of changes in the hemodynamic outputs and how they affect the anesthesia subsystem, especially BIS values, is performed. The simulation results during induction and maintenance phases show the efficiency of the proposed computer-controlled solution.

This chapter is structured as follows. The performance specifications for the induction and maintenance phases of the anesthesia and hemodynamic outputs are indicated in the next section. The design principles for FO-PIDs are detailed in the following section, while the fourth section presents the simulation results. Concluding remarks and further research ideas are included at the end of this chapter in the fifth section.

11.2 Performance Specifications for Designing the FO-PIDs

In this chapter, a benchmark patient model is considered [11]. To design the controllers, a nominal model out of the 24 patient database is considered. The characteristics of the 24 patients are indicated in Table 11.1, where C_{50} is the drug concentration necessary to reach the half-maximal effect and γ is the Hill-coefficient of sigmodicity of the pharmacodynamic component in the patient model [11].

Decentralized control is the preferred approach due to the simplicity in implementing and further adjustment options of the controller parameters. As such, nominal transfer functions that describe the relation between propofol and BIS, remifentanil and RASS, atracurium and NMB, dopamine and MAP, and sodium nitroprusside and CO are independently used to design 5 FO-PID controllers, one for each specific output signal.

Specific performance criteria for each of these 5 output signals are determined according to clinical requirements.

For the MAP and CO outputs, a maximum overshoot/undershoot of 10% is acceptable, with a settling time of less than 18–20 min [20, 30]. Cardiac output and mean arterial pressure variation ranges are restricted to 4–8 l/min and 65–110 mmHg, respectively. For the NMB and RASS signals, maximum overshoot and settling time requirements are 5% and 4 min [23, 24], while a 10–13% range is considered acceptable for NMB and −2.6, −2.4 for the RASS, respectively.

To control the BIS signal, the two phases of anesthesia are considered separately, as previous research has led to the conclusion that two separate controllers achieve

Table 11.1 Patient characteristics used in this study [11]

Index	Age (yrs)	Height (cm)	Weight (kg)	C_{50} (mg/ml)	γ (−)
1	74	164	88	2.5	3
2	67	161	69	4.6	2
3	75	176	101	5	1.6
4	69	173	97	1.8	2.5
5	45	171	64	6.8	1.78
6	57	182	80	2.7	2.8
7	74	155	55	1.7	3.5
8	71	172	78	7.8	2.9
9	65	176	77	2.9	1.88
10	72	192	73	3.9	3.1
11	69	168	84	2.3	3.1
12	60	190	92	4.8	2.1
13	61	177	81	2.5	3
14	54	173	86	2.5	3
15	71	172	83	4.3	1.9
16	53	186	114	2.7	1.6
17	72	162	87	4.5	2.9
18	61	182	93	2.7	1.78
19	70	167	77	6.8	3.1
20	69	168	82	9.8	1.6
21	69	158	81	3.2	2.1
22	60	165	85	5.1	2.51
23	70	173	69	3.67	3.1
24	56	186	99	5.8	2.3

better results. The performance criteria are also specific to the anesthesia phase. For the induction phase, according to clinical requirements [14], a settling time of 3 to 5 min is desirable, with a maximum undershoot of approximately 5%. The setpoint value for the BIS output should be 50%. In the maintenance phase, disturbance rejection is vital. Disturbances are caused by incisions performed by the surgeon, which affect directly BIS levels. Additional disturbances are caused by the anesthesiologist supplying propofol boluses. In control engineering terms, these are output and input disturbances that should be properly tackled by a FO-PID designed to control the BIS signal. At the same time, interactions coming from the hemodynamic subsystem should be kept minimum. The designed controller should be able to maintain BIS values between 40 and 60 during the maintenance phase.

Table 11.2 details the measuring units of all input and output signals.

Table 11.2 Description of input and output signals

Inputs	Units	Outputs	Units
Propofol	[mg/kg*min]	BIS	[%]
Remifentanil	[μg/kg*min]	RASS	[–]
Atracurium	[μg/ml]	NMB	[%]
Dopamine (DP)	[μg/kg*min]	CO	[l/min]
Sodium nitroprusside (SNP)	[μg/kg*min]	MAP	[mmHg]

11.3 Design Principles for Multivariable Robust FO-PID Controllers

Various design methods for fractional order PIDs are available. The main characteristic of such controllers is the increased number of parameters compared to the traditional PID controller. The transfer function of the generalized FO-PID is indicated next:

$$H_{\mathrm{c}}(s) = k_{\mathrm{p}}\left(1 + \frac{k_{\mathrm{i}}}{s^{\lambda}} + k_{\mathrm{d}}s^{\mu}\right) \tag{11.1}$$

where s stands for the Laplace variable, while k_{p}, k_{i}, and k_{d} are the proportional, integral and derivative gains. These are the traditional tuning parameters in a PID controller. Apart from these, the FO-PID in (11.1) has two additional parameters, the fractional orders of integration and differentiation, $\lambda \in [0, 2]$ and $\mu \in [0, 2]$, respectively. For $\lambda = \mu = 1$, the classical PID controller transfer function is obtained. Thus, the FO-PID in (11.1) is a generalization of the integer order PID. As many researchers have argued and demonstrated, the additional parameters of the FO-PID lead to an increased flexibility and allow for better closed loop results, even in terms of robustness. This is simply because with more parameters to tune, more performance specifications can be addressed.

Fractional order controllers are easier to represent in the frequency domain, which is why most design techniques use the frequency response and performance specification for determining the parameters of the FO-PID. In this chapter, this approach is preferred. The frequency response, for a given frequency ω [rad/s], of the FO-PID in (11.1) is given next:

$$H_{\mathrm{c}}(j\omega) = k_{\mathrm{p}}\left(1 + k_{\mathrm{i}}\omega^{-\lambda}\left(\cos\frac{\lambda\pi}{2} - j\sin\frac{\lambda\pi}{2}\right) + k_{\mathrm{d}}\omega^{\mu}\left(\cos\frac{\mu\pi}{2} + j\sin\frac{\mu\pi}{2}\right)\right) \tag{11.2}$$

Nevertheless, frequency domain specifications offer an adequate estimation of the most common time domain specifications, such as overshoot, settling time, disturbance rejection, or noise cancellation abilities.

The gain crossover frequency, ω_c, is frequently used to address settling time requirements. The equation attached to the gain crossover frequency specification is the magnitude requirement mathematically described as follows:

$$|H_{OL}(j\omega_c)| = 1 \tag{11.3}$$

where $H_{OL}(s) = H_c(s)H_p(s)$ is the open loop transfer function and $H_p(s)$ stands for the Laplace representation of each component in the anesthesia and hemodynamic subsystems. For example, $H_p(s)$ is the transfer function that relates the BIS variation with respect to the propofol or the cardiac output variation to changes in the sodium nitroprusside.

The overshoot requirement can be indirectly addressed through the phase margin, PM, specification:

$$\angle H_{OL}(j\omega_c) = -\pi + \text{PM} \tag{11.4}$$

Robustness is tackled using the iso-damping requirement in (11.5), a specification that requests that the overshoot remains the same despite gain variations in the $H_p(s)$ transfer function. Since such variations occur frequently due to patient intra- and inter-variability, the iso-damping property is frequently used when designing FO-PIDs:

$$\left.\frac{d\angle H_{OL}(j\omega)}{d\omega}\right|_{\omega=\omega_c} = 0 \tag{11.5}$$

Disturbance rejection and noise cancellation are common performance specifications that have to be addressed in the anesthesia and hemodynamic system, as well. In control engineering terms, this requirement is specified through a constraint on the complementary sensitivity $T(s)$, as follows:

$$\left|T(j\omega) = \frac{H_{OL}(j\omega)}{1 + H_{OL}(j\omega)}\right|_{dB} \leq A_{dB} \tag{11.6}$$

where A is a design parameter.

Equally important is the ability of the designed controller to deal with load disturbances. This is mathematically expressed using a constraint on the sensitivity function $S(s)$, as indicated next:

$$\left|S(j\omega) = \frac{1}{1 + H_{OL}(j\omega)}\right|_{dB} \leq B_{dB} \tag{11.7}$$

where B is a design specification.

A decentralized approach is considered and 5 controllers similar to (11.1) are tuned for each output that needs to be maintained in a prespecified range. The tuning of the controller parameters is done specifically in each case using graphical approaches or optimization routines, depending on the complexity of the controller that needs to be designed and on the number of parameters that need to be determined.

A supplementary performance specification is added to the previous five. According to the already mentioned performance specifications, maintaining all outputs at strict setpoints is not mandatory. Rather, ensuring that the outputs do not exceed the safe limits is desirable. Thus, the integral of absolute error (IAE) is used as a means to ensure that the outputs remain as close as possible to their reference values. The IAE is mathematically expressed as:

$$\mathrm{IAE} = \int_0^{\infty} |e(t)|\, dt \tag{11.8}$$

where $e(t)$ is the error signal, the difference between the reference value and the measured output signal. To ensure that the outputs remain within the specified minimum and maximum admissible limits, a minimization of the IAE is desired.

11.4 FO-PIDs for Hemodynamic and Anesthesia Control

A general two-input-two-output hemodynamic model is used in this chapter as described in [20]:

$$\begin{pmatrix} \mathrm{CO}\,(\mathrm{s}) \\ \mathrm{MAP}(s) \end{pmatrix} = \begin{pmatrix} \frac{K_{11}\, e^{-T_{11}s}}{\tau_{11}s+1} & \frac{K_{12}\, e^{-T_{12}s}}{\tau_{12}s+1} \\ \frac{K_{21}\, e^{-T_{21}s}}{\tau_{21}s+1} & \frac{K_{22}\, e^{-T_{22}s}}{\tau_{22}s+1} \end{pmatrix} \begin{pmatrix} \mathrm{DP}(s) \\ \mathrm{SNP}(s) \end{pmatrix} \tag{11.9}$$

The possible variation ranges for the hemodynamic parameters in (11.9) are included in Table 11.3 [4]. A nominal patient model is obtained using the typical values listed in Table 11.3.

From both clinical and control engineering point of views, the cardiac output is maintained by properly adjusting dopamine values, whereas sodium nitroprusside is used to regulate the mean arterial pressure. Previous research regarding fractional order multivariable control of the hemodynamic system included the design of FO-PI controllers in various configurations [20], including time delay compensation approaches to diminish their effect upon closed loop performance. However, this previous research has considered a hemodynamic system independently from the anesthesia components. Recent findings [4, 8, 11] have proven that this is a simplistic approach since the anesthesia and hemodynamic systems are strongly coupled. Due to the effect that both dopamine and sodium nitroprusside have upon the BIS signal and the vice-versa effect of propofol and remifentanil on cardiac output and MAP, these previously designed FO-PI controllers produced an unstable closed loop response. Thus, a derivative effect has been added to enhance the overall

Table 11.3 Hemodynamic parameters ranges

Parameter	Typical	Range/units
K_{11}	5	1–12 / [ml/μg]
τ_{11}	300	70–600 / [s]
T_{11}	60	15–60 / [s]
K_{12}	3	0–9 / [mmHg·kg·min/μg]
τ_{12}	40	30–60 / [s]
T_{12}	60	15–60 / [s]
K_{21}	12	(−15)-25 / [ml/μg]
τ_{21}	150	70–600 / [s]
T_{21}	50	15–60 / [s]
K_{22}	−15	(−50)-(−1) / [mmHg·kg·min/μg]
τ_{22}	40	30–60 / [s]
T_{22}	50	15–60 / [s]

stability. One fractional order PID has been designed for each of the hemodynamic subsystem output. The nominal patient model is used in the tuning.

To meet the performance specifications, a phase margin PM $= 75°$ and a gain crossover frequency of $\omega_c = 0.0038$ rad/s are used to tune the FO-PID for the CO output. To account for the possible gain variations, as indicated in Table 11.3, the iso-damping condition is also used as a third performance specification, to enhance the robustness of the controller. To keep errors as small as possible, despite the possible parameter variations in the patient model, an IAE requirement is added. An optimization routine is used to determine the parameters of the FO-PID that meet all of the design specifications:

$$H_{\mathrm{CO}}(s) = 0.1378\left(1 + \frac{0.0071}{s^{0.35}} + 4.7742 s^{1.003}\right) \tag{11.10}$$

For the MAP output, the same performance specifications are used to tune the controller: robustness to gain variations, minimization of IAE, $\omega_c = 0.0044$ rad/s, and a large phase margin of PM $= 80°$. The parameters are computed using standard optimization routines, with the FO-PID controller transfer function being given as:

$$H_{\mathrm{MAP}}(s) = 0.0244\left(1 + \frac{0.0073}{s^{0.70}} + 0.69 s^{1.09}\right) \tag{11.11}$$

For the anesthesia subsystem, a patient nominal model is considered to tune three FO-PIDs for the three output signals: neuromuscular blockade, BIS, and RASS. The characteristics for this nominal patient are listed in Table 11.1, under index 1. Remifentanil has little influence upon the neuromuscular blockade; thus, this is completely decoupled from the rest of the anesthesia components. The hemodynamic variables do not affect the neuromuscular blockade. Then, a simple controller that is able to meet reference tracking without steady-state errors is sufficient. Thus, a simplified robust FO-PI controller is designed to meet the previously mentioned performance specifications. To ensure the settling time and overshoot requirements, as well as to increase the robustness, a gain crossover frequency of $\omega_c = 0.01$ rad/s is imposed, along with PM = 85° and iso-damping. The resulting FO-PI controller is:

$$H_{\mathrm{NMB}} = 0.08\left(1 + \frac{0.0227}{s^{1.05}}\right) \tag{11.12}$$

To meet the performance specifications for the RASS signal, the FO-PID controller is tuned for a phase margin of PM = 78°, a gain crossover frequency of $\omega_c = 0.021$ rad/s and the iso-damping property. Matlab optimization routines are used to determine the parameters of the controller. For simplification purposes, $\lambda = \mu$, and the transfer function of the controller is obtained as:

$$H_{\mathrm{RASS}}(s) = 0.0753\left(1 + \frac{0.1553}{s^{0.94}} + 0.0388s^{0.94}\right) \tag{11.13}$$

The BIS signal is affected by the synergistic effect of both propofol and remifentanil [11, 23]. Additionally, previous research regarding fractional order control applied to the anesthesia system has led to the conclusion that the best solution consists in separate controllers for the induction and maintenance phases [24]. Recent findings [11, 23] demonstrated that gain variations occur in the BIS due to patient variability. These requirements, along with different performance specifications for the anesthesia phases, have led to the need for designing two fractional order controllers, one for the induction phase and one for the maintenance phase. For the latter one, a FO-PI is sufficient since disturbance rejection is the main task of the controller, along with maintaining the BIS value within the 40–60% range. To simplify the tuning routine, $\lambda = \mu$. For the induction phase, according to the performance specifications mentioned in the previous section, the following design specifications are imposed: $\omega_c = 0.017$ rad/s, PM = 84° and robustness to gain variations. The system of nonlinear Eqs. (11.3)–(11.5) is solved using optimization routines and led to the following transfer function of the controller:

$$H_{\mathrm{BIS\_I}}(s) = 0.0033\left(1 + \frac{0.0537}{s^{0.97}} + 0.0134s^{0.97}\right) \tag{11.14}$$

For the maintenance phase, a FO-PI controller is tuned to meet the following design specifications, according to the performance criteria indicated in the previous section: minimization of IAE, with constraint related to load and output disturbance rejection. To solve the minimization problem (11.8), with constraints on (11.6) and

(11.7), the Matlab "fmincon" function was used. The transfer function with the resulting parameters of the controller was determined as:

$$H_{\mathrm{BIS\_M}}(s) = 0.0053\left(1 + \frac{0.0532}{s^{1.04}}\right) \tag{11.15}$$

A bumpless transfer is required, which has led to the necessity of designing a synchronization function [24]. Usually, a BIS value of 50% is considered the setpoint for the induction phase. Once this value has been reached, the maintenance phase that corresponds to the start of surgical procedures is initiated. Thus, the first time the induction controller reaches the 50% feedback mark, the switching to the maintenance controller is performed. This switching occurs only once.

11.5 Robustness Analysis

A detailed analysis of the designed fractional order control strategy, during both induction and maintenance phases, has been presented already, with a focus on the anesthesia subsystem [24]. As such, this chapter focuses on the hemodynamic stabilization. Variations in the CO and MAP values are considered, and their effect upon BIS levels is analyzed. The study includes the 24 patients as described in Table 11.1. A robustness analysis is also performed, considering also hemodynamic parameter variations as indicated in Table 11.3. Two case scenarios are presented. The first one refers to variations in the MAP values and how these affect the depth of hypnosis as reflected by BIS values. The target is to maintain these within the 40–60% range. A similar test is performed by varying the CO values, and a similar analysis upon BIS values is performed.

Figure 11.1 shows the closed loop results with the designed controllers, considering a MAP setpoint variation 85 mmHg → 80 mmHg → 83 mmHg → 85 mmHg. The simulation results are presented for all 24 patients. The maximum overshoot varies between 7% and 9%, while the settling time varies in the 14–15 min range. The design then meets the performance specifications as they have been previously mentioned. Due to the strong interactions in the hemodynamic subsystem, large variations are present in the CO, as indicated in Fig. 11.2. These interactions can be further diminished by introducing a decoupler [20]. The corresponding dopamine and sodium nitroprusside are indicated in Figs. 11.3 and 11.4, and these represent variations of the administered drugs around a nominal value.

An analysis of the NMB and RASS signals shows that these are not affected or slightly by changes in the MAP setpoint, as indicated in Figs. 11.5 and 11.6, with corresponding atracurium and remifentanil input drugs given in Figs. 11.7 and 11.8. The controllers meet the design specifications, as in both cases there is no overshoot and the settling time is less than 4 min.

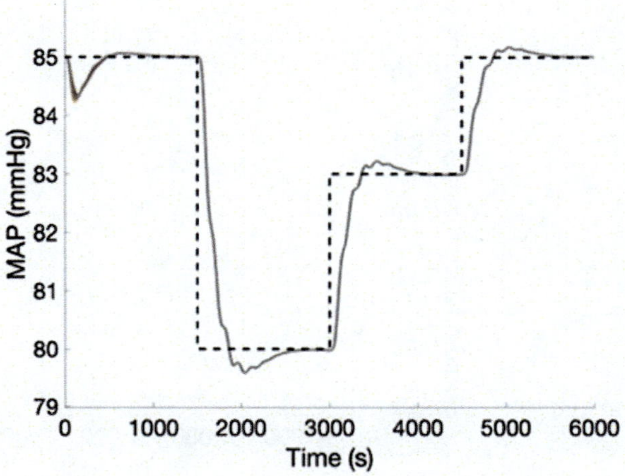

Fig. 11.1 MAP variation for setpoint tracking

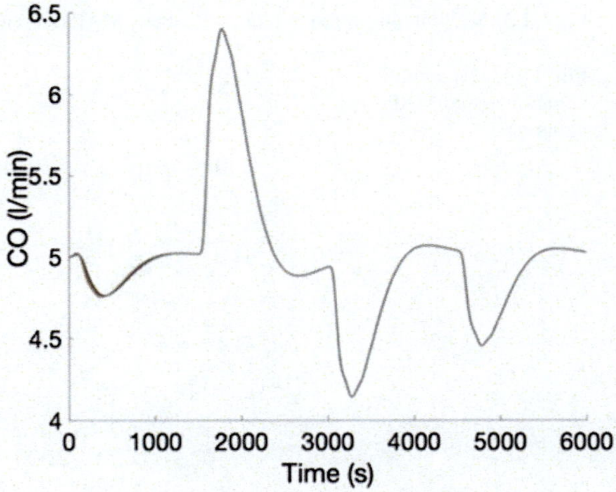

Fig. 11.2 CO variation due to changes in the MAP setpoint

A final analysis regarding the effect that MAP setpoint changes had on BIS values is performed based on the closed loop results in Fig. 11.9, with the associated control input, propofol, given in Fig. 11.10. To evaluate the changes, the time to target (TT) is computed. TT refers to the time required for the controller to bring BIS for the first time in the target range of 45–55%. Apart from the time to target, the BIS-NADIR value is also evaluated as the lowest/highest observed BIS value. The lowest BIS value is referred to as the BIS-NADIRn, while the highest one is referred to as the BIS-NADIRp. For the induction phase, the quantitative closed loop results are given in Table 11.4. Notice that in two cases, the undershoot exceeds but remains close to the 5% value defined as a performance criteria. The time to target varies between 97 and 177 seconds (or 1.6 min to approximately 3 min), with an average value of 128 seconds (or approximately 2 min). The maximum TT value is 3 min, which

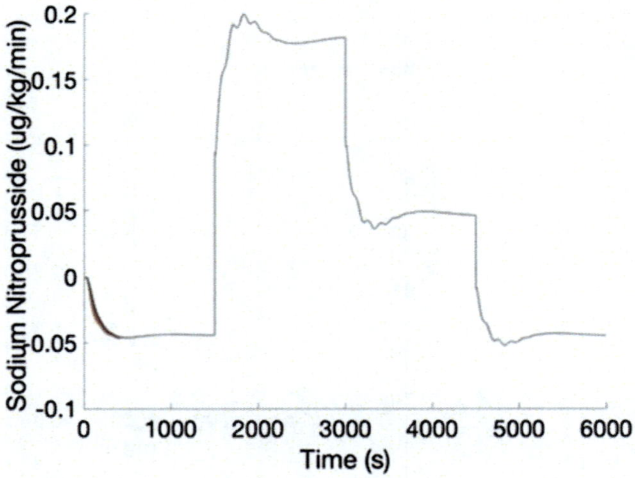

Fig. 11.3 Sodium nitroprusside required to meet MAP reference tracking

Fig. 11.4 Dopamine variation due to MAP variation

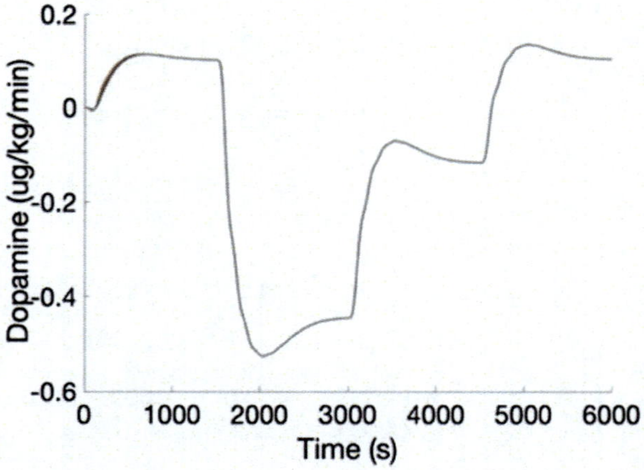

also meets the design specifications. The BIS value does not go below the minimum safety value of 40%.

The results regarding BIS variations due to hemodynamic changes in the MAP setpoint are given in Table 11.5. Notice the time to target varies between 140 seconds (or 2.3 min) and 173 seconds (or 2.9 min), with an average value of 156.5 seconds (or 2.6 min). The maximum TT value is 2.9 min, which also meets the design specifications. The smallest BIS-NADIRn value corresponds to 41.56, whereas the largest BIS-NADIRp value corresponds to 55.08. The BIS-NADIR values show that the BIS signal is kept within the safe operating range of 40–60%. The designed control strategy is able to meet the performance specifications, both in the induction and maintenance phases.

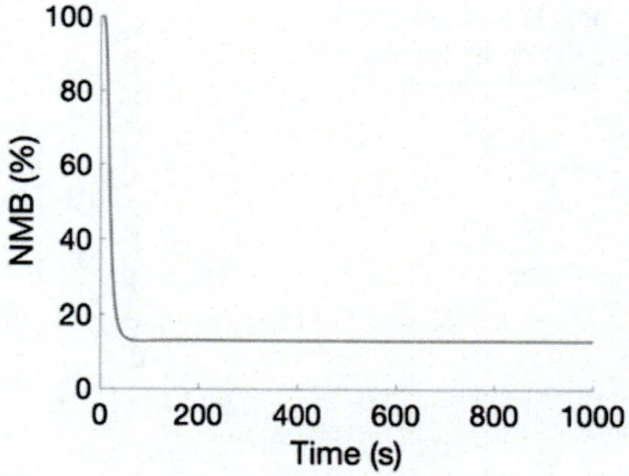

Fig. 11.5 NMB variation due to changes in the MAP setpoint

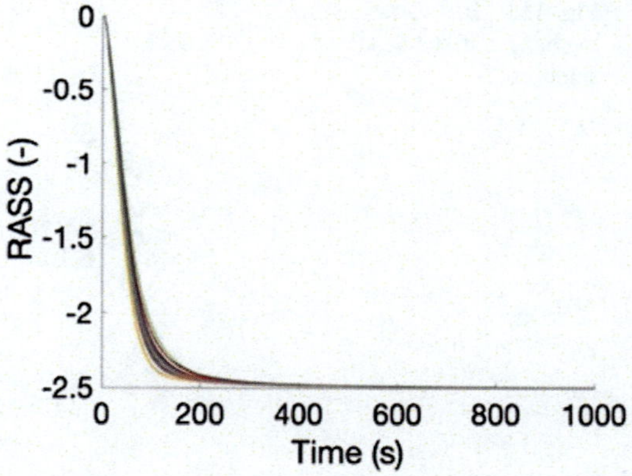

Fig. 11.6 RASS variation due to changes in the MAP setpoint

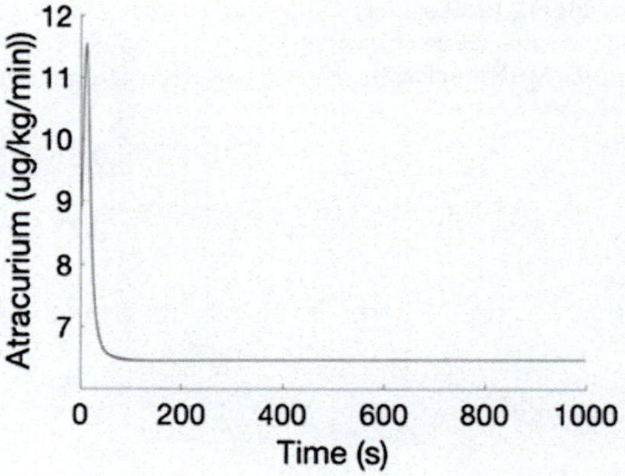

Fig. 11.7 Atracurium variation due to changes in the MAP setpoint

The second case scenario tackles the effect of CO variations upon MAP and anesthesia components. The closed loop reference tracking results for CO are given in Fig. 11.11, considering a setpoint change for CO as follows: 5 l/min → 4.5 l/min → 5 l/min → 5.5 l/min. The simulation results are presented for all 24 patients. The maximum overshoot varies between 7.5% and 8%, while the settling time varies

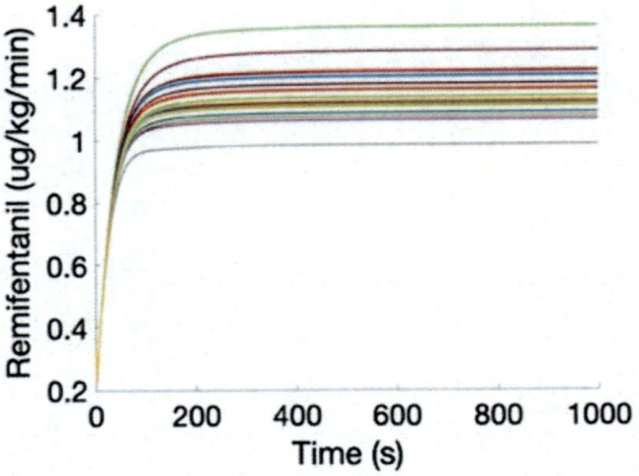

Fig. 11.8 Remifentanil variation due to changes in the MAP setpoint

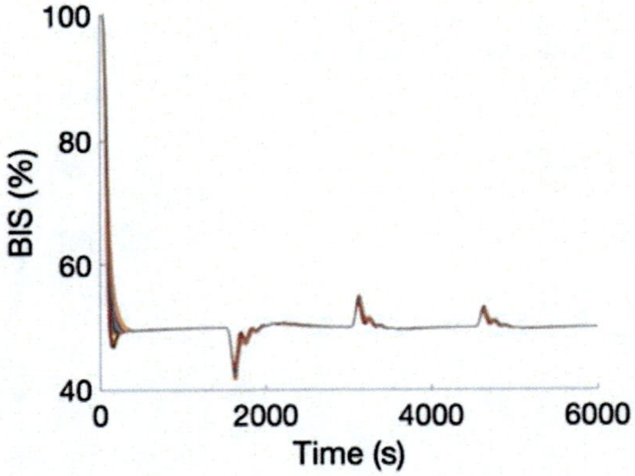

Fig. 11.9 BIS variation due to changes in the MAP setpoint

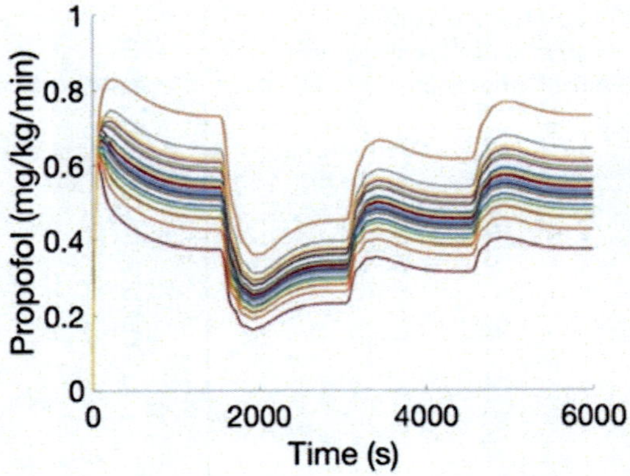

Fig. 11.10 Propofol variation due to changes in the MAP setpoint

in the 16–17 min range. The design then meets the performance specifications as they have been previously mentioned. The designed multivariable control algorithm is able to tackle better interactions on this side; thus, only small variations occur in the MAP, as indicated in Fig. 11.12. The corresponding dopamine and sodium

Table 11.4 Quantitative analysis for depth of hypnosis (BIS) in the induction phase

Patient	TT (seconds)	BIS-NADIR (%)	Undershoot (%)
1	135	49.43	0
2	104	47.86	4.3
3	167	49.52	0
4	147	49.52	0
5	112	46.68	6.64
6	128	49.28	0
7	97	46.88	6.24
8	116	49.27	0
9	119	49.26	0
10	129	49.37	0
11	127	49.33	0
12	140	49.5	0
13	123	49.27	0
14	126	49.18	0
15	123	49.35	0
16	177	49.6	0
17	131	49.4	0
18	137	49.44	0
19	114	49.25	0
20	121	49.3	0
21	123	49.26	0
22	123	49.17	0
23	109	48.85	2.3
24	144	49.51	0
Mean	128	49.06	0.81
Min	97	46.68	0
Max	177	49.6	6.64

nitroprusside are indicated in Figs. 11.13 and 11.14, and these represent variations of the administered drugs around a nominal value.

An analysis of the NMB and RASS signals shows that, in this case as well, these are not affected or slightly affected by changes in the CO setpoint. The controllers meet the design specifications, as in both cases there is no overshoot and the settling time is less than 4 min. For simplicity, the corresponding figures are not included in this chapter.

A final analysis regarding the effect that CO setpoint changes have upon BIS values is performed based on the closed loop results in Fig. 11.15, with the associated control input, propofol, given in Fig. 11.16. To evaluate the changes, the time to target (TT) is computed and the BIS-NADIRn and BIS-NADIRp values are estimated.

The results regarding BIS variations due to hemodynamic changes in the CO setpoint are given in Table 11.6. Notice a very fast time to target that varies between 34 and 39 seconds, less than 1 min, with an average value of 35.75 seconds.

Table 11.5 Quantitative analysis for depth of hypnosis (BIS) due to variations in the MAP setpoint

Patient	TT (seconds)	BIS-NADIRn (%)	BIS-NADIRp (%)
1	158	42.74	54.35
2	147	43.43	53.92
3	165	42.28	54.63
4	162	42.39	54.56
5	152	42.91	54.25
6	159	42.54	54.39
7	140	43.9	53.65
8	152	43.11	54.13
9	155	42.91	54.25
10	156	42.84	54.29
11	157	42.79	54.32
12	162	42.36	54.6
13	157	42.73	54.36
14	158	42.54	54.47
15	154	42.95	54.23
16	173	41.56	55.08
17	158	42.76	54.34
18	161	42.41	54.55
19	151	43.16	54.1
20	154	42.95	54.23
21	155	42.89	54.26
22	157	42.71	54.37
23	149	43.28	54.02
24	164	42.18	54.7
Mean	156.5	42.76	54.33
Min	140	41.56	53.65
Max	173	43.9	55.08

The drawback to such a fast time to target consists in larger variation of the BIS signal, with the smallest BIS-NADIRn value being 40.38 and the largest BIS-NADIRp = 59.69. The designed control strategy meets the performance criteria related to the TT values and is able to maintain the BIS signal within 40–60%.

The hemodynamic parameters variations in Table 11.3 are considered hereafter. For simplicity, only the case study considering MAP setpoint variations is included, and the robustness analysis refers to this situation only. Similar results were however obtained for the case when CO setpoint variations were considered. The designed control algorithm has been tested for gain variations, time delay variations, and time constant variations. Since NMB and RASS outputs were not affected significantly, the closed loop simulation results considering these two signals are not presented in this chapter.

The same setpoint variation of the MAP signal has been analyzed. Figures 11.17 and 11.18 show the simulation result for the MAP and BIS values considering

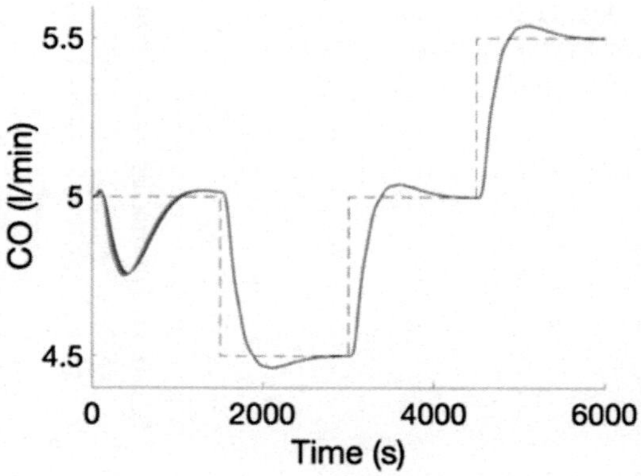

Fig. 11.11 CO variation for setpoint tracking

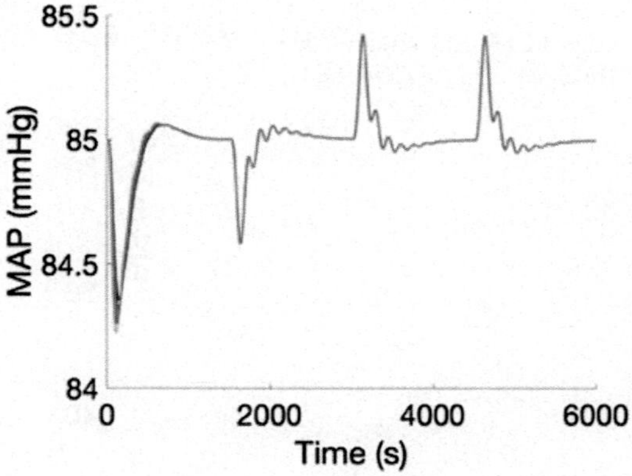

Fig. 11.12 MAP variation due to changes in the CO setpoint

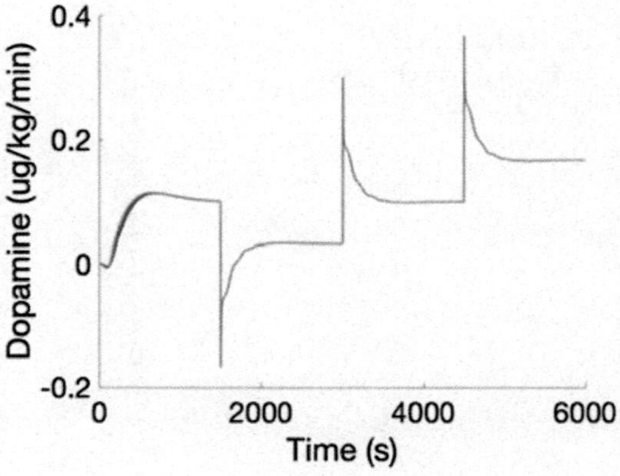

Fig. 11.13 Dopamine required to meet CO reference tracking

±30% gain variation. The controllers have been tuned to be robust to this type of inter-/intra-patient variability. In the case of MAP signal, the maximum overshoot is below 10% and remains quasi constant as expected due to the tuning approach. The settling time varies between 13.2 and 17.1 min, meeting the performance criteria as well.

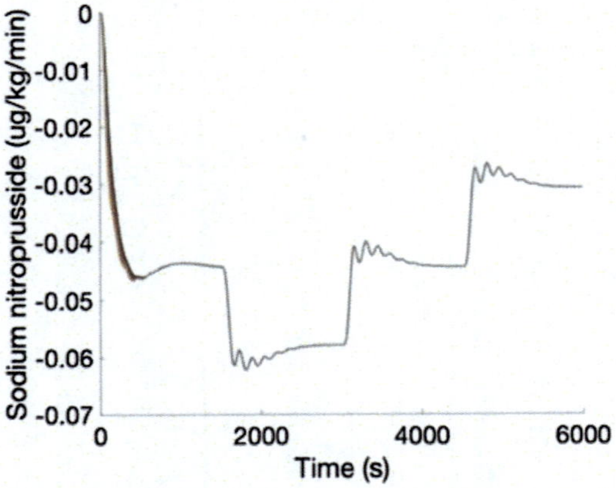

Fig. 11.14 Sodium nitroprusside variation due to CO variation

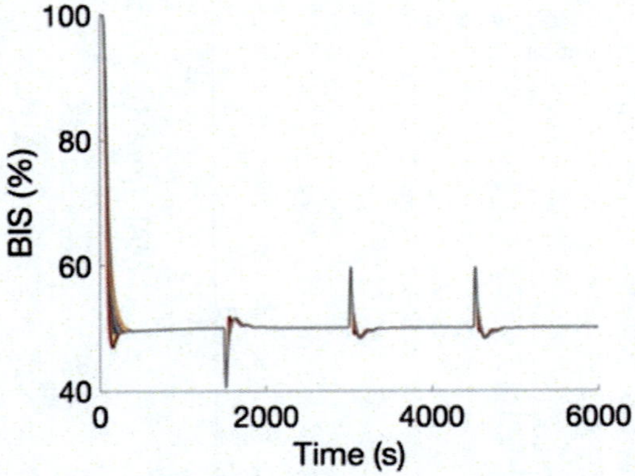

Fig. 11.15 BIS variation due to changes in the CO setpoint

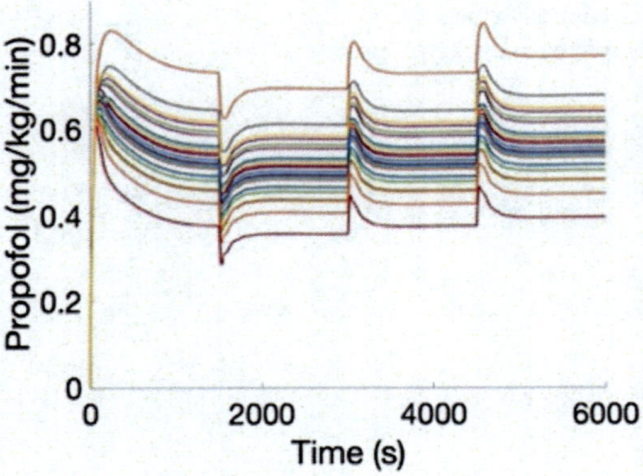

Fig. 11.16 Propofol variation due to changes in the CO setpoint

In the case of the BIS signal, the time to target ranges between 85 and 170] seconds, or 1.42–2.83 min. The minimum BIS-NADIRn value is 40.5%, while the maxim BIS-NADIRp = 55.8%. The controller is robust and able to maintain the values withing the safe range of 40–60%.

Table 11.6 Quantitative analysis for depth of hypnosis (BIS) due to variations in the CO setpoint

Patient	TT (seconds)	BIS-NADIRn (%)	BIS-NADIRp (%)
1	35	40.42	59.65
2	34	40.40	59.67
3	37	40.44	59.64
4	37	40.43	59.64
5	36	40.44	59.63
6	36	40.44	59.63
7	34	40.38	59.69
8	35	40.41	59.66
9	35	40.42	59.65
10	35	40.42	59.66
11	35	40.42	59.65
12	37	40.44	59.63
13	36	40.43	59.64
14	37	40.44	59.63
15	35	40.42	59.65
16	39	40.47	59.61
17	36	40.42	59.65
18	37	40.44	59.63
19	35	40.41	59.66
20	35	40.41	59.65
21	34	40.42	59.65
22	36	40.43	59.64
23	35	40.41	59.67
24	37	40.45	59.62
Mean	35.75	40.43	59.65
Min	34	40.38	59.61
Max	39	40.47	59.69

Fig. 11.17 MAP variation for setpoint tracking considering gain variations

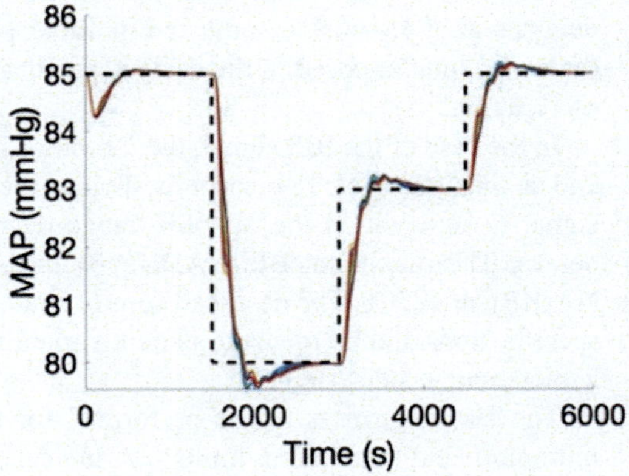

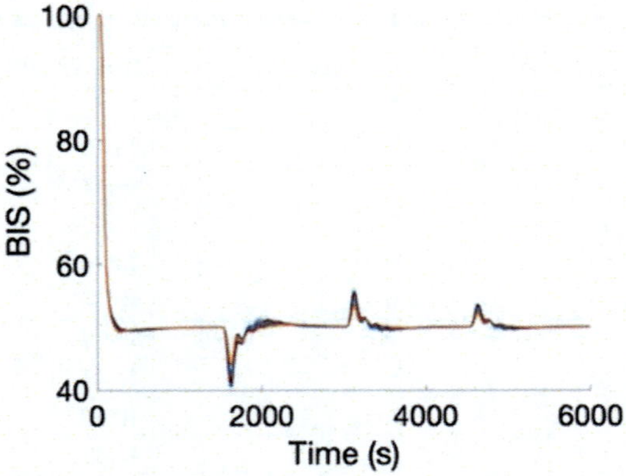

Fig. 11.18 BIS variation due to changes in the MAP setpoint considering gain variations

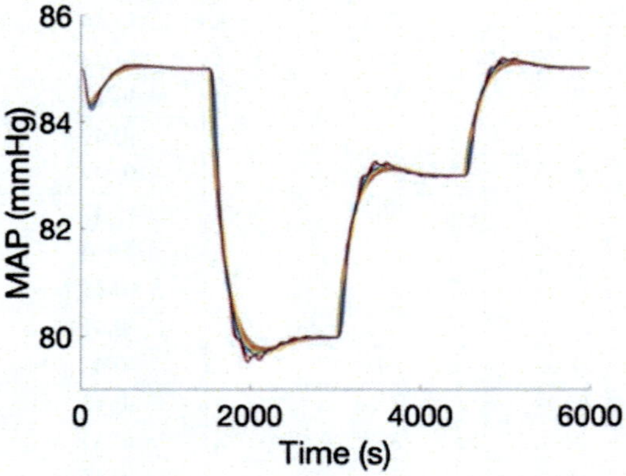

Fig. 11.19 MAP variation for setpoint tracking considering time delay variations

The control strategy is further evaluated for various possible time delay variations. The closed loop simulation results are presented in Figs. 11.19 and 11.20. In the case of MAP signal, the maximum overshoot is less than 10% for time delays within 15–60s, as indicated in Table 11.3. The settling time is kept within the constraints imposed in the design specifications, with a maximum settling time of 17.3 min.

In the case of the BIS signal, the TT varies in between a minimum of 50 seconds and a maximum of 181 seconds, below the required 3–5 min limit. The BIS signal is kept within the 40–60% range despite the large variations in the time delays. The maximum BIS-NADIRp value is 54.5%, and the minimum is BIS-NADIRn = 42.7%. The designed control strategy manages to meet the performance specifications and be robust to patient variability that affects the time delays of the hemodynamic subsystem.

The last robustness test is performed for time constant variations, within the minimum and maximum limits, as indicated in Table 11.3. The closed loop simulation results are presented in Figs. 11.21 and 11.22. In the case of MAP signal, the maximum overshoot is less than 10%, except for three extreme cases,

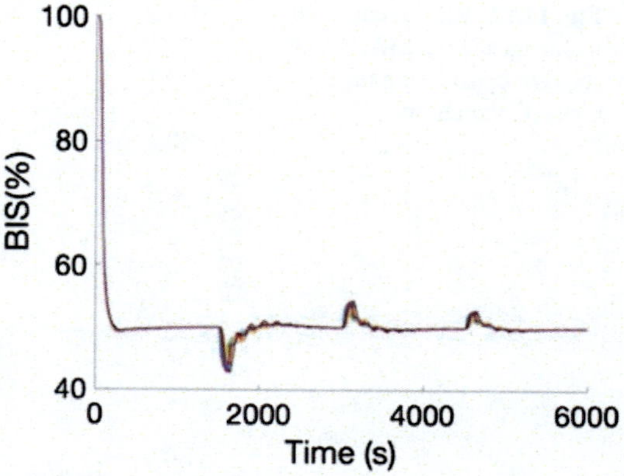

Fig. 11.20 BIS variation due to changes in the MAP setpoint considering time delay variations

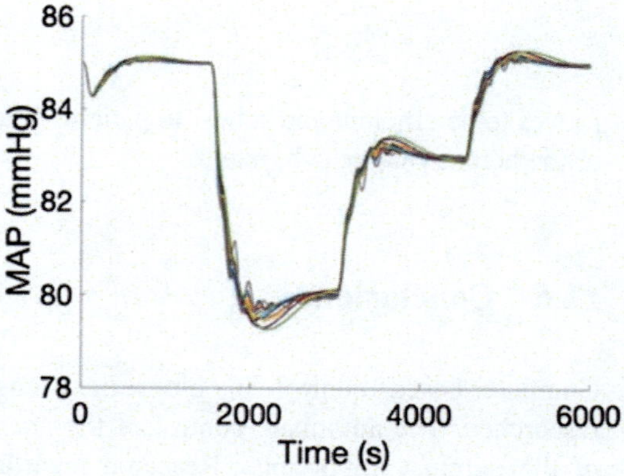

Fig. 11.21 MAP variation for setpoint tracking considering time constant variations

where the time constants of the hemodynamic system in (11.9) are close to the maximum limit, $\tau_{11} > 500$s and $\tau_{12} > 450$s, while τ_{21} and τ_{22} are maximum. In these extreme cases, the maximum overshoot is still small 13%. The settling time is kept within the constraints imposed in the design specifications, except for the case when the time delays approach the minimum values, as specified in Table 11.3. For $\tau_{11} = \tau_{12} = 70$s and $\tau_{21} = \tau_{22} = 30$s oscillatory dynamics appear on the MAP output, leading to an increased settling time of more than 20 min.

In the case of the BIS signal, a maximum TT = 170 seconds = 2.83 min is obtained. In the case of $\tau_{11} = \tau_{12} = 70$s and $\tau_{21} = \tau_{22} = 30$s, the BIS signal is no longer kept within the 40–60% range and the time to target increases to TT = 324 seconds = 5.4 min. The resulting TT in this case is slightly larger than the imposed performance constraint. In all other situations, the BIS output is maintained within the safe range. In this case, the maximum BIS-NADIRp value is 56%, and the minimum is BIS-NADIRn = 41%.

Even though poorer performance is obtained for extreme values of the time constants (either at the lower or upper limit), overall the designed control strategy

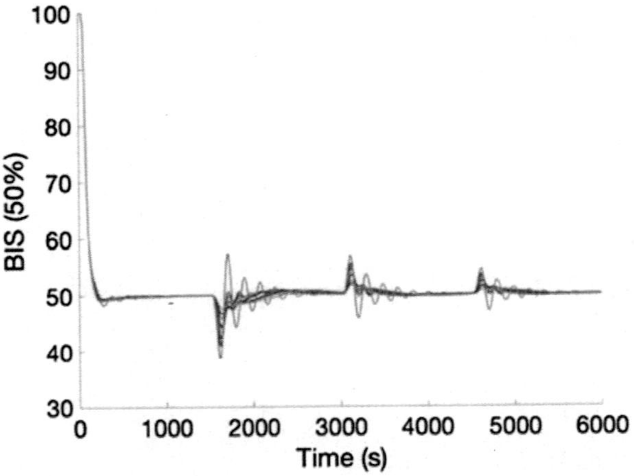

Fig. 11.22 BIS variation due to changes in the MAP setpoint considering time constant variations

proves to be efficient and robust to patient variability that affects the time constants of the hemodynamic subsystem.

11.6 Conclusion

Computer-based control in biomedical engineering is nowadays intensively researched. The automatic control of the anesthesia and hemodynamic outputs is of the outmost importance. Research regarding a multivariable control strategy for the anesthesia and hemodynamic subsystems is limited, even though the two subsystems interact, mainly due to the synergistic and antagonistic effects of the various administered drugs. The control of such a system is even more challenging due to parameter variations occurring because of patient intra- and inter-variability. Such a control system is presented and evaluated in this chapter. A recently developed benchmark patient model encompasses more accurately the interaction between the anesthesia and hemodynamic components. For a nominal patient model, a multivariable fractional order control algorithm is developed to meet a set of performance specifications. The closed loop performance of the designed controllers is then analyzed for a set of 24 patients. The focus is on the hemodynamic stabilization and how variations in the hemodynamic outputs affect the depth of hypnosis. A robustness analysis is included with a wide range of parameter variations in the hemodynamic system. The results show that the proposed solution is a suitable option for controlling the combined anesthesia and hemodynamic subsystems. Further research includes an improved minimization of interactions on the hemodynamic system using a decoupled design.

Acknowledgment This work was supported by a grant of the Romanian Ministry of Research, Innovation and Digitization, CPNRR-III-C9-2022 – I9, grant number 760018/27.01.2023 and by a grant of the Ministry of Research, Innovation and Digitization, CNCS-UEFISCDI, project number PN-III-P1-1.1-PD-2021-0204.

References

1. L. Kovacs, Linear parameter varying (lpv) based robust control of type 1 diabetes driven for real patient data. Knowl.-Based Syst. **122**, 199–213 (2017)
2. B. Kiss, J. Sapi, L. Kovacs, Imaging method for model-based control of tumor diseases, in *Proc. IEEE 11th Int. Symp. on Intelligent Systems and Informatics (SISY)*, (2013), pp. 271–275
3. C.M. Ionescu, R. De Keyser, B. Torrico, T. De Smet, M. Struys, J. Normey-Rico, Robust predictive control strategy applied for Propofol dosing using BIS as a controlled variable during anesthesia. I.E.E.E. Trans. Biomed. Eng. **55**(9), 2161–2170 (2008)
4. D. Copot, *Automated Drug Delivery in Anesthesia* (Elsevier, Amsterdam, 2020)
5. J.K. Popovic, D. Spasic, J. Tovic, J. Kolarovic, R. Malti, I.M. Mitic, S. Pilipovic, T. Atanackovic, Fractional model for pharmacokinetics of high dose methotrexate in children with acute lymphoblastic leukaemia. Commun. Nonlinear Sci. Numer. Simul. **22**, 451–471 (2015)
6. A. Joosten, J. Rinehart, A. Bardaji, P. Van der Linden, et al., Anesthetic management using multiple closed-loop systems and delayed neurocognitive recovery: A randomized controlled trial. Anesthesiology **132**, 253–266 (2020)
7. C. Zaouter, A. Joosten, S. Rinehart, S. Michel, T. Hemmerling, Autonomous systems in anesthesia where do we stand in 2020? A narrative review. Anesth. Analg. **130**, 1120–1132 (2020)
8. M. Ghita, M. Neckebroek, C. Muresan, D. Copot, Closed-loop control of anesthesia: Survey on actual trends, challenges and perspectives. IEEE Access **8**, 206264–206279 (2020). https://doi.org/10.1109/ACCESS.2020.3037725
9. J.L. Fellahi, E. Futier, C. Vaisse, et al., Perioperative hemodynamic optimization: From guidelines to implementation—An experts' opinion paper. Ann. Intensive Care **11**, 58 (2021). https://doi.org/10.1186/s13613-021-00845-1
10. M.J. Khodaei, N. Candelino, A. Mehrvarz, N. Jalili, Physiological closed-loop control (PCLC) systems: Review of a modern frontier in automation. IEEE Access **8**, 23965–24005 (2020)
11. C. Ionescu, M. Neckebroek, M. Ghita, D. Copot, An open source patient simulator for design and evaluation of computer based multiple drug dosing control for anesthetic and hemodynamic variables. IEEE Access (2021). https://doi.org/10.1109/ACCESS.2021.3049880
12. F. Padula, C. Ionescu, N. Latronico, M. Paltenghi, A. Visioli, G. Vivacqua, Optimized PID control of depth of hypnosis in anesthesia. Comput. Methods Prog. Biomed. **144**, 21–35 (2017). https://doi.org/10.1016/j.cmpb.2017.03.013
13. L. Merigo, F. Padula, N. Latronico, M. Paltenghi, A. Visioli, Event-based control tuning of propofol and remifentanil coadministration for general anaesthesia. IET Control Theory Appl. **14**(19), 2995–3008 (2020). https://doi.org/10.1049/iet-cta.2019.1067. IET Digital
14. A. Pawłowski, M. Schiavo, N. Latronico, M. Paltenghi, A. Visioli, Event-based MPC for propofol administration in anesthesia. Comput. Methods Prog. Biomed. **229**, 107289 (2023). https://doi.org/10.1016/j.cmpb.2022.107289
15. B.J. Patel, H.G. Patel, Design of CRONE-based fractional-order control scheme for BIS regulation in intravenous anesthesia, in *Advances in Control Instrumentation Systems*, Lecture Notes in Electrical Engineering, ed. by V. George, B. Roy, vol. 660, (Springer, 2020)
16. D. Copot, C. Muresan, R. De Keyser, C. Ionescu, Patient specific model based induction of hypnosis using fractional order control. IFAC- PapersOnLine **50**(1), 15097–15102 (2017)
17. N. Liu, T. Chazot, S. Hamada, A. Landais, N. Boichut, C. Dussaussoy, B. Trillat, L. Beydon, E. Samain, D.I. Sessler, M. Fischler, Closed-loop coadministration of propofol and remifentanil

guided by bispectral index: A randomized multicenter study. Anesth. Analg. **112**(3), 546–557 (2011)

18. T. Mendonça, P. Rocha, J. Silva, Chapter 6: Modeling and control of neuromuscular blockade level in general anesthesia: The neuromuscular blockade case, in *Automated Drug Delivery in Anesthesia*, ed. by D. Copot, (Academic, 2020), pp. 167–195. https://doi.org/10.1016/B978-0-12-815975-0.00011-4
19. R.R. Rao, B. Aufderheide, B.W. Bequette, Experimental studies on multiple-model predictive control for automated regulation of hemodynamic variables. IEEE Trans. Biomed. Eng. **50**(3), 277–288 (2003). https://doi.org/10.1109/TBME.2003.808813
20. D. Copot, C. Muresan, I. Birs, L. Kovacs, Robust hemodynamic control under general anesthesia conditions. IFAC-PapersOnLine **53**(2), 16179–16184 (2020). https://doi.org/10.1016/j.ifacol.2020.12.608
21. M. Gorges, N. West, S.M. Brodie, Hemodynamic changes during closed loop induction of anesthesia. Anesth. Analg. **122**(3) (2016)
22. A. Maxim, D. Copot, Closed-loop control of anesthesia and hemodynamic system: A model predictive control approach. IFAC-PapersOnLine **54**(15), 37–42 (2021)
23. E. Hegedus, I. Birs, C. Muresan, Fractional order control of the combined anaesthesia-hemodynamic system: A preliminary study. IFAC-PapersOnLine **54**(15), 19–24 (2021). https://doi.org/10.1016/j.ifacol.2021.10.225
24. E.T. Hegedus, I.R. Birs, M. Ghita, C.I. Muresan, Fractional-order control strategy for anesthesia–hemodynamic stabilization in patients undergoing surgical procedures. Fractal Fract. **6**, 614 (2022). https://doi.org/10.3390/fractalfract6100614
25. M. Ghita, D. Copot, I.R. Birs, C.I. Muresan, R. De Keyser, M. Neckebroek, M.M.R.F. Struys, C.M. Ionescu, Uncertainty minimization and feasibility study for managing the complex and interacting anesthesia-hemodynamic system, in *2022 IEEE 61st Conference on Decision and Control (CDC), Cancun, Mexico*, (2022), pp. 6064–6069. https://doi.org/10.1109/CDC51059.2022.9992350
26. C. Monje, Y.Q. Chen, B.M. Vinagre, D. Xue, V. Feliu, *Fractional-Order Systems and Controls. Fundamentals and Applications* (2010). isbn:978-1-84996-334-3
27. C.M. Ionescu, R. Hodrea, R. De Keyser, Variable time-delay estimation for anesthesia control during intensive care. I.E.E.E. Trans. Biomed. Eng. **58**(2), 363–369 (2011)
28. A. Tepljakov et al., Towards industrialization of FOPID controllers: A survey on milestones of fractional-order control and pathways for future developments. IEEE Access **9**, 21016–21042 (2021). https://doi.org/10.1109/ACCESS.2021.3055117
29. T.P. Agarkar, M.D. Patil, V.A. Vyawahare, Analysis of control effort for fractional-order PID controller, in *2021 International Conference on Advances in Computing, Communication, and Control (ICAC3), Mumbai, India*, (2021), pp. 1–7. https://doi.org/10.1109/ICAC353642.2021.9697329
30. C. Palerm, W. Bequette, Hemodynamic control using direct model reference adaptive control – experimental results. Eur. J. Control. **11**, 558–571 (2015)

Chapter 12
Fractional-Order Event-Based Control Meets Biomedical Applications

Isabela Birs and Cristina Muresan

12.1 Introduction

Fractional calculus is a branch of mathematics that generalizes differentiation and integration operations to any, non-real, arbitrary order. The limitation that differintegral operations should be of integer orders is lifted opens many doors for both modeling and control of systems with highly complex dynamics, such as the case of biomedical systems.

Fluid shifts, drug diffusion, and viscoelastic phenomena associated to blood, a non-Newtonian fluid, are a problem best characterized using fractional calculus. Going "in between" integer-order differentiation offers a better representation of physical phenomena, with complex dynamics encapsulated in a reduced number of parameters. Fractional-order calculus has been proven to be a better choice than integer-order operators for both identification and control in a manifold of applications concerning different domains, including biomedical engineering [1, 2].

Apart from modeling, fractional calculus also provides a generalization of the classical proportional integral Derivative (PID) controller through two extra parameters representing arbitrary orders of integration and differentiation, leading to a fractional-order PID (FOPID) controller. For the integral case, the $\frac{1}{s}$ is replaced

I. Birs
Automation Department, Technical University of Cluj-Napoca, Cluj-Napoca, Romania

Department of Electromechanical, Systems and Metal Engineering; Flanders Make EEDT Core Lab, Ghent University, Gent, Belgium
e-mail: isabela.birs@aut.utcluj.ro

C. Muresan (✉)
Automation Department, Technical University of Cluj-Napoca, Cluj-Napoca, Romania
e-mail: cristina.muresan@aut.utcluj.ro

C. M. A. Pinto, C. M. Ionescu (eds.), *Computational and Mathematical Models in Biology*, Nonlinear Systems and Complexity 38,
https://doi.org/10.1007/978-3-031-42689-6_12

by $\frac{1}{s^\alpha}$, where α can have any nonrational value. The same applies to the derivative action, s being generalized s^α [1]. This leads to a more advanced FOPID controller that offers more degrees of freedom than the classical PID controller, enabling the possibility to meet a more restricted set of performance specifications simultaneously [3]. Among the benefits introduced through the FOPID controller, we can enumerate increased stability, better time-domain performance such as settling time and overshoot, and an increased robustness of the closed loop system. The latter is of utmost importance in biomedical systems, where model variations (i.e., inter-/intra-patient variability) require a highly robust control loop that guarantees stability and performance when faced to system uncertainties.

Another control engineering branch gaining more focus in the last decade is event-based control [4, 5]. This type of control triggers the computation of a new control variable through intrinsic and extrinsic factors associated to the process' context. It has been shown that event-based control is a natural solution to control problems where the effects of various events can be tackled before affecting the process output [6, 7]. Various biomedical events are identified and studied in [8] to provide a clear overview with respect to biomedical entities. Integer-order event-based control strategies have been proposed for biomedical fields such as controlling of different anesthesia stages in [9–11].

The generalization of the fractional-order event-based control strategy has been only recently introduced in the control engineering field. Theoretical aspects and general implementation guidelines have been published in [12]. The paper proves that fractional-order event-based control obtains similar results to classical, discrete-time implementations, but with a drastically reduced computational effort, for a class of relevant processes from the control field. The first real-life validation of the proposed strategy has been performed on a circulatory system replica consisting of controlling a submersed miniature unmanned vehicle in a bloodlike environment with variable non-Newtonian fluid properties in [13].

Integer-order event-based control proved highly effective in controlling for biomedical system control such as general anesthesia. However, these control strategies have been explored from a restricted, integer-order, point of view. Since fractional calculus is a natural solution to modeling biomedical-related applications, a fractional-order control approach is more suitable [14–16].

Currently, fractional-order event-based control is presented at a theoretical level, lacking clear implementation strategies. The aim of this chapter is to present implementation possibilities of fractional-order event-based control, to enable its broader adoption for biomedical systems. Implementation possibilities for event-based fractional-order PID controllers as well as fractional-order internal model control are detailed.

The chapter is structured as follows: Sect. 12.2 details the suitability of event-based control for biomedical related applications; Sect. 12.3 briefly presents the concept of event-based control with its novel fractional-order generalization; Sect. 12.4 explores implementation strategies for both FOPID and FOIMC controller types.

Case studies of the implementation strategies on an experimental device are given in Sect. 12.5, while Sect. 12.6 concludes the chapter.

12.2 Suitability of Fractional Calculus and Event-Based Control in Biomedical Applications

Most properties of non-Newtonian fluids overlap with that of viscoelastic materials, such as polymers, lung tissue, gel-like substances, rubber, etc. [17]. Specific properties as memory, creep, and shear stress do not follow classical Newton's law of dynamic viscosity and have been manifold proven to be well characterized by combinations of power law with exponential functions [18]. These are nonrational expressions of combined nonlinear effects in material creep and strain, and they represent a generalization from integer-order differential equations to fractional-order differential equations. Their local solution is the Mittag-Leffler function [19–21]. Such generalized order equations are mathematical tools emerging from fractional calculus and successfully introduced in engineering, medical, and interdisciplinary application fields [22, 23]. Hitherto, the use of fractional calculus tools for modeling and control of non-Newtonian environments has been very scarce, while the majority of theoretical state of art relies on the well-known, mature, classical Newtonian approach.

Existing state of the art related to accurate non-Newtonian modeling based on the highly suitable fractional calculus paradigm is scarcely available. In [24], it is proposed a complex model to describe time-dependent flow in non-Newtonian fluids by means of a fractional-order differintegral equation for muddy clay. This is a generalization of classic Newtonian fluid flow theory following the fractionalization rationale. The idea has been extended in [25] for Maxwell elements in mechanical models of viscoelasticity. Further on, these mechanical models are then the basis for electrical model analogous [26]. The physical basis for this nonuniform velocity gradient can be the nonuniformity of fluid particles (e.g., mixtures of solid and liquid particles), molecular interaction, and biological and chemical effects.

Fluid shifts, drug diffusion, and viscoelastic phenomena associated to blood (a non-Newtonian fluid) are a problem best characterized using fractional calculus, a powerful tool that provides a generalization of differintegral operations to any arbitrary order. Going "in between" integer-order differentiation offers a better representation of physical phenomena, with complex dynamics encapsulated in a reduced number of parameters. Fractional-order calculus has been proven to be a better choice than integer-order operators for both identification and control in a manifold of applications concerning different domains, including biomedical engineering [2]. A general fractional-order model for non-Newtonian fluid interactions in changing environments (blood, tissue, etc.) has been already developed in [16]. The existing model can be adapted for the generalized drug diffusion problem to provide a better mathematical insight and a more accurate description of the real-life drug diffusion problem, with respect to patient inter-intra variability.

General anesthesia is the process of inducing a reversible loss of consciousness to allow for medical procedures. The goal of anesthesia control is to maintain a stable and safe level of anesthesia throughout the procedure while minimizing the amount of anesthetic used and reducing the risk of adverse effects. Traditional anesthesia control methods are based on continuous infusion of anesthetic drugs, which can lead to oversedation, under-sedation, or prolonged recovery times. Event-based control, on the other hand, triggers the administration of anesthetic drugs only when specific events occur, such as a change in the patient's heart rate or blood pressure. This approach can reduce the amount of anesthetic used, improve patient safety, and lead to faster recovery times.

The combination of event-based control and fractional-order control has shown promise in improving the performance and safety of anesthesia control. Fractional-order controllers can take into account the patient's past response to anesthesia, which can improve the accuracy of the control and reduce the risk of over- or underdosing. Event-based control can trigger a control value when something in the process changes, i.e., blood pressure, bolus, or surgical stimulus. However, there are still challenges in designing and implementing such control systems, including the identification of appropriate events and the selection of appropriate fractional-order control laws.

Anesthetized patients in vascular or cardiac surgery pose higher risks and challenges to the anesthesiologist. This is due to large fluid shifts taking place during the operation whose effects are not captured by available patient models. The anesthesiologist faces a prediction task with large model mismatch and uncertainty. His decision has to take into account the future actions of the surgeon and the expected effects on the body (i.e., a predictive regulatory action is required). For example, patients that take blood thinning medication pose extra challenges for modelling/control objectives and are higher risk for ensuring closed loop stability. Fractional calculus can aid to characterize and estimate (i.e., similar to observers) specific blood perfusion/drug diffusion changes which affect the uptake/clearance time constants (pharmacokinetics) and the degree of variability in drug effect (pharmacodynamics). This justifies the need of robust control, making the fractional-order PID controller the ideal candidate for the task with robustness that can be directly imposed in the tuning procedure [1].

The anesthesia maintenance phase has the objective to keep the patient in anesthetized states while surgical stimulation is ongoing. Surgery has predetermined events: intubation, incision, suction, etc. From the control point of view, these are disturbance signals. Additionally, the events in surgery are not continuous but obey a profile similar to that endorsed by event-based control, whereas only when some parameter of importance changes, the controller takes action; otherwise, it remains running of a piecewise steady-state value [27, 28]. Works such as [10, 29] use (integer-order) event-based control for depth of hypnosis. Some preliminary simulation studies show that event-based control leads to a fast induction phase with bounded overshoot and acceptable disturbance rejection. Compared to a classical PID controller, the event-based technique mimics the behavior of the anesthesiologist by providing a significant decrement of the total variation of the manipulated variable [9].

Essentially, each surgical stimulus profile is an event, justifying the choice of event-based fractional-order control. This can be either manually triggered (by anesthesiologist) as to activate the next in line surgical activity sequence for which a disturbance model profile exists. Or it can be detected through digitalization of surgery protocol signal. For example, a ramp-step form disturbance (e.g., intubation, internal organ handling) will require lower interval limit set point values than a light step form disturbance (e.g., suture, skin closure).

Event-based fractional-order control is a strategy that is dedicated to the anesthesia process and is highly customizable with respect to real-life conditions of general anesthesia. A breakthrough novel aspect is the possible adaptation of the novel fractional-order event-based strategy to control the anesthesia and hemodynamic system as a whole, customizing the triggered events with respect to real-life scenarios such as surgical stimulus. This approach solves the limitation related to disturbance signals causing high variability of drug dosing, since the control strategy shall "see" the disturbance as a custom event, determining the best course of action with respect to the type of stimulus and predefined anesthetic/hemodynamic thresholds. Event-based control is an elegant solution for controlling the anesthesia and hemodynamic systems due to its intrinsic nature to focus the control on actual, real-life, stimuli.

12.3 Event-Based Control

The fourth industrial revolution, known as Industry 4.0, integrates the latest scientific advances of the twenty-firstst century into the architecture of traditional manufacturing plants [30]. State-of-the-art smart manufacturing systems and technologies take the industrial setting a step further into the future. Human intervention is minimized by add-on automation technologies featuring state machines, self-monitoring, and abilities to perform machine to machine communication. The emerging Internet of Things (IOT) protocol is the principal constituent in the smart manufacturing trend, creating connected networks of distributed systems [31].

The new industrial cloud architectures are equipped with a wide range of sensors and actuators creating a widespread network that gathers and shares data. Decisions are (semi-)autonomously taken by interpreting the available information, sometimes through semantic reasoning. However, an overflow of unnecessary, unusable, or corrupted data leads to an overburden of the bandwidth and wireless network sharing. Introducing the context awareness modules into the smart manufacturing process enables the interpretation and filtering of the gathered information as well as performing decisions regarding its relevance [32]. The main idea is to process data in the lower levels of the system and forwarding the contextual information to the upper layers of the architecture - known as RAMI 4.0 (Reference Architectural

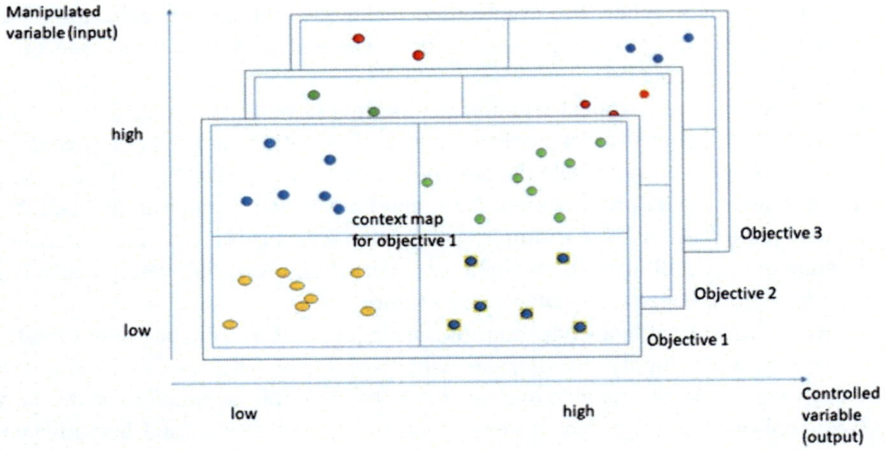

Fig. 12.1 Conceptual representation of layer-based context awareness management using context maps

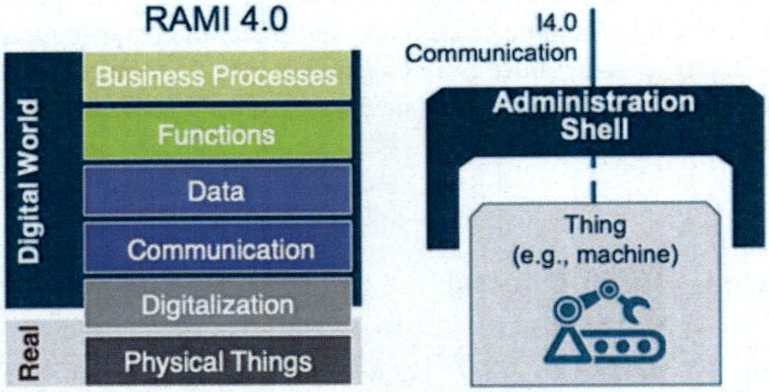

Fig. 12.2 RAMI4.0 layered architecture and element integration via the administration shell

Model Industry). There are four main layers of the context management, as presented in Fig. 12.1. Furthermore, the layers of the RAMI architecture together with the communication and administration shells are shown in Fig. 12.2. The connection takes place over the Industry 4.0 communication protocol, where the administration shell performs the digital part and the physical "things" perform the real part.

In the conceptual Fig. 12.1, at determined or triggered time instances, each objective of the system is analyzed, and a context map is created based on regression data. The context map determines various sets of input and output pairs that are particularly useful in reaching the objective (with green), pairs that are layabout for reaching the goal (with orange) or the pairs that hinder the main objective (with red). The next step involves an analysis of the process with respect to the previously defined sets and identifying events. These events will ultimately lead to a new context, with different solution pairs than the previous, which in turn can trigger the computation of a new set of control parameters in order to reach the main objective.

From the control engineering point of view, the natural solution of such context awareness paradigm lies in the event-based control methodology. The system's context is mapped based on some predefined conditions for reaching a certain objective, specific to the controlled process.

Even if traditional sampled data control, known as Riemann sampling, is the standard tool for implementing computer control, there are severe difficulties when dealing with systems having multiple sampling rates or systems with distributed computing [4]. With multi-rate sampling, the complexity of the system depends critically on the ratio of the sampling rates. For distributed systems, the theory requires that the clocks are synchronized. However, in networked distributed systems, current research revolves around solving problems such as sampling jitter, lost samples, and delays on computer-controlled systems [33]. Event-based sampling, also known as Lebesgue sampling, is a context aware alternative to periodic sampling [34]. Signals are then sampled only when significant events occurs, such as when a measured signal exceeds a limit or when an encoder signal changes. Event-based control has many conceptual advantages. Control is not executed unless it is required, control by exception. For a camera-based sensor, it could be natural to read off the signal when sufficient exposure is obtained. Event-based control is also useful in situations when control actions are expensive, for example, when controlling manufacturing complexes, and when it is expensive to acquire information like in computer networks [35].

All sampled systems, periodic as well as event based, share a common property that the feedback is intermittent and that control is open loop between the samples. After an event, the control signal is generated in an open loop manner and applied to the process. In traditional sampled data theory, the control signal is simply kept constant between the sampling instants, a scheme that is commonly called a zero-order hold (ZOH). In event-based systems, the generation of the open loop signal is an important issue, and the properties of the closed loop system depend critically on how the signal is generated [36]. There has not been much development of theory for systems with event-based control, compared to the widely available strategies developed for periodic sampled control [37].

Mathematical analysis related to the efficiency of the Lebesgue sampling algorithms has been realized by Astrom HÄgglund [36]. The study compares the performance of Riemann sampling to event-based sampling and shows that the latter gives a better performance, even if the variable sampling theory is still in its infancy.

Event-based or event-driven control is a strategy based on variable sampling of the control action. The concept is best explained in contrast to the classical, non-event-based, discrete-time implementation. The latter implies the realization of the discrete-time controller using a control signal computed every T_s seconds, without exceptions. For step references, the control signal varies while the system is in the transient regime; afterward, the control variable is constant for most processes. The only exception to this is the presence of unwanted disturbances, when the

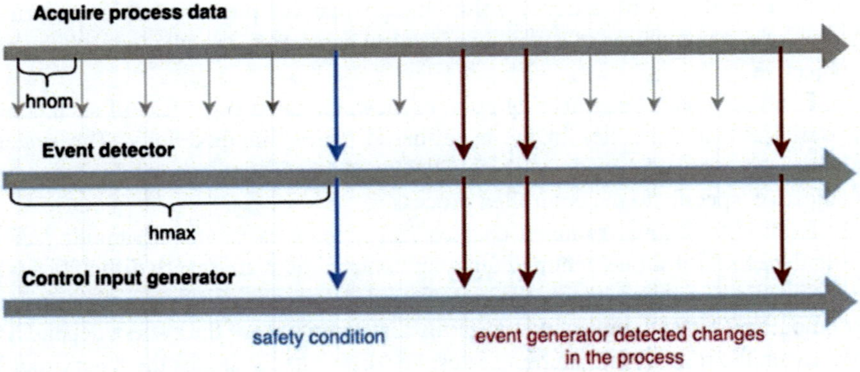

Fig. 12.3 Event-based operating principles

control action tries to bring the system back to the steady-state regime. Hence, the control variable changes during the transient regime until the process variable reaches the desired reference or when rejecting disturbances, while for the other time slots, it is constant. However, discrete-time control implementations compute a new value for the control signal every T_s seconds. From a computational point of view, the approach is inefficient and that the control effort can be reduced to the periods of time when the process variable is outside the steady-state boundaries. The efficiency problem is solved in an elegant manner by the introduction of event-based control approaches that allow the computation of a new control signal with variable sampling, removing the burden on the controlling board/circuitry for unnecessary computations. This is of particular interest in autonomous systems, which have limited energy resources (battery).

A conceptual approach to the event-based control paradigm is presented in Fig. 12.3.

h_{nom} is the nominal sampling time to acquire the data.The value is chosen similarly to the classical sampling period in discrete-time control. The *event detector* receives the process data and determined whether an event should be triggered. This component is highly customizable with respect to the main control objective. For example, in anesthesia control processes, the event detector can be programmed using biomedical events that can appear during the operating process.

Apart from process-specific events (i.e., surgical stimulus, bolus), it is advisable to include intrinsic control-specific events. The most widely used control-specific event is triggered by the process' output error variation:

$$|e(t) - e(t - h_{act})| \geq \Delta_e, \tag{12.1}$$

where $[-\Delta_e \ \Delta_e]$ is a predefined acceptable threshold interval, h_{act} is the elapsed time since the last event, $e(t - h_{act})$ is the error value when the last event was triggered, and $e(t)$ is the current error value.

Another "must-have" triggering rule in any event-based implementation approach is the safety condition. This represents the maximum time between two consecutive events, s denoted by h_{max}, where

$$h_{act} \geq h_{max}. \tag{12.2}$$

After an event is triggered, regardless of its type, the *control input generator* computes a new control value. The controller is included in the control input generator, as a discrete-time representation with variable sampling time.

As such, the challenge one needs to overcome in event-based implementations is the handling of a variable sampling for the fractional-order control domain.

12.4 Implementation Strategies of Event-Based Fractional-Order Control

The fractional operator is denoted by ${}_aD_t^\alpha f(t)$, with $t, a \in \mathbb{R}$, where $(t > a)$ are the upper and lower limits of the *differintegral*, representing a generalization of integral and derivative operations to any arbitrary order, $\alpha \in \mathbb{R}$.

The most widely used definitions of ${}_aD_t^\alpha f(t)$ have been introduced by Riemann-Liouville (RL), Grünwald-Letnikov (GL), and Caputo (C) [38].

Fractional-order control strategies are mainly developed based on the Riemann-Liouville definition given as

$${}_a^{RL}D_t^\alpha f(t) = \frac{1}{\Gamma(m'-\alpha)} \frac{d^{m'}}{dt^{m'}} \int_a^t \frac{f(\tau)}{(t-\tau)^{\alpha-m'+1}} d\tau, \tag{12.3}$$

where $m' - 1 \leq \alpha < m'$, with m' being the smallest integer greater than α and $\Gamma(m' - \alpha)$ is the Euler gamma function. The upper and lower bounds, t and a, need to be established in the case of the RL definition.

Another popular representation of the fractional differintegral operation is the Caputo fractional derivative, introduced in 1967 by Michele Caputo [39]. The advantage of this definition is that the fractional-order initial conditions do not have to be defined as in the RL case. Discussions on the issue of initial conditions are ongoing in the applied mathematics and physics community but are out of the scope of this work:

$${}_a^{C}D_t^\alpha f(t) = \frac{1}{\Gamma(m'-\alpha)} \int_a^t \frac{f^{(m')}(\tau)}{(t-\tau)^{\alpha-m'+1}} d\tau, \tag{12.4}$$

with $m' - 1 < \alpha < m'$ and $\alpha > 0$.

Grünwald-Letnikov (GL) defines ${}_aD_t^\alpha f(t)$ as

$$ {}_{a}^{GL}D_t^{\alpha} f(t) = \lim_{h \to 0^+} h^{-\alpha} \sum_{m'=0}^{\left[\frac{t-a}{h}\right]} (-1)^{m'} \binom{\alpha}{m'} f(t - m'h), \tag{12.5} $$

where $\alpha \in \mathbb{R}$ and $\alpha > 0$.

A numeric approximation of the GL definition from Eq. (12.5) has been proposed by Dorcak [40] as

$$ \begin{aligned} {}_{a}^{GL}D_t^{\alpha} f(t) &\approx T_s^{-\alpha} \sum_{m'=0}^{M(t)} (-1)^{m'} \binom{\alpha}{m'} f(t - m'T_s) \\ &\approx T_s^{-\alpha} \sum_{m'=0}^{M(t)} c_{m'}^{(\alpha)} \binom{\alpha}{m'} f(t - m'T_s), \end{aligned} \tag{12.6} $$

where T_s denotes the sampling time and L is a memory length chosen to satisfy

$$ \begin{aligned} L &\geq \frac{1}{\delta_0^2 \Gamma(\alpha)} \\ \delta_0 &= \frac{|{}_{a}^{GL}D_t^{\alpha} f(t) - {}_{t-L}^{GL}D_t^{\alpha} f(t)|}{P}. \\ P &= \max_{[0,\infty]} |f(t)| \end{aligned} \tag{12.7} $$

$M(t)$ is the minimum between t/h and L/h, while $c_{m'}^{(\alpha)}$ are binomial coefficients defined as

$$ c_{m'}^{(\alpha)} = \left(1 - \frac{1+\alpha}{m'}\right) c_{m'-1}^{(\alpha)} \tag{12.8} $$

with $c_0^{(\alpha)} = 1$.

Another important piece in fractional calculus theory is the Mittag-Leffler (ML) function:

$$ E_{\alpha}(t) = \sum_{m'=0}^{\infty} \frac{t^{m'}}{\Gamma(\alpha m' + 1)} \tag{12.9} $$

that connects the pure exponential and power law behavior, characterizing both integer- and fractional-order phenomena, resulting in

$$ \mathscr{L}\{E_{\alpha}(\pm a t^{\alpha})\} = \frac{s^{\alpha-1}}{s^{\alpha} \mp a}. \tag{12.10} $$

12.4.1 The FOPID Controller

The FOPID controller uses fractional calculus to control complex and nonlinear systems. Unlike a traditional PID controller, which uses integer-order calculus, a fractional-order PID controller uses fractional calculus to determine the appropriate control action to take based on the system's past behavior.

An integer-order PID controller consists of three terms: a proportional term k_p, an integral term k_i, and a derivative term k_d. The proportional term provides an immediate response to the system's current state, the integral term integrates the error over time to reduce steady-state error, and the derivative term anticipates future changes in the system and provides a damping effect to reduce overshoot and oscillations.

In a FOPID controller, the derivative term is a fractional-order one, denoted by s^{μ}. The fractional-order integral term is expressed as s^{λ}, and it is expressed as the integral of the error signal raised to a fractional order. The fractional orders of differentiation and integration provide a more accurate and flexible way of dealing with systems that exhibit nonlinear and time-varying behavior, as it can account for memory effects and long-range dependencies in the system.

The transfer function that characterizes an FOPID controller is

$$H_{FOPID}(s) = k_p + \frac{k_i}{s^{\lambda}} + k_d s^{\mu} \tag{12.11}$$

that gives the control signal

$$U_{FOPID}(s) = u_p(s) + u_i(s) + u_d(s), \tag{12.12}$$

with $u_p(s)$, $u_i(s)$, and $u_d(s)$ being the proportional, integral, and derivative effects control signals

$$\begin{aligned} u_p(s) &= k_p e_{\beta}(s), \\ u_i(s) &= \frac{k_p k_i}{s^{\lambda}} e(s), \\ u_d(s) &= k_p k_d \, s^{\mu} \, e_{\gamma}(s). \end{aligned} \tag{12.13}$$

The tuning of the controller can be performed using manifold of methodologies. An in-depth review of various tuning possibilities has been published by Birs et al. [2].

A direct discrete-time approximation based on the Muir recursion from [41] is used further to approximate the fractional-order term s^{α}, $\alpha > 0$ to a discrete transfer function:

$$s^{\alpha} \approx \left(\frac{2}{T_s}\right)^{\alpha} \frac{a_n z^{-n} + a_{n-1} z^{-n+1} + \ldots + a_0}{b_n z^{-n} + b_{n-1} z^{-n+1} + \ldots + b_0} = \left(\frac{2}{T_s}\right)^{\alpha} \frac{N_n(z^{-1})}{D_n(z^{-1})}. \tag{12.14}$$

For the Muir recursion mapper of s^α, the polynomials are $N_n(z^{-1}) = A_n(z^{-1}, \alpha)$ and $D_n(z^{-1}) = A_n(z^{-1}, -\alpha)$.

The model has isolated the sampling time and will treat it as a variable in the event-based implementation. The values of $N_n(z^{-1})$ and $D_n(z^{-1})$ can be calculated only once and will remain constant for each iteration through the control signal generator.

Additionally, the fractional-order terms from (12.13) will be handled independently and approximated to the discrete-time domain using the generalized formula from Eq. (12.14). As it doesn't introduce any fractional-order operation, the fractional-order proportional component is equivalent to the integer-order PID proportional term:

$$u_{fp}(k) = k_p e_\beta(k). \tag{12.15}$$

It is possible to express the fractional-order integral term as $\frac{1}{s^\lambda} = \frac{s^{1-\lambda}}{s}$, which satisfies the condition of $\alpha > 0$ in the direct discretization of s^α. As a result, the discrete-time transfer function of the fractional-order integral component is determined by utilizing the Tustin formula for the $1/s$ term and a direct nth discrete-time formula for the $s^{1-\lambda}$ term, given that $1 - \lambda > 0$ and λ belongs to the range of values between 0 and 1:

$$\frac{1}{s^\lambda} = \frac{1}{s} s^{1-\lambda} = \frac{T_s}{2} \frac{1 + z^{-1}}{1 - z^{-1}} \left(\frac{2}{T_s}\right)^{1-\lambda} \frac{N_n(z^{-1})}{D_n(z^{-1})}. \tag{12.16}$$

By expanding Eq. (12.16), we arrive at the recurrence formula for the fractional-order integral component:

$$\begin{aligned} u_{fi}(k) &= \left(\frac{2}{h_{act}}\right)^{-\lambda} \frac{k_p k_i}{d_0} \mathcal{E}_{fi_n}(k) - \frac{1}{d_0} \mathcal{U}_{fi_n}(k), \\ \mathcal{E}_{fi_n}(k) &= n_0 e(k) + (n_0 + n_1) e(k-1) + (n_1 + n_2) e(k-2) + \\ &\quad + \ldots + (n_{n-1} + n_n) e(k-n) + n_n e(k-(n+1)), \\ \mathcal{U}_{fi_n}(k) &= (d_1 - d_0) u_{fi}(k-1) + (d_2 - d_1) u_{fi}(k-2) + \\ &\quad + \ldots + (d_n - d_{n-1}) u_{fi}(k-n) - d_n u_{fi}(k-(n+1)). \end{aligned} \tag{12.17}$$

$e(k-n)$ refers to the error signal at the nth previous step, while $u_{i_n}(k-n)$ represents the fractional-order integral control action at the nth previous step.

The fractional-order derivative component is directly obtained from (12.14) , leading to

$$
\begin{aligned}
u_d(k) &= \left(\frac{2}{h_{act}}\right)^{\mu} \frac{k_p T_d}{d_0} \mathcal{E}_{d_n}(k) - \frac{1}{d_0} \mathcal{U}_{d_n}(k), \\
\mathcal{E}_{d_n}(k) &= n_0 e_\gamma(k) + n_1 e_\gamma(k-1) + \ldots + n_n e_\gamma(k-n), \\
\mathcal{U}_{d_n}(k) &= d_1 u_d(k-1) + d_2 u_d(k-2) + \ldots + d_n u_d(k-n).
\end{aligned} \tag{12.18}
$$

In the above equation, $e_\gamma(k-n)$ denotes the previous nth error value that has been adjusted by a weight factor γ, based on its set point value. Meanwhile, $u_d(k-n)$ represents the previous nth fractional-order derivative control value.

The fractional-order PID introduced in this section provides a foundation for implementing fractional-order controllers in an event-based context. For instance, a similar approach can be utilized for a fractional-order internal model control (IMC) implementation. However, treating all the effects of the fractional-order controller as a single entity is only a dependable approach when the controller contains a single fractional-order operator.

It is essential to emphasize that the process of tuning a fractional-order controller is entirely separate from its event-based implementation. Any available method can be used to tune fractional-order controllers. The algorithm presented here provides a direct discretization option and a fractional-order event-based implementation, thereby completing the implementation of the fractional-order controller. However, the controller's performance should be tuned in line with any performance specifications, just as it would be for a non-event-based implementation.

12.4.2 The FOIMC Controller

Unlike the PID control method that utilizes the process model implicitly and relies on pre-imposed specifications for the closed loop system, the IMC strategy explicitly incorporates the model into the design procedure. One of the major drawbacks of the PID approach is that it cannot accurately predict how the process model will impact the tuning decision. On the other hand, the IMC methodology is centered on the process model, with the incorporation of a tuning parameter that represents the closed loop time constant. Therefore, while the PID approach requires determination of three parameters, the IMC method only necessitates the determination of a single parameter. This makes IMC a more efficient approach compared to the PID method [42].

Fractional-order internal model control (FOIMC) is a variant of IMC that utilizes fractional calculus to enhance the controller's performance. In recent years, FOIMC has been proposed as a potential control strategy for anesthesia systems [43].

FOIMC works by incorporating a fractional-order integrator into the controller's structure, which allows for more precise modeling of the system being controlled. The fractional-order integrator is used to capture the system's long-term memory,

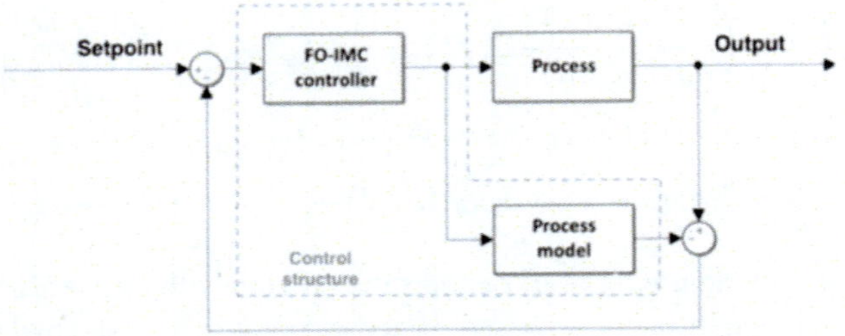

Fig. 12.4 FOIMC control diagram

which can be critical in anesthesia systems, where the patient's response to anesthesia can change over time.

In FOIMC, the fractional-order integrator is used to estimate the patient's blood propofol concentration. The controller then adjusts the infusion rate of the propofol to maintain the desired blood concentration. This approach can help to minimize the risk of over- or underdosing the patient, which can be especially important in complex surgeries where the patient's condition can change rapidly [44, 45].

One advantage of FOIMC is that it can improve the controller's response time, which can be crucial in anesthesia systems where the delivery of anesthesia must be precise and responsive to changes in the patient's condition. FOIMC can also help to reduce the risk of side effects associated with propofol, such as hypotension and respiratory depression, as shown in [46].

Research in FOIMC for anesthesia systems is ongoing, and there is still much to be learned about its effectiveness and safety. However, the early results are promising [45], and FOIMC is an area of active investigation in the field of anesthesia control.

The FOIMC controller can be obtained through two methods: by inverting a fractional-order model or by incorporating a fractional-order filter to a PID-type controller. This controller has been explored for both integer-order and fractional-order processes. In cases where fractional-order systems are used, the FOIMC can be obtained directly as a fractional controller, as demonstrated in [20, 47]. The fractional-order filter tuning approach typically involves adjusting two parameters: the closed loop time constant and an additional fractional element. The resultant controller is viewed as a PID-type controller with a fractional-order filter, according to [48]. A wide range of tuning techniques can be employed to identify the parameters of the FOIMC, including optimization methods and solving a system of frequency domain specifications, as noted in [49].

According to [50], the fundamental assumption of the IMC strategy is that accurate modeling of the process is essential for achieving perfect control. Figure 12.4 shows the operating principle of the FOIMC controller.

The process represents the physical process on which the control is applied. Furthermore, the process' model is a mathematical representation, generalized as

$$H_m(s) = \frac{a_m s^{\alpha_m} + a_{m-1} s^{\alpha_{m-1}} + \ldots + a_0}{b_n s^{\beta_n} + b_{n-1} s^{\beta_{n-1}} + \ldots + b_0}, \tag{12.19}$$

with $\{a_m, a_{m-1}, \ldots, a_0, b_n, b_{n-1}, \ldots, b_0\}$ model coefficients and $\{\alpha_m, \alpha_{m-1}, \ldots, \beta_n, \beta_{n-1}, \ldots\} \in \mathbb{R}_+$ representing fractional orders of differentiation, $\alpha_m \leq \beta_m$.

Inverting the process model and adding a fractional-order component gives the FOIMC controller as

$$H_{FOIMC}(s) = \frac{b_n s^{\beta_n} + b_{n-1} s^{\beta_{n-1}} + \ldots + b_0}{a_m s^{\alpha_m} + a_{m-1} s^{\alpha_{m-1}} + \ldots + a_0} \frac{1}{\lambda s^{\gamma} + 1}, \tag{12.20}$$

with λ is the time constant of the closed loop system, γ denoting the fractional order, and $\gamma > 0$.

The FOIMC loop employs feedback by comparing the process output with the output of an internal model, taking into account both model inaccuracies and disturbances in the output signal. Research has demonstrated that IMC strategies are resilient to the effects of model uncertainties and disturbances [51].

The control structure shown in Fig. 12.4 can be written as

$$H_c(s) = \frac{H_{FOIMC}(s)}{1 - H_{FOIMC}(s) H_m(s)}. \tag{12.21}$$

Replacing (12.19) and (12.20) in (12.21) obtains

$$H_c(s) = \frac{b_n s^{\beta_n} + b_{n-1} s^{\beta_{n-1}} + \ldots + b_0}{a_m s^{\alpha_m} + a_{m-1} s^{\alpha_{m-1}} + \ldots + a_0} \frac{1}{\lambda s^{\gamma}}, \tag{12.22}$$

leading to the following open loop system

$$H_{ol}(s) = H_c(s) H_m(s) = \frac{1}{\lambda s^{\gamma}}. \tag{12.23}$$

One can assert that the complexity of the process model does not affect the simplicity of the resulting open loop system with the FOIMC controller. To tune the controller, it is necessary to identify two parameters, λ and γ, based on the model's parameters, and to impose open loop characteristics.

The controller parameters can be determined by imposing two frequency domain specifications, namely, the gain crossover frequency ω_{gc} and the phase margin ϕ_m. In addition to these specifications, FOPID-type controllers typically utilize a third robustness requirement called isodamping to compute the parameters. The isodamping property guarantees a robust performance for a restricted range of gain uncertainties [1].

When replacing s with $j\omega$, the Laplace plane can be mapped into the frequency domain. If we want the magnitude of the open loop system from (12.23) to be equal to 0dB, we can impose the gain crossover frequency. The magnitude specification can be expressed as follows:

$$|H_{ol}(j\omega)|_{\omega=\omega_{gc}} = 1. \tag{12.24}$$

with the desired phase margin

$$\angle H_{ol}(j\omega)|_{\omega=\omega_{gc}} = -pi + \phi_m. \tag{12.25}$$

The frequency domain expression of Eq. (12.23) is derived by utilizing the trigonometric representation of the complex number $s = j\omega = \omega\ (cos\frac{\pi}{2} + j\sin\frac{\pi}{2})$, using De Moivre's formula $s^\gamma = (j\omega)^\gamma = \omega^\gamma\ (cos\frac{\gamma\pi}{2} + j\sin\frac{\gamma\pi}{2})$

$$H_{ol}(j\omega) = \frac{1}{\lambda\omega^\gamma\ (cos\frac{\gamma\pi}{2} + j\sin\frac{\gamma\pi}{2})}. \tag{12.26}$$

Replacing (12.26) in (12.24) gives

$$|H_{ol}(j\omega_{gc})| = \frac{1}{\sqrt{\lambda^2\omega_{gc}^{2\gamma}\ (cos^2\frac{\gamma\pi}{2} + \sin^2\frac{\gamma\pi}{2})}} = \frac{1}{\lambda\omega_{gc}^\gamma}. \tag{12.27}$$

The phase equation is obtained by replacing (12.26) in (12.25) as

$$\angle H_{ol}(j\omega_{gc}) = -\arctan\left(\frac{\lambda\omega_{gc}^\gamma \sin\frac{\gamma\pi}{2}}{\lambda\omega_{gc}^\gamma \cos\frac{\gamma\pi}{1}}\right) = -\frac{\gamma\pi}{2}. \tag{12.28}$$

The FOIMC tuning is ultimately the solving of the system:

$$\begin{cases} \frac{1}{\lambda\omega_{gc}^\gamma} = 1 \\ -\frac{\gamma\pi}{2} = -\pi + \phi_m \end{cases}. \tag{12.29}$$

Solving the system using simple methods is straightforward. However, not all solutions produced by this method correspond to proper FOIMC controllers, and the main difficulty lies in selecting appropriate values for ω_{gc} and ϕ_m that ensure the relevance of the resulting controller for the process.

There are two possibilities for an event-based implementation of the FOIMC controller. The first one relies on separating the controller into a PID with a fractional-order filter:

$$H_{FOIMC}(s) = k_p\left(1 + \frac{1}{T_i s} + T_d s\right)\frac{1}{s^\lambda}. \tag{12.30}$$

The parameters from (12.30) denote an integer-order PID controller with k_p proportional gain, T_i integral time, and T_d derivative time. Note that the integer-order PID controller is a particular case of the FOPID from (12.11) with orders of integration and differentiation being equal to 1. The $\frac{1}{s^\lambda}$ term can be implemented in series with the integer-order PID controller using the Muir approximation from (12.14).

Another implementation strategy is to separate the control signal computed using the equivalent structure in various fractional-order components and implement them individually using the fractional-order event-based paradigm of the PID controller.

Both implementation approaches shall be analyzed further based on a biomedical case study.

12.5 Case Study

During the administration of sedatives, the anesthetic substances tend to accumulate in the tissues, and the accumulation level depends on the infusion rate of the substance. In order to maintain controlled anesthesia, it is essential to measure the quantity of substance that has accumulated in the tissues to close the control loop. At present, the only available option to determine the amount of accumulated substance is by using predefined pharmacokinetic principles. Based on this information, the computer automatically doses the anesthetic substances, calculating the necessary amount of substance required to achieve a preset tissue concentration [52]. However, the dosage calculation is generalized based on patient characteristics such as age, weight, sex, height, and different biomarkers. As each individual is unique, the sedative substance will have a slightly different effect on each person.

A useful tool in automated anesthetic application is the ability to measure the tissue concentration instead of approximating it. This would be possible by joining targeted drug delivery concepts and automated anesthesia concepts [53].

The robots used for localized delivery can be used to sense the environment and therefore measure the intoxication levels, providing accurate feedback needed to close the control loop for automated substance administration. Using this approach ensures a higher accuracy in dosing the right quantities of sedatives instead of approximating them. The drawbacks caused by a patient's individuality to substance absorption are completely eliminated. Medical staff would also have accurate real-time data of the substance accumulation.

The case study combines the control needs of a carrier prototype with energy efficient solutions in a bloodlike environment. The developed fractional-order controller is an FOIMC that is implemented using event-based strategies discussed in Sect. 12.4.

Figures 12.5 and 12.6 illustrate the platform utilized for designing the FOIMC control structure. The non-Newtonian environment is composed of tubes with various diameters containing steel insertions that expand and contract as a result of the injection of non-Newtonian fluid by a variable flow pump. The carrier prototype

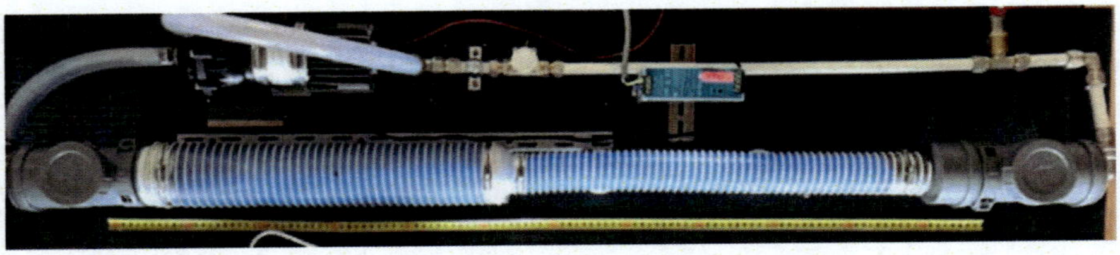

Fig. 12.5 Non-Newtonian environment

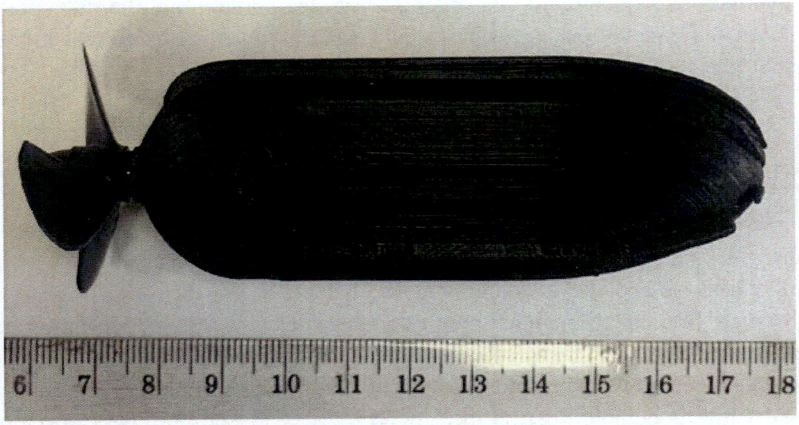

Fig. 12.6 Carrier prototype

is equipped with a propeller that serves as the thrust unit and is capable of navigating through the non-Newtonian environment. In a targeted drug delivery scenario, the carrier travels to the target area and administers the required substance. The main focus of this study is regulating the velocity of the submersible within the two tubes. The transition of the robot between the larger and smaller tubes, or vice versa, causes a disturbance to the output velocity owing to the different flow rates.

In [54], an analytical model has been created that links the velocity of the carrier to the pulse width modulation (PWM) duty ratio that drives the propeller:

$$H_m(s) = \frac{63.3580}{s^{1.7263} + 19.4139s^{0.8682} + 175.9943}. \tag{12.31}$$

A corresponding control structure with an FOIMC controller has been designed for the velocity model in [55] as follows:

$$H_c(s) = \frac{s^2 + 60.13s + 667.1}{232} \frac{1}{0.1013s^{1.2778}}. \tag{12.32}$$

In the first implementation approach of the control structure, the transfer function in Eq. (12.32) is divided into a PID controller of the form $k_p\left(1+\frac{1}{T_i s}+T_d s\right)$ multiplied by a differintegral of the form $\frac{1}{s^{\lambda}}$.

The PID component is implemented using its event-based form, while the fractional-order filter is discretized using the 5th order Muir recursion:

$$s^{\alpha}=\left(\frac{2}{T_s}\right)^{\alpha}\left(\frac{1-z^{-1}}{1+z^{-1}}\right)^{\alpha}=\left(\frac{2}{T_s}\right)^{\alpha}\lim_{n\to\infty}\frac{A_n(z^{-1},\alpha)}{A_n(z^{-1},-\alpha)}, \tag{12.33}$$

where $A_n(z^{-1},\alpha)$ is a polynomial computed as

$$\begin{aligned} &A_0(z^{-1},\alpha)=1,\\ &A_n(z^{-1},\alpha)=A_{n-1}(z^{-1},\alpha)-c^n z^n A_n(z^{-1},\alpha) \end{aligned} \tag{12.34}$$

and $c_n=\frac{\alpha}{n}$ if n is odd or $c_n=0$ if n is even. Equation (12.34) computes the fifth-order polynomial $A_5(z^{-1},\alpha)$ as

$$\begin{aligned} A_5(z^{-1},\alpha)=&-\frac{1}{5}\alpha z^{-5}+\frac{1}{5}\alpha^2 z^{-4}-\left(\frac{1}{3}\alpha+\frac{1}{15}\alpha^3\right)z^3+\\ &+\frac{2}{5}\alpha^2 z^{-2}-\alpha z^{-1}+1. \end{aligned} \tag{12.35}$$

The effectiveness of this implementation strategy is demonstrated in Fig. 12.7 for reference tracking and Fig. 12.8 for disturbance rejection scenarios.

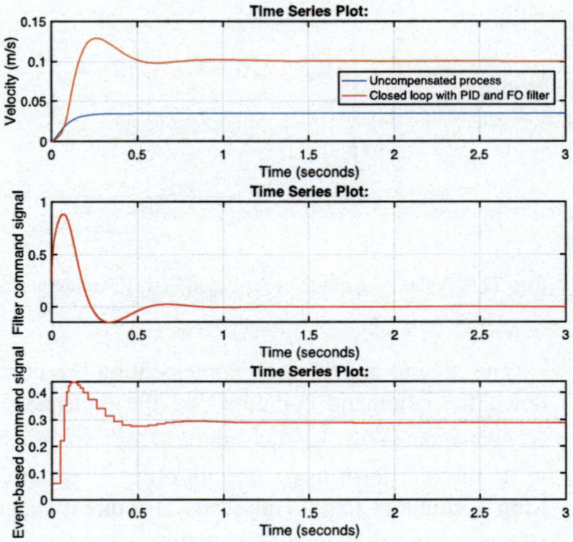

Fig. 12.7 Validation of the PID and FO filter implementation approach – reference tracking

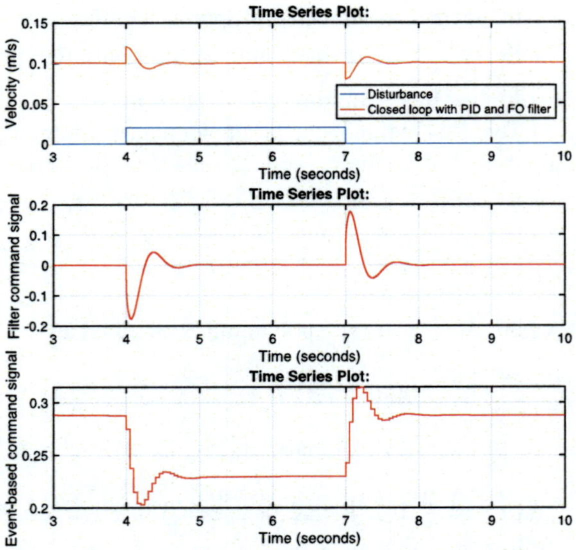

Fig. 12.8 Validation of the PID and FO filter implementation approach – disturbance rejection

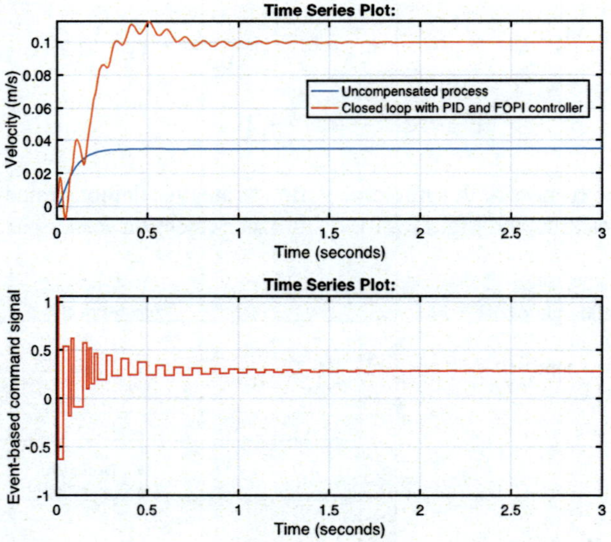

Fig. 12.9 Validation of the event-based FOIMC implementation approach – reference tracking

The second approach to implementing the control structure involves breaking down the command computed by the equivalent structure into fractional-order components: $\frac{k_p}{s^{0.2778}}$, $\frac{k_p}{T_i s^{1.2778}}$, and $k_p T_d s^{0.7222}$. Each differintegral is approximated to its discrete form using the 5th order direct discrete-time mapper based on the Muir formula (12.35). This leads to a direct event-based fractional-order control strategy. The validation of this method can be seen in Figs. 12.9 and 12.10.

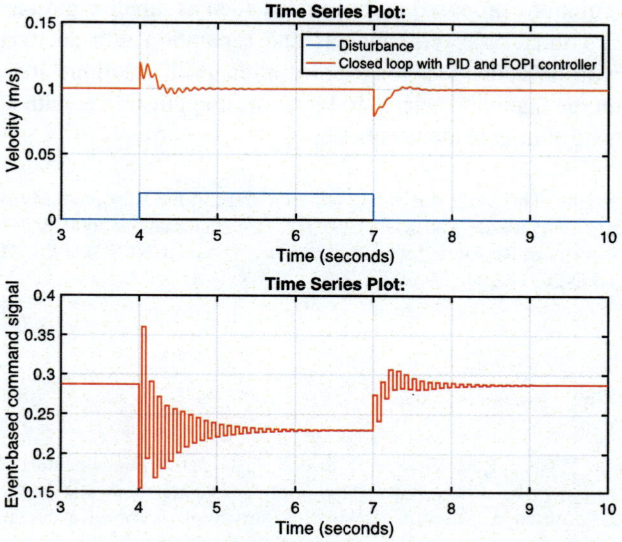

Fig. 12.10 Validation of the event-based FOIMC implementation approach – disturbance rejection

As can be seen in the validations, both event-based strategies are viable implementation solutions of event-based fractional-order controllers.

12.6 Conclusions and Future Directions

The events in surgery are not continuous but obey a profile similar to that endorsed by event-based control, whereas only when some parameter of importance changes, the controller takes action; otherwise, it remains running of a piecewise steady-state value. Some of our initial simulation studies show that event-based control leads to a fast induction phase with bounded overshoot and acceptable disturbance rejection. Compared to a classical PID controller, the event-based technique mimics the behavior of the anesthesiologist by providing a significant decrement of the total variation of the manipulated variable.

At the present time, there are scarce studies featuring real-life applications of the combined benefits of event-based PID control and fractional calculus. The generalization of the fractional-order event-based control strategy has been only recently introduced in the control engineering field. Theoretical aspects and general implementation guidelines have been published, but clear implementation strategies are missing.

This chapter explores implementation possibilities of event-based fractional-order control from different perspectives for FOPID and FOIMC controller. There are several solutions proposed and validated such as direct event-based implementations of fractional-order components and separation into an integer-order PID controller with an added fractional-order filter. Both solutions are validated on a carrier prototype that is associated to biomedical applications, with an emphasis on automatic control of general anesthesia.

Acknowledgments This work was supported by a grant of the Romanian Ministry of Research, Innovation and Digitization, CPNRR-III-C9-2022 – I9, grant number 760018/27.01.2023 and by a grant of the Ministry of Research, Innovation and Digitization, CNCS-UEFISCDI, project number PN-III-P1-1.1-PD-2021-0204.

References

1. C.A. Monje, Y. Chen, B.M. Vinagre, D. Xue, V. Feliu-Batlle, Fundamentals of fractional-order systems, in *Fractional-Order Systems and Controls Fundamentals and Applications* (2010)
2. I. Birs, C. Muresan, I. Nascu, C. Ionescu, A survey of recent advances in fractional order control for time delay systems. IEEE Access **7**, 30951–30965 (2019)
3. C.A. Monje, B.M. Vinagre, V. Feliu, Y. Chen, Tuning and auto-tuning of fractional order controllers for industry applications. Control Eng. Practice **16**(7), 798–812 (2008). http://www.sciencedirect.com/science/article/pii/S0967066107001566
4. K.J. Åström, B. Bernhardsson, *Systems with Lebesgue Sampling* (Springer, Berlin, 2007), pp. 1–13
5. K.J. Aström, *Event Based Control* (Springer, Cham, 2021), pp. 1–13
6. L. Grüne, S. Hirche, O. Junge, P. Koltai, D. Lehmann, J. Lunze, A. Molin, R. Sailer, M. Sigurani, C. Stöcker, F. Wirth, Event-based control, in *Control Theory of Digitally Networked Dynamic Systems* (2014)
7. J. Lunze, D. Lehmann, A state-feedback approach to event-based control. Automatica **46**(1), 211–215 (2010)
8. R.V.S.S. Patchigolla, S. Sahu, A. Anand, Biomedical event trigger identification using bidirectional recurrent neural network based models, in *BioNLP 2017* (Association for Computational Linguistics, Vancouver, 2017), pp. 316–321. https://aclanthology.org/W17-2340
9. L. Merigo, M. Beschi, F. Padula, N. Latronico, M. Paltenghi, A. Visioli, Event-based control of depth of hypnosis in anesthesia. Comput. Methods Programs Biomed. **147**, 63–83 (2017). https://www.sciencedirect.com/science/article/pii/S0169260716306988
10. L. Merigo, F. Padula, N. Latronico, M. Paltenghi, A. Visioli, Event-based control tuning of propofol and remifentanil coadministration for general anaesthesia. IET Control Theory Appl. **14**, 211–215 (2020)
11. M. Schiavo, F. Padula, N. Latronico, L. Merigo, M. Paltenghi, A. Visioli, Performance evaluation of an optimized pid controller for propofol and remifentanil coadministration in general anesthesia. IFAC J. Syst. Control **15**, 100121 (2021). https://www.sciencedirect.com/science/article/pii/S2468601820300250
12. I. Birs, I. Nascu, C. Ionescu, C. Muresan, Event-based fractional order control. J. Adv. Res. **25**, 191–203 (2020). Recent Advances in the Fractional-Order Circuits and Systems: Theory, Design and Applications. https://www.sciencedirect.com/science/article/pii/S2090123220301338
13. I. Birs, C. Muresan, C. Ionescu, An event based implementation of a fractional order controller on a non-newtonian transiting robot, in *2020 European Control Conference (ECC)* (2020), pp. 1436–1441

14. C. Ionescu, A memory-based model for blood viscosity. Commun. Nonlinear Sci. Numer. Simul. **45**, 29–34 (2017)
15. C.M. Ionescu, Phase constancy in a ladder model of neural dynamics. IEEE Trans. Syst. Man Cybern. A: Syst. Hum. **42**, 1543–1551 (2012)
16. I. Birs, C. Muresan, D. Copot, I. Nascu, C. Ionescu, Identification for control of suspended objects in non-newtonian fluids. Fract. Calc. Appl. Anal. **22**, 1378–1394 (2020)
17. R. Lakes, *Viscoelastic Materials* (Cambridge University Press, Cambridge, 2009)
18. C. Ionescu, J.F. Kelly, Fractional calculus for respiratory mechanics: power law impedance, viscoelasticity, and tissue heterogeneity. Chaos Solitons Fractals **102**, 433–440 (2017)
19. Y. Li, Y.Q. Chen, I. Podlubny, Mittag-Leffler stability of fractional order nonlinear dynamic systems. Automatica **45**, 1965–1969 (2009)
20. Y. Li, Y.Q. Chen, I. Podlubny, Stability of fractional-order nonlinear dynamic systems: Lyapunov direct method and generalized Mittag-Leffler stability. Comput. Math. Appl. **59**, 1810–1821 (2010)
21. C.M. Ionescu, A memory-based model for blood viscosity. Commun. Nonlinear Sci. Numer. Simul. **45**, 29–34 (2017)
22. R. Magin, B. Vinagre, I. Podlubny, Can cybernetics and fractional calculus be partners? Searching for new ways to solve complex problems. IEEE Systems Man Cybern. Mag. **4**(3), 23–28 (2018)
23. C. Ionescu, A. Lopes, D. Copot, J.A.T. Machado, J.H. Bates, The role of fractional calculus in modeling biological phenomena: a review. Commun. Nonlinear Sci. Numer. Simul. **51**, 141–151 (2017)
24. D. Yin, W. Zhang, C. Cheng, Y. Li, 'Fractional time-dependent Bingham model for muddy clay. J. Non-Newtonian Fluid Mech. **187–188**, 32–35 (2012)
25. J.F. Kelly, R.J. McGough, Fractal ladder models and power law wave equations. J. Acoust. Soc. Am. **126**, 2072–2081 (2009)
26. G. Ala, M.D. Paola, E. Francomano, Y. Li, F.P. Pinnola, Electrical analogous in viscoelasticity. Commun. Nonlinear Sci. Numer. Simul. **19**, 2513–2527 (2014)
27. J.A.R. Navarro V, Eventos crÃ-ticos en anestesia. Colomb. J. Anestesiol. **39**, 573–586 (2011)
28. J. Liberman, J. Slagle, G. Whitney, M. Shotwell, A. Lorinc, E. Porterfield, M. Weinger, Incidence and classification of nonroutine events during anesthesia care. Anesthesiology **133**, 1 (2020)
29. A. Pawlowski, L. Merigo, J.L. Guzman, A. Visioli, S. Dormido, Event-based GPC for depth of hypnosis in anesthesia for efficient use of propofol, in *2017 3rd International Conference on Event-Based Control, Communication and Signal Processing (EBCCSP)* (2017), pp. 1–7
30. E. Oztemel, S. Gursev, Literature review of industry 4.0 and related technologies. J. Intell. Manuf. **31**, 127–182 (2020)
31. A. Mohamed, M.A. Imran, P. Xiao, R. Tafazolli, Memory-full context-aware predictive mobility management in dual connectivity 5g networks. IEEE Access **6**, 9655–9666 (2018)
32. J. Sun, J. Yang, S. Li, Reduced-order GPIO based dynamic event-Triggered tracking control of a networked one-DOF link manipulator without velocity measurement. IEEE/CAA J. Autom. Sin. **7**(3), 725–734 (2020)
33. L. Merigo, M. Beschi, F. Padula, A. Visioli, A noise-filtering event generator for PIDPlus controllers. J. Franklin Inst. **355**(2), 774–802 (2018)
34. W. Dunham, *The Calculus Gallery: Masterpieces from Newton to Lebesgue* (Princeton University Press, Princeton, 2015)
35. X. Xu, Y. Lu, B. Vogel-Heuser, L. Wang, Industry 4.0 and industry 5.0-inception, conception and perception. J. Manuf. Syst. **61**, 530–535 (2021)
36. K.J. Astrom, T. HÄgglund, Advanced PID control (2006)
37. E. Aranda-Escolastico, M. Guinaldo, R. Heradio, J. Chacon, H. Vargas, J. Sanchez, S. Dormido, Event-based control: a bibliometric analysis of twenty years of research. IEEE Access **8**, 47188–47208 (2020)
38. D. Valério, J. Trujillo, M. Rivero, J. Machado, D. Baleanu, Fractional calculus: a survey of useful formulas. Eur. Phys. J. Special Topics **222**, 1827–1846 (2013)

39. M. Caputo, Linear models of dissipation whose Q is almost frequency independent-II. Geophys. J. Int. **13**(5), 529–539 (1967)
40. L. Dorcak, Numerical models for the simulation of the fractional-order control systems. arXiv: Optimization and Control (2002), pp. 1–12
41. B.M. Vinagre, Y.Q. Chen, I. Petráš, Two direct Tustin discretization methods for fractional-order differentiator/integrator. J. Franklin Inst. **340**(5), 349–362 (2003)
42. J.C. Hung, Internal model control, in *Control and Mechatronics* (2016)
43. D. Copot, C. Muresan, I. Birs, L. Kovacs, Robust Hemodynamic Control Under General Anesthesia Condition—Dana Copot is the holder of the prestigious post-doctoral scholarship awarded by the Flanders Research Centre, grant number 12X6819N. Isabela Birs acknowledges the Flanders Research Centre for the indivisual doctoral fellowship, grant number 3504719. Levente Kovacs is the holder of the European Research Council (ERC)—European Union's Horizon 2020 Research and Innovation Program grant agreement no. 679681. IFAC-PapersOnLine **53**(2), 16179–16184 (2020), 21st IFAC World Congress. https://www.sciencedirect.com/science/article/pii/S2405896320309095
44. C.L. Beck, Modeling and control of pharmacodynamics. Eur. J. Control **24**, 33–49 (2015), SI: ECC15. https://www.sciencedirect.com/science/article/pii/S0947358015000618
45. M. Ghita, M. Ghita, D. Copot, 2—an overview of computer-guided total intravenous anesthesia and monitoring devices—drug infusion control strategies and analgesia assessment in clinical use and research, in *Automated Drug Delivery in Anesthesia*, ed. by D. Copot (Academic Press, New York, 2020), pp. 7–50. https://www.sciencedirect.com/science/article/pii/B9780128159750000072
46. J.-O. Hahn, G.A. Dumont, J.M. Ansermino, Robust closed-loop control of hypnosis with propofol using wavcns index as the controlled variable. Biomed. Signal Process. Control **7**(5), 517–524 (2012). https://www.sciencedirect.com/science/article/pii/S1746809411001078
47. M. Bettayeb, R. Mansouri, Fractional IMC-PID-filter controllers design for non integer order systems. J. Process Control **24**(4), 261–271 (2014)
48. B. Hanane, A. Charef, IMC based fractional order control design for automatic voltage regulator system, in *Proceedings of 2015 7th International Conference on Modelling, Identification and Control, ICMIC 2015* (2016)
49. T. Vinopraba, N. Sivakumaran, S. Narayanan, T.K. Radhakrishnan, Design of internal model control based fractional order PID controller. J. Control Theory Appl. **10**, 297–302 (2012)
50. F. Shinskey, *Process Control Systems: Application, Design, and Tuning* (1996)
51. T.L. Chia, I. Lefkowitz, Internal model-based control for integrating processes. ISA Trans. **49**(4), 519–527 (2010)
52. M. Struys, T. De Smet, J.I.B. Glen, H.E.M. Vereecke, A.R. Absalom, T.W. Schnider, The history of target-controlled infusion. Anaesth. Analg. **122**(1), 56–69 (2016)
53. I.R. Birs, C. Muresan, A non-Newtonian impedance measurement experimental framework: modeling and control inside bloodlike environments—fractional-order modeling and control of a targeted drug delivery prototype with impedance measurement capabilities, in *Automated Drug Delivery in Anesthesia* ed. by D. Copot (Academic Press, New York, 2020), pp. 51–90
54. I. Birs, C. Muresan, O. Prodan, S. Folea, C. Ionescu, Analytical modeling and preliminary fractional order velocity control of a small scale submersible, in *SICE ISCS 2018—2018 SICE International Symposium on Control Systems* (2018)
55. I. Birs, I. Nascu, C.I. Muresan, Fractional order internal model control strategies for a submerged nanorobot, in *2020 International SAUPEC/RobMech/PRASA Conference* (2020), pp. 1–6

Chapter 13
Numerical Simulation and Validation of a Nonlinear Differential System for Drug Release Boosted by Light

J. A. Ferreira, H. Gómez, and L. Pinto

13.1 Introduction

Let Ω be the square $(0,1)^2$ and $\partial\Omega$ its boundary. We assume that $\partial\Omega = \overline{\Gamma}_l \cup \overline{\Gamma}_u \cup \overline{\Gamma}_r \cup \overline{\Gamma}_d$ (see Fig. 13.1). In this work, we consider a numerical scheme for the system of partial differential equations:

$$\begin{cases} \dfrac{1}{\beta}\dfrac{\partial I}{\partial t} = \nabla\cdot(D_I\nabla I) - \mu_a I, & (13.1)\\ \dfrac{\partial c_f}{\partial t} = \nabla\cdot(D_d\nabla c_f) + \gamma c_b I, & (13.2)\\ \dfrac{\partial c_b}{\partial t} = -\gamma c_b I, & (13.3)\end{cases}$$

for $x \in \Omega, t \in (0,T]$. In (13.1)–(13.3), D_I, D_d are diagonal matrices, with nonnegative entries defined in $\Omega \times (0,T]$, given by

$$D_I = \begin{pmatrix} D_{I,11} & 0 \\ 0 & D_{I,22}\end{pmatrix} \quad \text{and} \quad D_d = \begin{pmatrix} D_{d,11} & 0 \\ 0 & D_{d,22}\end{pmatrix},$$

where the diagonal entries are bounded from below by a positive constant.

To close the system (13.1)–(13.3), we assume the initial conditions:

$$I(x,0) = 0, \quad c_f(x,0) = 0, \quad c_b(x,0) = c_{b,0}(x), \quad x \in \Omega, \qquad (13.4)$$

J. A. Ferreira (✉) · H. Gómez · L. Pinto
Department of Mathematics, University of Coimbra, CMUC, Coimbra, Portugal
e-mail: ferreira@mat.uc.pt; hugopena@mat.uc.pt; luisp@mat.uc.pt

C. M. A. Pinto, C. M. Ionescu (eds.), *Computational and Mathematical Models in Biology*, Nonlinear Systems and Complexity 38,
https://doi.org/10.1007/978-3-031-42689-6_13

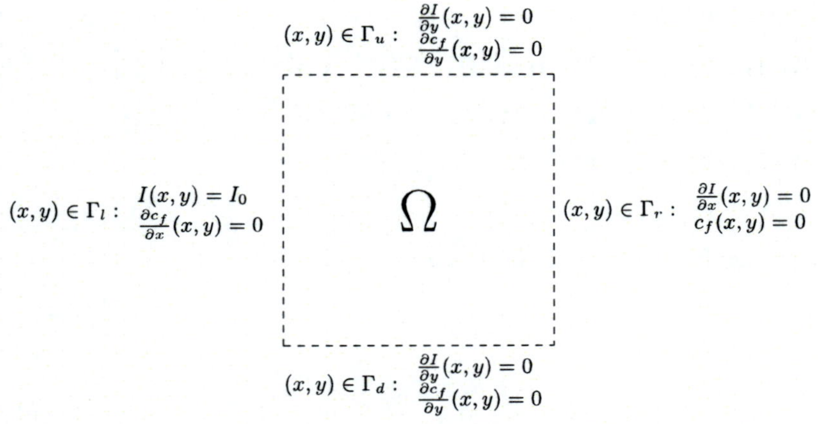

Fig. 13.1 Domain Ω and boundary conditions

and the boundary conditions

$$I(x,t) = I_0(t), \quad x \in \overline{\Gamma}_l, t \in (0,T], \tag{13.5}$$

$$\nabla I(x,t) \cdot \eta = 0, \quad x \in \partial\Omega - \overline{\Gamma}_l - \mathcal{C}_I, t \in (0,T], \tag{13.6}$$

$$\nabla c_f(x,t) \cdot \eta = 0, \quad x \in \partial\Omega - \overline{\Gamma}_r - \mathcal{C}_f, t \in (0,T], \tag{13.7}$$

$$c_f(x,t) = 0, \quad x \in \overline{\Gamma}_r, \ t \in (0,T], \tag{13.8}$$

where $\mathcal{C}_I = \{(1,0),(1,1)\}$, $\mathcal{C}_f = \{(0,0),(0,1)\}$ and η denotes the unitary exterior normal.

Local drug delivery systems (DDS) have emerged as a solution to protect healthy tissues from eventual drug side effects, particularly severe in chemotherapy cancer treatment. This technique relies on nanocarriers that transport the drugs and release them at the target tissues. The most common nanocarriers are liposomes, dendrimers, micelles, and nanoparticles. The drug release can be triggered by endogenous (e.g., pH and enzymes) or exogenous (e.g., light, heat, ultrasound, electric fields, and magnetic fields) stimuli. Due to their spatiotemporal control, reduced toxicity, noninvasive nature, ease of production, simplicity of operation, and good controllability (wavelength and intensity), light-based stimuli are particularly appealing. Near-infrared light has particularly attracted the researcher's attention since it allows deeper tissue penetration [1, 2].

The differential system (13.1)–(13.8) can describe a local DDS where a polymeric drug carrier releases the drug by action of a light stimulus. In this case, in Eq. (13.1), I denotes the light intensity. Light propagation through a scattering and absorbing medium can be modeled by the radiative transfer equation (RTE) [3]. One can consider different RTE simplifications depending on the magnitude of the absorption and scattering coefficients, μ_a and μ_s, respectively. Following [4], we use the diffusion approach to describe light intensity (13.1), where D_I is the scalar light diffusion coefficient $D_I = 1/(3(\mu_a + \mu_s))$ and β is the speed propagation of light in the medium.

System (13.1)–(13.8) also assumes that the drug molecules trapped into the polymeric matrix are cleaved due to energy absorption, i.e., the cleavage of the drug-polymer linkage occurs by light irradiation. Equation (13.2) describes the evolution of the free drug concentration c_f, and the term $\gamma c_b I$ represents the conversion of the bound drug into a free drug that is allowed to diffuse through the polymeric structure. Equation (13.3) describes the evolution of the bound drug concentration c_b.

At the initial time $t = 0$, we assume that the light intensity I and the free drug c_f concentration are zero in the domain and that the bound drug c_b distribution is known (13.4). We impose Dirichlet boundary conditions for light intensity on $\overline{\Gamma}_l$, i.e., we place a known source light at this boundary. At the remaining boundary $\partial\Omega - \overline{\Gamma}_l - \mathcal{C}_I$, we assume homogeneous Neumann boundary conditions (13.5)–(13.6). Finally, for free drug concentration, we assume a zero flux on $\partial\Omega - \overline{\Gamma}_r - \mathcal{C}_f$ and a homogeneous Dirichlet boundary condition on $\overline{\Gamma}_r$ (13.7)–(13.8), i.e., all the drug reaching this boundary is immediately removed.

The authors studied systems similar to (13.2)–(13.3) in previous works. For a one-dimensional setting and considering the simplified Beer-Lambert law for light intensity complemented with Dirichlet boundary conditions, see [5]. For a two-dimensional setting but with an emphasis on numerical simulation of local DDS and assuming Dirichlet boundary conditions for light intensity, see [5].

In this work, we propose a numerical scheme for the initial boundary value problem (IBVP) (13.1)–(13.8) that consists of a nonuniform finite difference method in space and an implicit-explicit (IMEX) linearized method in time. The finite difference scheme for the light intensity and free drug concentration has a first-order truncation error in the norm $\|.\|_\infty$. Nevertheless, we will numerically show that the scheme is second-order convergent in a discrete H^1-norm. This nonstandard phenomenon, known in the literature as supraconvergence, has been observed by other authors in several contexts over the last two decades of the last century; see, for instance, [6–12]. Traditionally, the theoretical supraconvergence results rely on the assumption of smooth solutions. In the first decade of this century, in [13] and [14], the authors provided a novel approach for supraconvergence analysis for nonsmooth solutions based on the Bramble-Hilbert Lemma [15]. More recently, this approach was used in [16–19]. We also note that we discretize the boundary conditions by adding fiction grid nodes and resorting to centered finite difference operators that lead to second-order truncation errors. For the time discretization, we will numerically show that the scheme is first-order convergent in a discrete L^2-norm.

The authors obtained supraconvergence results for systems like (13.1)–(13.8) in [5] and [20]. In [5], for a one-dimensional version of the coupled Eqs. (13.2)–(13.3), the authors obtain a supraconvergent result for a finite difference scheme assuming nonsmooth solutions, i.e., for $c_f(t) \in H^3(a,b) \cap H^1_0(a,b)$. In [20], for the full two-dimensional system (13.1)–(13.8), but with Dirichlet boundary conditions for light intensity, the authors obtain a supraconvergent result for a finite difference scheme assuming smooth solutions, i.e., $I(t) \in C^4(\overline{\Omega}_1^*)$ and $c_f(t) \in C^4(\overline{\Omega}_2^*)$ where $\overline{\Omega}_1^*$ and $\overline{\Omega}_2^*$ are convenient domains containing $\overline{\Omega}$.

In what follows, we numerically illustrate that the numerical scheme proposed for system (13.1)–(13.8) is supraconvergent in space in the sense that the numerical approximations for light intensity and free drug, as well as for their gradients, are second-order convergent. We also investigate the regularity requirements the theoretical solution must satisfy for the superconvergence result to hold. Moreover, unlike the previous works, we also analyze the discretization in time.

We organize the paper as follows: Sect. (13.2) introduces some definitions and notations. Section (13.3) presents the numerical scheme for the IBVP (13.1)–(13.8). The numerical results illustrating the convergence properties of the scheme are included in Sect. (13.4). In Sect. (13.5), we validate the IBVP (13.1)–(13.8) in the drug release context, comparing our simulation results with experimental data from [21].

13.2 Preliminaries

In what follows, we introduce the definitions and notations that define the functional scenario used to build the numerical scheme for the IBVP (13.1)–(13.8).

In $\overline{\Omega}$, we introduce a nonuniform partition $\overline{\Omega}_H$ as follows. Let Λ be a sequence of vectors $H = (h, k)$ with $h = (h_1, \ldots, h_{N_1})$, $k = (k_1, \ldots, k_{N_2})$ with positive entries such that

$$\sum_{i=1}^{N_1} h_i = \sum_{i=1}^{N_2} k_i = 1,$$

with $H_{max} = \max\{h_{max}, k_{max}\} \to 0$, where $h_{max} = \max_{i=1,\ldots,N_1} h_i$ and $k_{max} = \max_{i=1,\ldots,N_2} k_i$. Let $\overline{\Omega}_H$ be the nonuniform grid:

$$\overline{\Omega}_H = \{(x_i, y_j) : x_i = x_{i-1} + h_i, y_j = y_{j-1} + k_j, i = 1, \ldots, N_1, j = 1, \ldots, N_2\}, \tag{13.9}$$

with $x_0 = y_0 = 0$, $x_{N_1} = y_{N_2} = 1$. Let $\Omega_H = \overline{\Omega}_H \cap \Omega$, $\partial\Omega_H = \overline{\Omega}_H \cap \partial\Omega$, and $\Gamma_{i,H} = \Gamma_i \cap \partial\Omega_H, i = l, u, r, d$.

By $W(\overline{\Omega}_H)$, we represent the space of grid functions defined in $\overline{\Omega}_H$; $W_{0,l}(\overline{\Omega}_H)$ and $W_{0,r}(\overline{\Omega}_H)$ denote the spaces of grid functions in $W(\overline{\Omega}_H)$ that are null on $\overline{\Gamma}_{l,H}$ and $\overline{\Gamma}_{r,H}$, respectively. We also need to consider the space $W_b(\overline{\Omega}_H - \overline{\Gamma}_{r,H})$ of grid functions defined in $\overline{\Omega}_H - \overline{\Gamma}_{r,H}$.

We introduce now the following finite difference operators:

$$D_{-x}u_H(x_i, y_j) = \frac{u_H(x_i, y_j) - u_H(x_{i-1}, y_j)}{h_i},$$

$$D_x^* u_H(x_i, y_j) = \frac{u_H(x_{i+1}, y_j) - u_H(x_i, y_j)}{h_{i+\frac{1}{2}}},$$

where $h_{i+\frac{1}{2}} = \frac{1}{2}(h_{i+1} + h_i)$,

$$D_{c,x}u_H(x_i, y_j) = \frac{u_H(x_{i+1}, y_j) - u_H(x_{i-1}, y_j)}{h_i + h_{i+1}},$$

and

$$\nabla_H u_H = (D_{-x}u_H, D_{-y}u_H), \qquad \nabla_H^* u_H = (D_x^* u_H, D_y^* u_H),$$

$$\nabla_{c,H} u_H = (D_{c,x}u_H, D_{c,y}u_H),$$

where D_{-y}, $D_{c,y}$, and D_y^* are the finite difference operators defined analogously to D_{-x}, $D_{c,x}$, and D_x^*, respectively.

In $W(\overline{\Omega}_H)$, we consider the following inner product:

$$(u_H, v_H)_H = \sum_{(x_i, y_j) \in \overline{\Omega}_H} |\square_{ij}| u_H(x_i, y_j) v_H(x_i, y_j),$$

where $\square_{ij} = [x_{i-\frac{1}{2}}, x_{i+\frac{1}{2}}] \times [y_{j-\frac{1}{2}}, y_{j+\frac{1}{2}}] \cap \overline{\Omega}$, and $u_H, v_H \in W(\overline{\Omega}_H)$. The norm induced by this inner product is denoted by $\| \cdot \|_H$. We also use the following notations:

$$(u_H, v_H)_{x,+} = \sum_{(x_i, y_j) \in \overline{\Omega}_H - \overline{\Gamma}_{l,H}} |\square_{x,ij}| u_H(x_i, y_j) v_H(x_i, y_j),$$

$$\|u_H\|_{x,+} = \sqrt{(u_H, u_H)_{x,+}},$$

where $\square_{x,ij} = [x_{i-1}, x_i] \times [y_{j-\frac{1}{2}}, y_{j+\frac{1}{2}}] \cap \overline{\Omega}$, and $u_H, v_H \in W(\overline{\Omega}_H - \overline{\Gamma}_{l,H})$. Analogously, we also use the notations:

$$(u_H, v_H)_{y,+} = \sum_{(x_i, y_j) \in \overline{\Omega}_H - \overline{\Gamma}_{d,H}} |\square_{y,ij}| u_H(x_i, y_j) v_H(x_i, y_h),$$

$$\|u_H\|_{y,+} = \sqrt{(u_H, u_H)_{y,+}},$$

where $\square_{y,ij} = [x_{i-\frac{1}{2}}, x_{i+\frac{1}{2}}] \times [y_{j-1}, y_j] \cap \overline{\Omega}$, and $u_H, v_H \in W(\overline{\Omega}_H - \overline{\Gamma}_{d,H})$.

For $u_H = (u_{H,1}, u_{H,2})$, $v_H = (v_{H,1}, v_{H,2})$, where $u_{H,1}, v_{H,1} \in W(\overline{\Omega}_H - \overline{\Gamma}_{l,H})$, $u_{H,2}, v_{H,2} \in W(\overline{\Omega}_H - \overline{\Gamma}_{d,H})$, we take

$$(u_H, v_H)_{H,+} = (u_{H,1}, v_{H,1})_{x,+} + (u_{H,2}, v_{H,2})_{y,+} \quad \text{and} \quad \|u_H\|_+ = \sqrt{(u_H, u_H)_{H,+}}.$$

We introduce the average operators:

$$M_h u_H(x_i, y_j) = \frac{1}{2}\big(u_H(x_i, y_j) + u_H(x_{i-1}, y_j)\big),$$

being M_k defined analogously. By $D_I(M_H u_H)$, we denote the diagonal matrix with diagonal entries $D_{I,11}(M_h u_H)$ and $D_{I,22}(M_k u_H)$. The diagonal matrix $D_d(M_H u_H)$ is defined analogously.

To discretize the light intensity equation (13.1) and the boundary conditions (13.5)–(13.6), we need to consider the auxiliary point $x_{N_1+1} = x_{N_1} + h_{N_1}$, $y_{N_2+1} = y_{N_2} + k_{N_2}$, and $y_{-1} = -y_1$ and the fictitious points $\Gamma_{i,H}^{(I)}$, for $i = d, u, r$, defined by

$$\Gamma_{d,H}^{(I)} = \{(x_i, y_{-1}), i = 1, \ldots, N_1\}, \tag{13.10}$$

$$\Gamma_{u,H}^{(I)} = \{(x_i, y_{N_2+1}), i = 1, \ldots, N_1\}, \tag{13.11}$$

$$\Gamma_{r,H}^{(I)} = \{(x_{N_1+1}, y_j), j = 0, \ldots, N_2\}. \tag{13.12}$$

Similarly, to discretize the free drug concentration equation (13.2) and the boundary condition (13.7)–(13.8), we need to introduce the auxiliary point $x_{-1} = -x_1$ and the fictitious points $\Gamma_{i,H}^{(c)}$, for $i = l, d, u$, defied by

$$\Gamma_{l,H}^{(c)} = \{(x_{-1}, y_j), j = 0, \ldots, N_2\}, \tag{13.13}$$

$$\Gamma_{d,H}^{(c)} = \{(x_i, y_{-1}), i = 0, \ldots, N_1 - 1\}, \tag{13.14}$$

$$\Gamma_{u,H}^{(c)} = \{(x_i, y_{N_2+1}), i = 0, \ldots, N_1 - 1\}. \tag{13.15}$$

Let $W_{I,H}^*$ and $W_{c,H}^*$ be the space of grid functions defined in $\overline{\Omega}_H \cup (\cup_{i=d,r,u} \Gamma_{i,H}^{(I)})$ and $\overline{\Omega}_H \cup (\cup_{i=l,d,u} \Gamma_{i,H}^{(c)})$, respectively (Fig. 13.2).

13.3 Numerical Scheme

Here, we present a numerical method for a generalized version of system (13.1)–(13.8). We focus on the differential system:

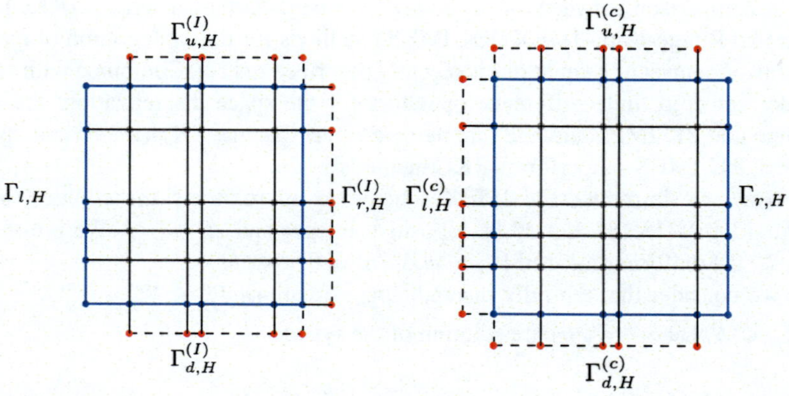

Fig. 13.2 Illustration of the ghost points $\Gamma^{(I)}_{i,H}$, for $i = d, u, r$ (on the left) and $\Gamma^{(c)}_{i,H}$, for $i = l, d, u$ (on the right)

$$\begin{cases} \frac{1}{\beta}\frac{\partial I}{\partial t} = \nabla \cdot (D_I \nabla I) - G(I), & (13.16) \\ \frac{\partial c_f}{\partial t} = \nabla \cdot (D_d \nabla c_f) + F(I, c_f, c_b), & (13.17) \\ \frac{\partial c_b}{\partial t} = S(I, c_f, c_b), & (13.18) \end{cases}$$

where $G : \mathbb{R} \to \mathbb{R}$ and $F, S : \mathbb{R}^3 \to \mathbb{R}$. Notice that system (13.1)–(13.3) is obtained from (13.16)–(13.18) by making a particular choice of the reaction terms. In what follows, we present an IMEX method to approximate the solution of the system (13.16)–(13.18) with initial and boundary conditions (13.4)–(13.8).

Ideally, an IMEX approach should have better stability properties than a fully explicit scheme and reduce the computational cost associated with a fully implicit scheme, which requires the solution of nonlinear systems at each time level.

IMEX methods have been recently used in [22, 23] to compute the numerical solutions of wave equations containing terms with different timescales (fast and slow). The results show that IMEX methods where the slow terms are approximate explicitly and the fast ones are approximate implicitly have similar accuracy to a fully implicit method but with much faster performance.

In [24, 25], IMEX methods are used in the numerical approximation for reaction-diffusion problems. The stiff term (the diffusion term) is discretized implicitly, and the nonlinear reaction terms are discretized explicitly. Compared to a fully discrete scheme, this approach avoids a restrictive time step condition associated with the diffusion term. On the other hand, compared to a fully implicit scheme, it avoids the solution of nonlinear systems associated with the reaction terms. The advantage of IMEX methods to deal with reaction-diffusion equations is discussed in [26].

A comparison between the second-order Runge-Kutta-Chebyshev (RKC) methods and Runge-Kutta-IMEX (RK-IMEX) methods for time integration of reaction-diffusion-convection equations is recently described in [27]. Combined with second-order centered finite difference operators for the space discretization, the author found that RK-IMEX methods are more accurate than the RK methods and, in many cases, RK-IMEX outperforms RKC methods.

Now, we introduce the IMEX scheme to approximate the solution of system (13.16)–(13.18). Let M be a positive integer and $\Delta t = \frac{T}{M}$. We introduce in $[0, T]$ the uniform time grid $\{t_m = m\Delta t, m = 0, \cdots, M\}$.

We consider that the fully discrete approximations $I_H^m \in W_{I,H}^*$, $c_{f,H}^m \in W_{c,H}^*$, $c_{b,H}^m \in W_b(\overline{\Omega}_H - \overline{\Gamma}_{r,H})$ are solution of the system

$$\begin{cases} \frac{1}{\beta} D_{-t} I_H^m = \nabla_H^* \cdot ((D_I(M_H I_H^m)\nabla_H I_H^{m+1})) \\ \quad +G(I_H^m) & \text{in } \overline{\Omega}_H - \overline{\Gamma}_{l,H} \quad (13.19) \\ D_{-t} c_{f,H}^m = \nabla_H^* \cdot ((D_d(M_H c_{f,H}^m)\nabla_H c_{f,h}^{m+1})) \\ \quad +F(I_H^m, c_{f,H}^m, c_{b,H}^m) & \text{in } \overline{\Omega}_H - \overline{\Gamma}_{r,H} \quad (13.20) \\ D_{-t} c_{b,H}^m = S(I_H^m, c_{f,H}^m, c_{b,H}^m) & \text{in } \overline{\Omega}_H - \overline{\Gamma}_{r,H} \quad (13.21) \end{cases}$$

complemented by the following initial and boundary conditions:

$$I_H^m = R_H I_i(t_m) \text{ on } \overline{\Gamma}_{H,l}, \; \nabla_{c,H} I_H^m \cdot \eta = 0 \text{ on } (\partial\Omega_H - \overline{\Gamma}_{H,l}) \times (0, T], \tag{13.22}$$

$$c_{f,H}^m = 0 \text{ on } \overline{\Gamma}_{r,H} \times (0, T], \; \nabla_{c,H} c_{f,H}^m \cdot \eta = 0 \text{ on } (\partial\Omega_H - \overline{\Gamma}_{r,H}) \times (0, T]. \tag{13.23}$$

Using the fictitious points, we can state that $\nabla_H^* \cdot ((D_d(M_H c_{f,H}^m)\nabla_H c_{f,h}^{m+1}))$ is well defined in $\overline{\Omega}_H - \overline{\Gamma}_{H,l}$. Moreover, $\nabla_{c,H} I_H^m . \eta$ is well defined in $\Gamma_{H,i}$, $i = u, r, d$ and in the corner points $(1, 0)$ and $(1, 1)$ where we consider two unitary normals: the vectors $e_1, -e_2$ and e_1, e_2, respectively, where $\{e_1, e_2\}$ is the canonical basis of $\mathbb{R}^2$. The same remarks hold for the discretizations of the free drug concentration.

System (13.19)–(13.23) can be established using the so-called method of lines approach: spatial discretization using finite difference operators, which allows us to replace the IBVP (13.16)–(13.18) by an ordinary differential problem, followed by a time integration using an IMEX approach to deal with the nonlinear terms. We discretize the nonlinear reaction terms ($G(\cdot)$, $S(\cdot)$, and $F(\cdot)$) and the nonlinear diffusion terms ($D_I(\cdot)$ and $D_d(\cdot)$) explicitly and the remaining terms implicitly. Following this approach, we can obtain the numerical approximation by solving a linear system at each time level. For instance, for the intensity light approximation I_H^{m+1}, we get the following matrix equation:

$$\frac{1}{\beta}\Big(I_d - \Delta t A_H(I_H^m)\Big) I_H^{m+1} = I_H^m + \Delta t (F_H(I_H^m) + G(I_H^m)), \tag{13.24}$$

where I_d is the identity matrix of order $N_1(N_2 + 1) \times N_1(N_2 + 1)$ and $A_H(I_H^m)$ is a tridiagonal block matrix.

We observe that we can fix Δt such that $\Delta t \|A_H\|_\infty < 1$. Consequently, $I_d - \Delta t A_H(I_H^m)$ is nonsingular, and then, for each time level, there exists a unique solution of the linear system (13.24). Using similar arguments, we can guarantee that, for each time level, there exists a unique numerical free drug concentration.

In the next section, we use several examples to demonstrate the convergence behavior of our method.

13.4 Numerical Examples: Convergence Behavior

This section has two main objectives: to investigate the convergence properties of the IMEX method (13.19)–(13.23) for smooth and nonsmooth solutions in space and to investigate the convergence in time, including a comparison with the fully implicit and second-order Crank-Nicolson (CN) method:

$$D_{-t} I_H^m = \frac{1}{2}\nabla_H^* \cdot ((D_I(M_H I_H^{m+1})\nabla_H I_H^{m+1}) + (D_I(M_H I_H^m)\nabla_H I_H^m)) + G_H^{m+1/2}$$
$$\text{in } \overline{\Omega}_H - \overline{\Gamma}_{H,l} \tag{13.25}$$

$$D_{-t} c_{f,H}^m = \frac{1}{2}\nabla_H^* \cdot ((D_d(M_H c_{f,H}^{m+1})\nabla_H c_{f,H}^{m+1}) + (D_d(M_H c_{f,H}^m)\nabla_H c_{f,h}^m)) + F_H^{m+1/2}$$
$$\text{in } \overline{\Omega}_H - \overline{\Gamma}_{H,r} \tag{13.26}$$

$$D_{-t} c_{b,H}^m = S_H^{m+1/2} \text{ in } \overline{\Omega}_H - \overline{\Gamma}_{H,r} \tag{13.27}$$

for $m = 0, \ldots, M-1$, complemented by the initial and boundary conditions (13.22)–(13.23), where $G_H^{m+1/2} = \frac{1}{2}(G(I^m) + G(I^{m+1}))$ being $F_H^{m+1/2}$, $S_H^{m+1/2}$ defined analogously. Unlike the IMEX method, the CN method requires the solution of a nonlinear system at each time step, which we solve by a trust-region-dogleg method. Therefore, the expected higher accuracy of the CN method comes with an increased computational cost.

To illustrate the convergence behavior, we introduce the following errors:

$$\text{Error}_I^2 = \max_{m=1,\ldots,M} \|E_I^m\|_H^2 + \Delta t \sum_{k=1}^{m} \|\nabla_H E_I^k\|_+^2, \tag{13.28}$$

$$\text{Error}^2_{c_f} = \max_{m=1,\ldots,M} \|E^m_{c_f}\|^2_H + \Delta t \sum_{k=1}^{m} \|\nabla_H E^k_{c_f}\|^2_+, \tag{13.29}$$

$$\text{Error}^2_{c_b} = \max_{m=1,\cdots,M} \|E^m_{c_b}\|_H, \tag{13.30}$$

where $E^m_I = R_H I(t_m) - I^m_H$ and R_H is the restriction operator. Analogously, $E^m_{c_f} = R_H c_f(t_m) - c^m_{f,H}$ and $E^m_{c_b} = R_H c_b(t_m) - c^m_{b,H}$.

The numerical convergence rates are estimated by

$$\text{R}_{\ell,H} = \log_2\left(\frac{\text{E}_{\ell,H}}{\text{E}_{\ell,H/2}}\right),$$

for $\ell = I, c_f, c_b$. Here, $E_{\ell,H/2}$ represent the errors associated with the spatial mesh $H/2$ that we obtain from a mesh H by introducing in the intervals (x_i, x_{i+1}) and (y_j, y_{j+1}) the corresponding midpoints $x_{i+1/2}$ and $y_{j+1/2}$.

To illustrate the convergence behavior of our method in time, we introduce $Error_{\Delta t,\ell}$, for $\ell = I, f, b$ defined by

$$Error^2_{\Delta t,I} = \max_{m=1,\cdots,M} \|E^m_I\|_H, \tag{13.31}$$

$$Error^2_{\Delta t,f} = \max_{m=1,\cdots,M} \|E^m_{c_f}\|_H, \tag{13.32}$$

$$Error^2_{\Delta t,b} = \max_{m=1,\cdots,M} \|E^m_{c_b}\|_H, \tag{13.33}$$

and the corresponding convergence rates

$$\text{R}^{\Delta t}_{\ell,H} = \log_2\left(\frac{\text{E}_{\Delta t,\ell}}{\text{E}_{\Delta t/2,\ell}}\right), \tag{13.34}$$

for $\ell = I, f, b$.

Example 1 Let us consider system (13.16)–(13.18) with initial boundary conditions given by (13.4)–(13.8) and diffusion coefficients and reaction functions defined by $D_{I,11} = D_{I,22} = I^2 + 0.1$, $D_{c,11} = D_{c,22} = 0.1c_f^2$, $G(I) = 2I^2 + I$, $S(I, c_f, c_b) = I^2 c_b^2 c_f$, and $F(I, c_f, c_b) = I^3 c_b c_f^2$. Let us also add suitable source functions to the right-hand side of Eqs. (13.16)–(13.18), such that the exact solution of this problem is

$$I(x, y, z) = 2t(2x - x^2)\frac{1}{\pi}\left(-(y-1)\cos(\pi y) + \frac{1}{\pi}\sin(\pi y)\right),$$

$$c_f(x, y, z) = \exp(t)\left(\frac{1}{x^2+1} - \frac{1}{2}\right)\left(\frac{y^2}{2} + \frac{1}{\pi^2}(\pi y\sin(\pi y) + \cos(\pi y))\right),$$

Table 13.1 Space errors and convergence rates

H_{max}	N_1	N_2	Error_I	Rate_I	Error_{c_f}	Rate_{c_f}	Error_{c_b}	Rate_{c_b}
1.3429E-1	12	10	1.3552E-3	–	3.4017E-6	–	1.2989E-4	–
6.7143E-2	24	20	5.6034E-6	3.9590	1.1923E-7	2.4172	9.1878E-6	1.9107
3.3572E-2	48	40	3.6037E-7	1.9794	8.0611E-9	1.9433	6.2452E-7	1.9395
1.6786E-2	96	80	2.2647E-8	1.9960	5.2261E-10	1.9736	4.0505E-8	1.9733
8.3929E-3	192	160	1.4182E-9	1.9986	3.3268E-11	1.9868	2.5775E-9	1.9870
4.1965E-3	384	320	8.8726E-11	1.9993	2.0987E-12	1.9933	1.6255E-10	1.9935

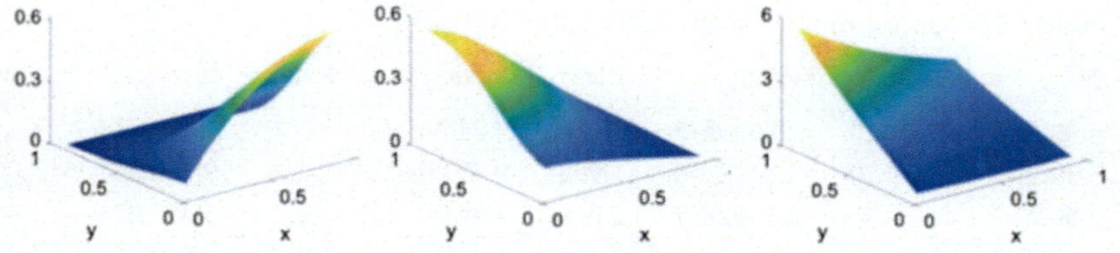

(a) I_H. (b) $c_{f,H}$. (c) $c_{b,H}$.

Fig. 13.3 Numerical solution for Example (1)

$$c_b(x, y, z) = 2\exp(t)\sinh(y)\frac{\cosh(10 - x)}{\cosh(10)}.$$

We solve the resulting system of differential equations using the method (13.19)–(13.23). We start with a random nonuniform spatial grid and define the sequence of grids by halving the step sizes. The time stepsize Δt is fixed, $\Delta t \leq H_{\max}^2$; consequently, the error induced by the time integration has a small contribution in the global error that will be dominated by the error in space. The total simulation time is $T = 1$. We give the errors and the estimated numerical convergence rates in Table 13.1.

The results presented in Table 13.1 suggest that the spatial errors of I, c_f, and c_b have second-order convergence. Note that by definition of the errors for I and c_f, Eqs. (13.28) and (13.29), respectively, the second-order accuracy holds not only for the numerical approximations I_H^m and $c_{f,H}^m$ but also for the numerical gradients $\nabla_H I_H^m$ and $\nabla_H c_{f,H}^m$. In Fig. 13.3, we plot I_H, $c_{f,H}$, and $c_{b,H}$ at final time.

Example 2 This example illustrates how the solution regularity can affect the convergence rates. Consider the system (13.16)–(13.18) with initial boundary conditions given by (13.4)–(13.8) and diffusion coefficients and reaction functions defined by $D_{I,11} = D_{I,22} = I + 1$, $D_{c,11} = D_{c,22} = c_f^2 + 1$, $G(I) = I$, $S(I, c_f, c_b) = Ic_b^2 c_f$, and $F(I, c_f, c_b) = Ic_b c_f$. For the fixed data, we consider (13.16)–(13.18) with convenient source terms such that this IBVP as the following solution:

Table 13.2 Space errors and convergence rates for the case $\alpha = 2.1$

H_{max}	N_1	N_2	$Error_I$	$Rate_I$	$Error_{c_f}$	$Rate_{c_f}$	$Error_{c_b}$	$Rate_{c_b}$
1.000E-1	10	10	2.8990E-5	–	5.2886E-8	–	1.4787E-6	–
5.000E-2	20	20	1.8312E-6	1.9924	1.3695E-8	0.97462	9.1431E-8	2.0077
2.500E-2	40	40	1.1192E-7	2.0161	3.7220E-9	0.93975	5.7390E-9	1.9969
1.250E-2	80	80	6.8539E-9	2.0147	9.3061E-10	0.99992	3.5845E-10	2.0005
6.250E-3	160	160	4.2312E-10	2.0089	2.1886E-10	1.0441	2.2328E-11	2.0024
3.125E-3	320	320	2.6269E-11	2.0048	4.9678E-11	1.0697	1.3861E-12	2.0049

Table 13.3 Space errors and convergence rates for $\alpha = 3.1$

H_{max}	N_1	N_2	$Error_I$	$Rate_I$	$Error_{c_f}$	$Rate_{c_f}$	$Error_{c_b}$	$Rate_{c_b}$
1.000E-1	10	10	2.8990E-5	–	9.2469E-9	–	1.4794E-6	–
5.000E-2	20	20	1.8312E-6	1.9924	9.1735E-10	1.6667	9.1528E-8	2.0073
2.500E-2	40	40	1.1192E-7	2.0161	7.3440E-11	1.8214	5.7511E-9	1.9961
1.250E-2	80	80	6.8539E-9	2.0147	5.1920E-12	1.9111	3.5992E-10	1.9990
6.250E-3	160	160	4.2312E-10	2.0089	3.4588E-13	1.9540	2.2503E-11	1.9998
3.125E-3	320	320	2.6269E-11	2.0048	1.4590E-14	1.9835	1.2813E-12	1.9999

$$I(x, y, z) = 2t(2x - x^2)\frac{1}{\pi}\left(-(y-1)\cos(\pi y) + \frac{1}{\pi}\sin(\pi y)\right),$$

$$c_f(x, y, z) = \exp(t)(x^3 - x^2)\left(\frac{y^3}{3} - \frac{y^2}{2}\right)|x - 0.5|^{\alpha},$$

$$c_b(x, y, z) = \exp(t)xy\sin(\pi y)\sin(\pi x),$$

for $x \in \Omega, t \in (0, T]$. We remark that, for $\alpha = 2.1$, $c_f \in H^2(\Omega)$, in contrast, for $\alpha = 3.1$, we have $c_f \in H^3(\Omega)$. Table 13.2 and 13.3 illustrate the convergence rates for each of the variables. Notice that for the case of the rate $\text{Rate}_{c_f,H}$, we have second-order convergence rate for the value $\alpha = 3.1$ (Table 13.3) and only first-order convergence rate for $\alpha = 2.1$ (Table 13.2).

This example suggests that the method's convergence order (in space) can depend on the theoretical solution regularity (Fig. 13.4).

Example 3 The goal of this example is to compare the convergence behavior of the two methods for the differential problem (13.16)–(13.18): the IMEX scheme (13.19)–(13.23) and the CN method, which is, theoretically, second-order convergent in time.

Let the diffusion coefficients and reaction functions defined by $D_{I,11} = D_{I,22} = I^4$, $D_{c,11} = D_{c,22} = c_f^2$, $G(I) = I^2$, $S(I, c_f, c_b) = I^2 c_b^2 c_f$, and $F(I, c_f, c_b) = I^3 c_f^2 c_b$. Let us also add suitable source functions to the right-hand side of Eqs. (13.16)–(13.18), such that the exact solution of this problem is

$$I(x, y, z) = \exp(t)\left(-\frac{x^3}{3} - \frac{x^2}{2}\right)(-\cos(\pi y)),$$

$$c_f(x, y, z) = \exp(t)(1 - x^2)\left(-\frac{y^3}{3} + \frac{y^2}{2}\right),$$

$$c_b(x, y, z) = \exp(t)x^2y^2\sin(\pi y^2)\sin(\pi x^2).$$

We fix $N_1 = 160$ and $N_2 = 192$ and successively halve the time step. To evaluate the performance in time of the mentioned methods, we use Eqs. (13.31)–(13.34) for the numerical errors and convergence rates. Table 13.4 and 13.5 illustrate the results obtained with the IMEX method and CN method, respectively. The results suggest that the IMEX method has first-order convergence, while the CN method has second-order convergence. In Fig. 13.5, we present the approximation I_H, $c_{f,H}$, and $c_{b,H}$ using the CN method.

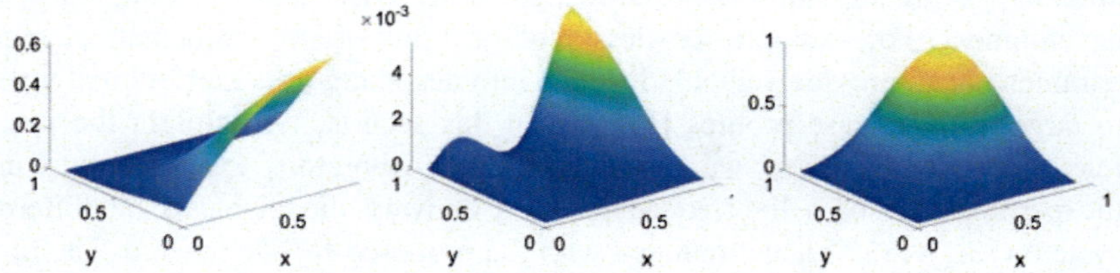

(a) I_H. (b) $c_{f,H}$. (c) $c_{b,H}$.

Fig. 13.4 Numerical solution for Example (2) for $alpha = 2.1$

Table 13.4 Error on time for IMEX method

Δt	$\text{Error}_{\Delta t,I}$	Rate_I	$\text{Error}_{\Delta t,c_f}$	Rate_{c_f}	$\text{Error}_{\Delta t,c_b}$	Rate_{c_b}
5.0E-1	3.5967E-2	–	4.1233E-2	–	4.0470E-2	–
2.50E-1	1.8542E-2	0.95587	2.1034E-2	0.97106	2.6208E-2	0.62684
1.250E-1	9.2611E-3	1.0015	1.0487E-2	1.0042	1.4217E-2	0.88236
6.25E-2	4.6192E-3	1.0036	5.2284E-3	1.0041	7.3463E-3	0.95255

Table 13.5 Error on time for CN method

Δt	$\text{Error}_{\Delta t,I}$	Rate_I	$\text{Error}_{\Delta t,c_f}$	rate_{c_f}	$\text{Error}_{\Delta t,c_b}$	Rate_{c_b}
5.0E-1	1.6047E-3	–	1.4969E-3	–	3.1500E-3	–
2.50E-1	5.4335E-4	1.5623	5.0486E-4	1.5680	1.0568E-3	1.5757
1.250E-1	1.4824E-4	1.8740	1.3881E-4	1.8628	2.8952E-4	1.8679
6.25E-2	3.2676E-5	2.1816	3.4938E-5	1.9902	6.9948E-5	2.0493

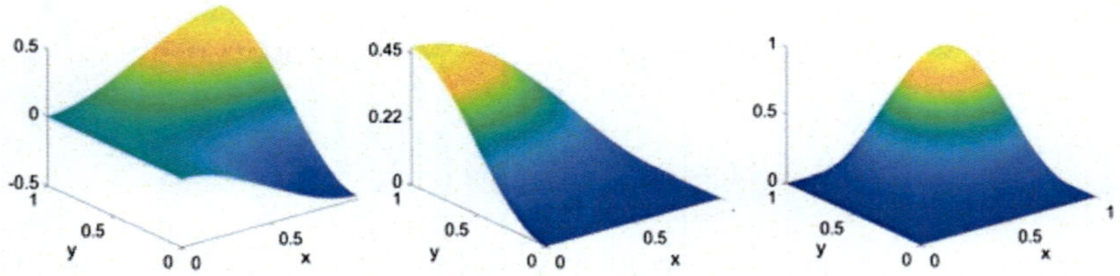

(a) I_H. (b) $c_{f,H}$. (c) $c_{b,H}$.

Fig. 13.5 Numerical solution for Example (3) using CN method

13.5 Validation: Simulation Results Versus Experimental Data

In drug delivery research, mathematical modeling and simulation allows rapid and efficient testing of multiple scenarios, lowering costs and speeding up product development. For example, costless modifications of the mathematical model's parameters can provide valuable insights into designing drug carriers that give rise to target drug release profiles [28–30]. In this section, to highlight the practical relevance of this paper, we use data from a laboratory experiment to assess the performance of a light-enhanced drug delivery model based on differential system (13.1)–(13.8) and simulated with the proposed IMEX scheme. The in vitro experiment is described in detail in [21]. Essentially, it consists of a polymeric structure loaded with drug molecules bound to the polymeric chains by photochemical links. By action of NIR light, the molecular bonds are break and the amount of drug diffused measured. To validate our model, we compare the experimental drug release to the simulated drug release obtained using the IMEX method (13.19)–(13.23).

13.5.1 Drug Release Enhanced by Light

Stimuli-responsive drug delivery systems have emerged as a solution for controlled local drug delivery. This technique relies on nanocarriers that transport the drugs to the target site, where they release it via a stimulus. Localized drug delivery is critical for minimizing undesirable side effects caused by high-toxicity drugs, while the controlled release is critical for keeping the drug concentration within its therapeutic window. This therapeutic approach is especially appealing in cancer treatment, where chemotherapy has serious side effects [28–30]. To increase the accumulation of nanocarriers in the target tissue, the transporters can be decorated with ligands that have an affinity with overexpressed receptors that usually arise in cancer tissues.

The development of controlled and localized drug delivery systems had a burst with the paradigm of nanomedicine based on nanotechnology. Some stimuli-responsive drug nanocarriers include dendrimers, liposomes, micelles, metal particles, polymeric nanoparticles, carbon nanotubes, and hydrogels [31–34]. To tune the drug release from the nanocarriers, the researchers are looking at endogenous (pH, redox, enzymes) and exogenous stimuli (temperature, ultrasound, light, electric fields, magnetic fields) [17, 34–38].

Let us note that local drug delivery involves some other challenging issues, namely, in solid tumors, where drug absorption is hampered by an abnormal vasculature (heterogeneous and chaotic vessel growth), a highly dense extracellular matrix (cross-linked structure), and high interstitial fluid pressure due to the high permeability of the tumor vasculature and absence of effective lymphatic drainage [29].

Many of the DDS are based on hydrogel reservoirs. Hydrogels are polymeric materials that can store large amounts of water or biological fluids, which makes them highly biocompatible and biodegradable. The physical and chemical properties of hydrogels are tunable. Properties like temperature and degradation rate can be controlled by an external stimulus. These properties make hydrogels an ideal candidate for controlled and localized drug delivery [31, 39, 40].

Due to the low adverse effects on human tissue and relatively deep tissue penetration, NIR light is considered a magic tool for controlled drug delivery [1]. NIR light is also highly tunable, and one can easily manipulate parameters like intensity, duration, and wavelength to fine-tune the drug release rate [1, 2, 29, 30, 32, 37, 39–42]. Furthermore, it is relatively easy to engineer polymeric drug carriers with photochemical or photothermal responsive links, i.e., responsive to photon energy absorption or to light-induced heating.

Once a NIR-light-responsive hydrogel is in contact with the target tissue, the drug entrapped in the polymeric matrix can be released by the stimulus of light radiation that breaks the links between the drug molecules and the polymeric chains [1]. A drug release of diffusion type then takes place, and it can be originated by different factors: temperature rise, hydrogel swelling due to increased osmotic pressure, or disintegration of the polymeric matrix (i.e., photocleavage) [43]. Furthermore, such processes are reversible, which means that drug diffusion can be controlled and regulated over time, and one can obtain the desired release rates by manipulating light parameters (e.g., intensity and duration) and hydrogel composition [1, 31, 39, 41].

13.5.2 *In Vitro Experiment*

In [21], the authors describe a laboratory experiment consisting of a NIR light-sensitive hydrogel designed for cancer treatment. The polymeric platform is an injectable cross-linked polyaspartic acid (PASP-SS) hydrogel encapsulating proteinase K (ProK) containing platinum nanoparticles (PtNPs) to control the polymer degradation, an oxidant, and lentinan, an anti-gastric cancer drug. The oxidant allows gel formation after the injection of the polymeric solution in the target.

In the experiment, $0.1g$ H-PtNPs-ProK-loaded lentinan was placed in a standard PBS solution, and NIR light stimulation was used to increase drug release. A chicken flesh specimen with a thickness of 30 mm was placed between the infrared laser (power density 1.5 W/cm^2) and the hydrogel to simulate muscle blockade. Samples from the PBS solution were collected at regular time intervals, and the amount of drug released was evaluated using a UV spectrometer.

For 23 hours, three NIR light protocols were tested: post-infrared, pre-infrared, and intermittent radiations. In the first scenario, NIR light was applied continuously during the last two hours, from hour 21 to hour 23. In the second scenario, NIR light was applied continuously during the first 2.5 hours. Finally, in the third scenario, NIR light was applied intermittently (20 minutes on, 20 minutes off).

Here, we focus on the pre-infrared radiation scenario, in which NIR radiation is applied during the first 2.5 hours. In Fig. 13.6, we present the experimental drug release profile. The amount of drug in the PBS solution rises quickly during the NIR irradiation period, and by the end of this 2.5 hours period, nearly 60% of the drug has been cleaved from the polymeric structure. Afterward, the drug release rate seems to reach a plateau, with the values oscillating around 60%. The authors state that the oscillations are due to measurement errors.

13.5.3 Simulation Versus Experimental Data

In this section, we try to replicate the experiment presented in [21] using the model (13.1)–(13.8). We measure the accuracy of our mathematical model (13.1)–(13.8), comparing the simulated release rate with the one presented in Fig. 13.6. For the simulation, we take $[0, T] = [0, 1400\text{min}]$, and we consider the initial conditions:

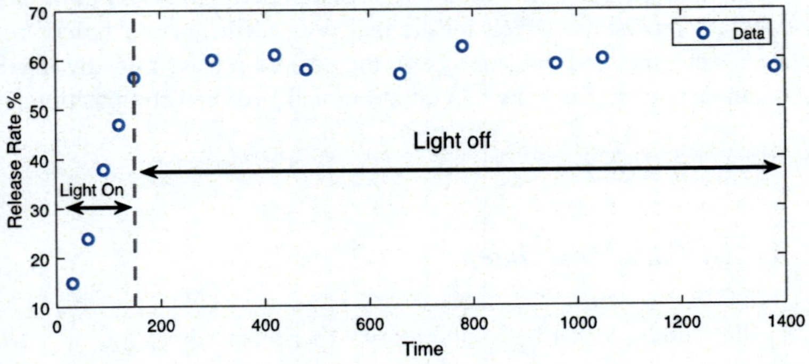

Fig. 13.6 Experimental data [21]

Table 13.6 Parameter values

Parameter	Value	Units
β	1.3324×10^{12}	cm/min
μ_a	0.025	cm^{-1}
μ_s	13	cm^{-1}
D_d	4×10^{-4}	cm^2/min
γ	4.4×10^{-3}	cm^2/min

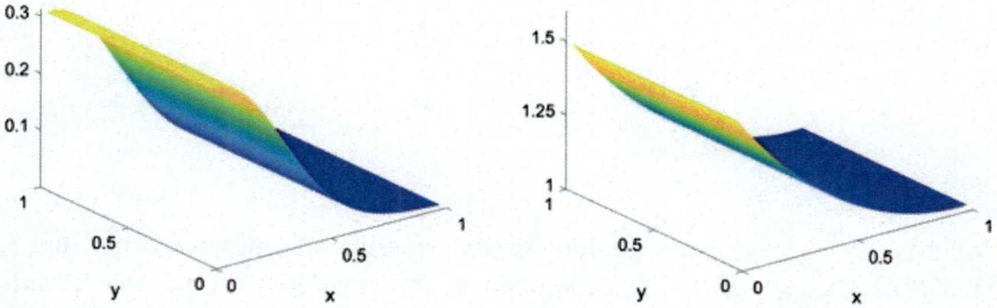

Fig. 13.7 Free drug concentration c_f (on the left) and light intensity I (on the right)

$$c_b(x, y, 0) = 1 \quad \text{for} \quad x \in [0, 0.25], \quad y \in [0, 1],$$

$$c_b(x, y, 0) = 0 \quad \text{for} \quad x \in (0.25, 1], \quad y \in [0, 1],$$

$$c_f(x, y, 0) = 0 \quad \text{for} \quad (x, y) \in \overline{\Omega},$$

$$I(x, y, 0) = 0 \quad \text{for} \quad (x, y) \in \overline{\Omega}.$$

These initial conditions reflect a bidimensional computational domain of size [0, 1] cm × [0, 1] cm, with the polymeric structure placed at [0, 0.25] cm × [0, 1] cm. The values of the parameters β, μ_a, μ_s, and D_d were obtained from the literature [44–46], and the value of γ was calculated using a best-fitting procedure. All the parameter values are given in Table 13.6.

The solution of the IBVP (13.1)–(13.8) is approximated with the fully discrete scheme (13.19)–(13.23) with the uniform space step 10^{-3} and the time step $\Delta t = 1$.

In Fig. 13.7, we plot the numerical approximations for c_f and I at the final time. In Fig. 13.8, we show the simulated free drug release rate. We see that there is a good agreement between simulation results (solid line) and experimental data (blue dots). The mathematical model accurately captures both the quick-release rate during NIR light irradiation and the plateau during the light-off period.

To quantify the accuracy of the model, we compute the mean absolute error (MAE) given by

$$\text{MAE} = \frac{1}{n} \sum_{k=1}^{n} |r_h(t_i) - r(t_i)|, \tag{13.35}$$

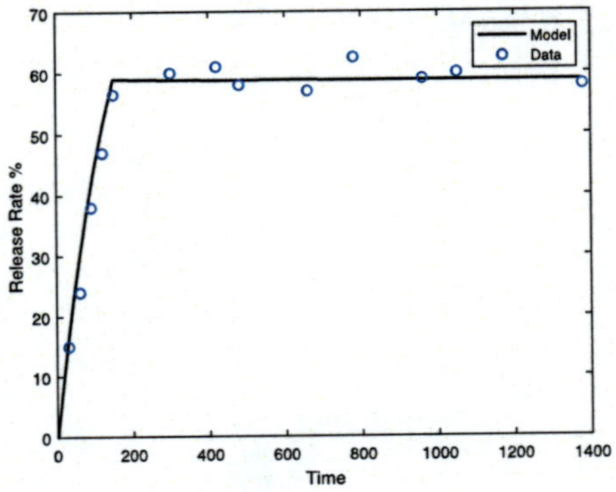

Fig. 13.8 c_f release rate

where t_i, $i = 1, \cdots, n$ is the time of the experimental measurements, and $r_h(t_i)$ and $r(t_i)$ correspond to the experimental and simulated release rate at time t_i, respectively. We obtain MAE $= 2.1254\%$, a relatively low value indicating that the model can be used to simulate light-enhanced drug delivery.

13.6 Conclusions and Future Work

Here, we consider an IMEX time scheme combined with a nonuniform finite difference scheme for the numerical solution of a nonlinear partial differential system. We numerically show that the approximations in space are second-order supraconvergent in a discrete H^1-norm. The numerical experiments also reveal that this supraconvergence phenomenon requires a certain smoothness of the theoretical solution. Regarding the time approximation, we compare the IMEX scheme with the Crank-Nicolson scheme. Unlike Crank-Nicolson, the IMEX scheme avoids the solution of nonlinear systems; however, the numerical convergence rate in a discrete L^2-norm decreases from two with Crank-Nicolson, to one with IMEX. In the experimental section, we also test the capability of the nonlinear system under analysis to model light-enhanced drug delivery. A comparison against experimental data shows that the model and the proposed scheme are reliable tools for the modeling and simulating specific drug delivery scenarios.

We left the formal analysis of the method's convergence and stability properties for future work. Some of the challenges arising in the theoretical analysis are the nonlinear nature of the associated differential system, the two-dimensional setting, and the presence of boundary conditions of the Neumann-Dirichlet type. Another critical point is the analysis of the impact of the smoothness of the theoretical solution on the method's order of convergence. For this, we need to replace the traditional error analysis techniques based on Taylor's decomposition with nonstandard ones based on Bramble-Hilbert lemma. Other research directions include expanding to a three-dimensional setting and developing and analyzing second-order IMEX schemes. We can also extend the nonlinear system to include temperature effects associated with light-enhanced drug delivery.

Acknowledgments This work was partially supported by the Centre for Mathematics of the University of Coimbra – UIDB/00324/2020, funded by the Portuguese government through FCT/MCTES. H.P. Gómez was also supported by the FCT PhD grant PD/BD/150551/2019, funded by the Portuguese government through FCT/MCTES.

References

1. Y. Tao, H. Chan, B. Shi, M. Li, H. Leong, Light: a magical tool for controlled drug delivery. Adv. Funct. Mater. **30**, 2005029 (2020). https://doi.org/10.1002/adfm.202005029
2. X. Wang, Z. Xuan, X. Zhu, H. Sun, J. Li, Z. Xie, Near-infrared photoresponsive drug delivery nanosystems for cancer photo-chemotherapy. J Nanobiotechnol. **18**, 108 (2020). https://doi.org/10.1186/s12951-020-00668-5
3. A. Ishimaru, *Wave Propagation and Scattering in Random Media* (Academic Press, New York, 1978)
4. L. Wang, H. WU, *Biomedical Optics—Principles and Imaging* (John Wiley and Songs, 2007)
5. J. Ferreira, H. Gómez, L.Pinto, A numerical scheme for a partial differential system motivated by light-triggered drug delivery. Appl. Numer. Math. **184**, 101–120 (2023). https://doi.org/10.1016/j.apnum.2022.09.005
6. P. Forsyth, P.H. Sammon, Quadratic convergence cell-centered grids. Appl. Numer. Math. **4**, 377–394 (1988). https://doi.org/10.1016/0168-9274(88)90016-5
7. R. Grigorieff, Some stability inequalities for compact finite difference operators. Mathematische Nachrichten **135**, 93–102 (1986). https://doi.org/10.1002/mana.19881350110
8. F. de Hoog, D. Jackett, On the rate of convergence of finite difference schemes on nonuniform grids. J. Aust. Math. Soc. B 377–394 (1988). https://doi.org/10.1017/S0334270000004495
9. H.O. Kreiss, T. Manteuffel, B. Swartz, B. Wendroff, J.A.B. White, Supraconvergent schemes on irregular grids. Math. Comput. **47**, 537–554 (1986). https://doi.org/10.2307/2008171
10. T. Manteuffel, J.A.B. White, The numerical solution of second order boundary value problems on nonuniform meshes. Math. Comput. **47**, 511–535 (1986). https://doi.org/10.2307/2008170
11. A. Weiser, M.F. Wheeler, On convergence of block-centered finite difference for elliptic problems. SIAM J. Numer. Anal. **25**, 351–375 (1988)
12. J.A. Ferreira, R. Grigorieff, On the supraconvergence of elliptic finite difference schemes. Appl. Numer. Math. **28**, 275–292 (1998). https://doi.org/10.1016/S0168-9274(98)00048-8
13. S. Barbeiro, J. Ferreira, R. Grigorieff, Supraconvergence of a finite difference scheme for solutions in $H^s(0, l)$. IMA J. Numer. Anal. **25**, 797–811 (2005). https://doi.org/10.1093/imanum/dri018
14. J. Ferreira, R. Grigorieff, Supraconvergence and supercloseness of a scheme for elliptic equations on nonuniform grids. Numer. Funct. Anal. Optim. **27**, 539–564 (2006)
15. J. Bramble, S. Hilbert, Estimation of linear functionals on Sobolev spaces with application to Fourier transforms and spline interpolation. SIAM J. Numer. Anal. **7**, 112–124 (1970)
16. S. Barbeiro, S. Bardeji, J. Ferreira, L. Pinto, Non-Fickian convection—diffusion models in porous media. Numerische Mathematik **138**, 869–904 (2017). https://doi.org/10.1007/s00211-017-0922-6
17. J. Ferreira, D. Jordão, L. Pinto, Approximating coupled hyperbolic-parabolic systems arising in enhanced drug delivery. Comput. Math. Appl. **76**, 81–97 (2018). https://doi.org/10.1016/j.camwa.2018.04.005
18. J. Ferreira, L. Pinto, G. Romanazzi, Supraconvergence and supercloseness in Volterra equations. Appl. Numer. Math. **62**, 1718–1739 (2012).

19. L. Pinto, Parabolic partial integro-differential equations: superconvergence estimates and applications, Ph.D. Thesis, University of Coimbra (2013)
20. J. Ferreira, H. Gómez, L. Pinto, A mathematical model for NIR light protocol optimization in controlled transdermal drug delivery. Appl. Math. Model. **112**, 1–17 (2022). https://doi.org/10.1016/j.apm.2022.07.014
21. J. Zhao, X. Liang, H. Cao, T. Tan, Preparation of injectable hydrogel with near-infrared light response and photo-controlled drug release. Bioresour. Bioprocess **7**, 1 (2020). https://doi.org/10.1186/s40643-019-0289-x
22. A. Zarzur, H.C. Velho, S. Freitas, S. Stephany, Combined explicit and implicit methods for time integration in partial differential equations. Am. Soc. Thermal Fluids Eng. 151–158 (2018). https://doi.org/10.1615/TFEC2018.cfd.021966
23. M. Hochbruck, J. Leibold, An implicit-explicit time discretization scheme for second-order semilinear wave equations with application to dynamic boundary conditions. Numerische Mathematik **147**, 869–899 (2021). https://doi.org/10.1007/s00211-021-01184-w
24. U. Ascher, S. Ruuth, B. Wetton, Implicit-explicit methods for time-dependent partial differential equations. SIAM J. Numer. Anal. **32**, 797–823 (1995). https://doi.org/10.1137/0732037
25. S. Ruuth, Implicit-explicit methods for reaction-diffusion problems in pattern formation. J. Math. Biol. **34**, 148–176 (1995). https://doi.org/10.1007/BF00178771
26. S. Ruuth, Implicit-explicit methods for reaction-diffusion problems in pattern formation. J. Math. Biol. **34**, 148–176 (1995)
27. A. Preuss, J. Lipoth, R.J. Spiteri, When and how to split? A comparison of two IMEX splitting techniques for solving advection–diffusion–reaction equations. J. Comput. Appl. Math. **414**, 114418 (2022). https://doi.org/10.1016/j.cam.2022.114418
28. P.N. Navya, A. Kaphle, S. Srinivas, S. Bhargava, V. Rotello, H. Daima, Current trends and challenges in cancer management and therapy using designer nanomaterials. Nano Convergence **6**, 23 (2019). https://doi.org/10.1186/s40580-019-0193-2
29. D. Rosenblum, N. Joshi, W. Tao, J. Karp, D. Peer, Progress and challenges towards targeted delivery of cancer therapeutics. Nat. Commun. **9**, 1410 (2018). https://doi.org/10.1038/s41467-018-03705-y
30. S. Senapati, A. Mahanta, S. Kumar, P. Maiti, Controlled drug delivery vehicles for cancer treatment and their performance. Signal Transduction Targeted Ther. **3**, 7 (2018). https://doi.org/10.1038/s41392-017-0004-3
31. A. Kasiński, E.O.M.Z.-Pisklak, M. Sobczak, Smart hydrogels—synthetic stimuli-responsive antitumor drug release systems. Int. J. Nanomed. **15**, 4541–4572 (2020). https://doi.org/10.2147/IJN.S248987
32. Y. Ji, J. Li, J. Zhao, S. Shand, C.-C. Chu, A light-facilitated drug delivery system from a pseudo-protein/hyaluronic acid nanocomplex with improved anti-tumor effects. Nanoscale **11**, 9987–10003 (2019). https://doi.org/10.1039/c9nr01909j
33. M. Tang, D. Svirskis, E. Leung, M. Kanamala, H. Wang, Z. Wu, Can intracellular drug delivery using hyaluronic acid functionalised pH-sensitive liposomes overcome gemcitabine resistance in pancreatic cancer? J. Controlled Release **305**, 89–100 (2019). https://doi.org/10.1016/j.jconrel.2019.05.018
34. C. Wells, M. Harris, L. Choi, V. Murali, F. Guerra, J. Jennings, Stimuli-responsive drug release from smart polymers. J. Funct. Biomater. **10**, 34 (2019). https://doi.org/10.3390/jfb10030034
35. R. Kalaydina, K. Bajwa, B. Qorri, A. Decarlo, M. Szewczuk, Recent advances in "smart" delivery systems for extended drug release in cancer therapy. Int. J. Nanomed. **13**, 4727–4745 (2018). https://doi.org/10.2147/IJN.S168053
36. S. Hossen, M. Hossain, M. Basher, M. Mia, M. Rahman, M. Uddin, Smart nanocarrier-based drug delivery systems for cancer therapy and toxicity studies: a review. J. Adv. Res. **15**, 1–18 (2019). https://doi.org/10.1016/j.jare.2018.06.005
37. Y. Shamay, A. Lily, A. Gonen, D. Ayelet, Light induced drug delivery into cancer cells. Biomaterials **32**, 1377–1386 (2011). https://doi.org/10.1016/j.biomaterials.2010.10.029
38. Y. Wang, D. Kohane, External triggering and triggered targeting strategies for drug delivery. Nat. Rev. Mater. **2**, 17020 (2017). https://doi.org/10.1038/natrevmats.2017.20

39. I. Tomatsu, K. Peng, A. Kros, Photoresponsive hydrogels for biomedical applications. Adv. Drug Delivery Rev. **63**, 1257–1266 (2011). https://doi.org/10.1016/j.addr.2011.06.009
40. H. Kim, H. Lee, K.-Y. Seong, E. Lee, S. Yang, J. Yoon, Visible light-triggered on-demand drug release from hybrid hydrogels and its application in transdermal patches. Adv. Funct. Mater. **4**, 2071–2077 (2015). https://doi.org/10.1002/adhm.201500323
41. M. Qiu, D. Wang, W. Liang, L. Liu, Y. Zhang, X. Chen, D. Sang, C. Xing, Z. Li, B. Dong, F. Xing, D. Fan, S. Bao, H. Zhang, Y. Cao, Novel concept of the smart NIR-light-controlled drug release of black phosphorus nanostructure for cancer therapy. Proc. Natl. Acad. Sci. **115**, 501–506 (2018). https://doi.org/10.1073/pnas.1714421115
42. T. Rappa, C. DeForest, Targeting drug delivery with light: a highly focused approach. Adv. Drug Delivery Rev. **171**, 94–107 (2021). https://doi.org/10.1016/j.addr.2021.01.009
43. W. Zhang, T. Ji, Y. Li, Y. Zheng, M. Mehta, C. Zhao, A. Liu, D. Kohane, Light-triggered release of conventional local anesthetics from a macromolecular prodrugfor on-demand local anesthesia. Nat. Commun. **11**, 2323 (2020). https://doi.org/10.1038/s41467-020-16177-w
44. J. Sandell, T. Zhu, A review of in-vivo optical properties of human tissues and its impact on PDT. J Biophoton. **4**, 773–787 (2011). https://doi.org/10.1002/jbio.201100062
45. S. Jacques, B. Pogue, Tutorial on diffuse light transport. J. Biomed. Opt. **13**(4) (2008). https://doi.org/10.1117/1.2967535
46. M. Grassi, I. Colombo, R. Lapasin, Drug release from an ensemble of swellable crosslinked polymer particles. J. Controlled Release **68**, 97–113 (2000) https://doi.org/10.1016/s0168-3659(00)00241-8

Index

C. M. A. Pinto, C. M. Ionescu (eds.), *Computational and Mathematical Models in Biology*, Nonlinear Systems and Complexity 38,
https://doi.org/10.1007/978-3-031-42689-6